Clean Energy Production Technologies

Series Editors

Neha Srivastava, Department of Chemical Engineering and Technology
IIT (BHU) Varanasi
Varanasi, India

P. K. Mishra, Department of Chemical Engineering and Technology
IIT (BHU) Varanasi
Varanasi, India

The consumption of fossil fuels has been continuously increasing around the globe and simultaneously becoming the primary cause of global warming as well as environmental pollution. Due to limited life span of fossil fuels and limited alternate energy options, energy crises is important concern faced by the world. Amidst these complex environmental and economic scenarios, renewable energy alternates such as biodiesel, hydrogen, wind, solar and bioenergy sources, which can produce energy with zero carbon residue are emerging as excellent clean energy source. For maximizing the efficiency and productivity of clean fuels via green & renewable methods, it's crucial to understand the configuration, sustainability and techno-economic feasibility of these promising energy alternates. The book series presents a comprehensive coverage combining the domains of exploring clean sources of energy and ensuring its production in an economical as well as ecologically feasible fashion. Series involves renowned experts and academicians as volume-editors and authors, from all the regions of the world. Series brings forth latest research, approaches and perspectives on clean energy production from both developed and developing parts of world under one umbrella. It is curated and developed by authoritative institutions and experts to serves global readership on this theme.

Mohammad Yaseen Mir • Javid Ahmad Parray
Sumanta Bhattacharya

Low-Carbon Hydrogen

Production Technologies and Emerging Applications

Springer

Mohammad Yaseen Mir
Centre of Research for Development
University of Kashmir
Naseem Bagh, Jammu and Kashmir, India

Javid Ahmad Parray
Department of Environmental Science
Govt. Degree College
Kupwara, Jammu and Kashmir, India

Sumanta Bhattacharya
Environmental Science
National Economic Research Bureau
Imphal, Manipur, India

ISSN 2662-6861 ISSN 2662-687X (electronic)
Clean Energy Production Technologies
ISBN 978-981-95-8141-2 ISBN 978-981-95-8142-9 (eBook)
https://doi.org/10.1007/978-981-95-8142-9

This Springer imprint is published by the registered company Springer Nature Singapore Pte Ltd.
The registered company address is: 152 Beach Road, #21-01/04 Gateway East, Singapore 189721, Singapore

Preface

The global transition toward low-carbon energy systems has become an urgent scientific, technological, and societal priority in response to climate change, energy security concerns, and the need for sustainable economic development. Within this transition, hydrogen has emerged as a pivotal energy carrier with the potential to decarbonize sectors that are difficult to electrify directly, including heavy industry, long-distance transport, chemicals, and seasonal energy storage. In particular, low-carbon hydrogen produced through pathways that significantly reduce or eliminate greenhouse gas emissions offers a versatile and scalable solution to support deep decarbonization while enabling greater integration of renewable energy sources.

This book, *Low-Carbon Hydrogen: Production Technologies and Emerging Applications*, provides a comprehensive and interdisciplinary overview of the current state, challenges, and future prospects of hydrogen technologies. It covers a broad spectrum of hydrogen production routes, including renewable-powered water electrolysis, biomass- and waste-derived hydrogen, methane reforming with carbon capture, and emerging bioinspired and hybrid systems. Beyond production, the book explores advances in hydrogen storage, transport, system integration, and end-use applications across industry, power systems, and mobility. Emphasis is placed on techno-economic performance, environmental impacts, and the role of policy, regulation, and innovation ecosystems in accelerating large-scale deployment. Designed for researchers, graduate students, engineers, policymakers, and industry practitioners, this volume aims to bridge fundamental research and real-world implementation. By bringing together contributions from leading experts across disciplines, the book seeks to foster a deeper understanding of hydrogen as a socio-technical system and to highlight pathways for its responsible, equitable, and sustainable adoption. We hope this work will serve as a valuable reference and inspire further research, collaboration, and innovation toward a low-carbon hydrogen future.

Naseem Bagh, Jammu and Kashmir, India — Mohammad Yaseen Mir
Kupwara, Jammu and Kashmir, India — Javid Ahmad Parray
Imphal, Manipur, India — Sumanta Bhattacharya

Competing Interests The authors have no competing interests to declare that are relevant to the content of this manuscript.

Contents

About the Authors

Mohammad Yaseen Mir is currently a Subject Expert Member at the Commission for Scientific and Technical Terminology (CSTT), Ministry of Education, Government of India, New Delhi, and teaches in the Department of Education, Jammu and Kashmir, India. Previously, he worked as a Senior Researcher on a Ministry of Environment, Forest & Climate Change (MoEF&CC)–funded project at G.B. Pant National Institute of Himalayan Environment & Sustainable Development, Uttarakhand, India, under the National Mission on Himalayan Studies. He also serves as a subject course expert for Environmental Science Massive Open Online Courses (MOOCs) approved by the Consortium for Educational Communication (CEC), Ministry of Education, Government of India, New Delhi. He earned his M.Phil. and Ph.D. in Botany with a specialization in Plant Biotechnology from the University of Kashmir, Jammu and Kashmir, India. His academic interests include plant biotechnology and environmental education. He is an editor and reviewer for several reputed journals and has authored multiple research articles, review papers, and book chapters. He has also authored and edited books published by internationally recognized publishers.

Javid Ahmad Parray is currently teaching in the Department of Environmental Science at Government Degree College, Kupwara, Jammu and Kashmir, India, affiliated with University of Kashmir Srinagar. He also serves as the National Course Coordinator for two Environmental Science MOOCs under CeC–SWAYAM, Ministry of Education, Government of India, New Delhi. He has completed his M.Phil. and Ph.D. in Environmental Science from the University of Kashmir, Jammu and Kashmir, India, after qualifying the JKSLET examination, and later pursued postdoctoral research at the same institution. His core research interests span ecological and agricultural microbiology, environmental microbiome, climate change, and microbial biotechnology. He has participated in several national and international conferences and courses across countries including Sri Lanka, Indonesia, and Malaysia. He has authored and edited numerous books, contributed extensively to high-impact journals and book chapters, and serves as editor, guest editor, and reviewer for reputed scientific journals. He is a member of various national and international scientific organizations and currently serves as Series Editor for the book series Microbiome Research in Plants and Soil and Microbial Genomics by an international publisher.

Sumanta Bhattacharya based in Salkia, Howrah, West Bengal, India, is currently an Expert Commissioner Member of International Union for Conservation of Nature (IUCN)'s Environmental, Economic, and Social Policy and an Individual Policy Analyst with the Asian Development Bank. He also holds advisory and membership roles with multiple Indian ministries, including Defence (USI, IDSA), Culture (AS), Personnel & Public Grievance (IIPA), and the think tank Centre for Joint Warfare Studies (CENJOWS). He holds a broad educational background with degrees in B.Tech and M.Tech in Textile Technology, M.Sc. in Environmental Science, MA in Development Studies, MA in Economics, MA in Security and Defence Law (National Academy of Legal Studies and Research (NALSAR), Telangana, India), LLB, MBA, multiple postgraduate diplomas, and an international certification in Public Policy (MPI, Oxford University, UK). He is a Ph.D. scholar at Asian International University, India, and a former Research Fellow at Maulana Abul Kalam Azad University of Technology (MAKAUT), West Bengal, India. His areas of expertise include climate change, sustainable development, biotechnology, nanotechnology, law, defense strategy, public policy, and environmental studies. He has presented at over 90 international and national conferences, including those in Russia, Germany, and Sweden, and has received 30 prestigious awards such as the Best Water Management Solution Award (2022), Mahatma Gandhi Ratna Award (2020), and Global Legend Award (2023). He is affiliated with organizations such as Institution of Engineers (India) (IEI), Indian Institute of Chemical Engineers (IICHE) Indian Institute of Public Administration (IIPA) Indian Direct Selling Association (IDSA) United Service Institution of India (USI) Centre for Land Warfare Studies (CLAWS) Youth non-governmental organization (YOUNGO), and the Asiatic Society.

Chapter 1
Hydrogen Production with Low- or Zero-Carbon Emissions from Renewable and Fossil Resources

Abstract The growing global energy demand driven by urbanization and population growth, coupled with the environmental impacts of fossil fuel use, has intensified the search for sustainable energy solutions. Hydrogen, as a clean and efficient energy carrier, offers a promising pathway toward low- or zero-carbon energy systems. This chapter provides a comprehensive review of hydrogen production technologies from renewable sources including biomass, solar, wind, geothermal and algae as well as from conventional fossil-based sources such as natural gas, coal, nuclear, and thermochemical processes. It evaluates the economic feasibility of each approach and examines the environmental implications, highlighting the potential of green hydrogen to support a sustainable energy future.

Keywords Hydrogen production · Fossil resources · Renewable resources · Greenhouse gas emissions · Environment

1.1 Introduction

The continuous rise in greenhouse gas (GHG) emissions has emerged as a major global concern, prompting world leaders, policymakers, and researchers to intensify efforts aimed at mitigating climate change and its associated environmental impacts. This challenge is further compounded by the steady increase in global energy demand, projected to grow at an average rate of approximately 1.3% annually until 2040. This growth is primarily driven by rapid technological advancements, population expansion, and sustained economic development, resulting in continued reliance on fossil fuels currently the dominant source of global energy and a major contributor to GHG emissions. Despite their adverse environmental consequences, fossil fuels are expected to retain a substantial share of the global energy mix until at least 2050 (Brockway et al. 2019; Mir 2025a). Additionally, declining fossil fuel reserves and volatile oil prices continue to exert significant pressure on industrial operations and end users alike (Thorbecke 2019). The COVID-19 pandemic caused an unprecedented disruption to the global energy system, leading to a temporary

D. M. Y. Mir et al., *Low-Carbon Hydrogen*, Clean Energy Production Technologies, https://doi.org/10.1007/978-981-95-8142-9_1

reduction of approximately 5% in global energy demand and a 7% decline in CO_2 emissions in 2020, alongside an 18% decrease in energy investments. However, renewable energy sectors—particularly electricity generation—proved relatively resilient during this period, experiencing minimal long-term impacts compared to conventional energy systems (Cozzi et al. 2020). In this context, the World Energy Outlook 2020 outlines a Vision 2050 that emphasizes the urgent need for clean energy–based sustainable development to achieve net-zero emissions. The roadmap highlights the critical role of the power sector in emissions reduction while underscoring the importance of low-carbon fuels, such as hydrogen, in enabling deep decarbonization across multiple sectors (Mir 2025b, c). Hydrogen is widely recognized as a promising energy carrier due to its abundance and environmentally benign characteristics. When utilized as a fuel, hydrogen produces only heat and water as by-products, thereby significantly reducing greenhouse gas emissions (Ruocco et al. 2019). Beyond energy applications, hydrogen is already extensively used in petroleum refining and chemical industries, underscoring its industrial relevance (Abdin et al. 2020). At present, the majority of hydrogen is produced via fossil fuel–based thermochemical pathways, including coal gasification, steam methane reforming, plasma reforming, and hydrocarbon pyrolysis. In contrast, increasing research efforts are being directed toward renewable hydrogen production using natural organic waste materials—such as rice, wheat, corn, and maize residues with or without photocatalysts (Sharma et al. 2020), as well as hydrogen generation through photoelectrochemical water splitting (Singla et al. 2020). These alternative pathways aim to enhance sustainability and assess the commercial viability of green hydrogen technologies. Global demand for high-purity hydrogen continues to rise, particularly for applications in petroleum refining, ammonia synthesis, and other industrial processes. Currently, global hydrogen production stands at approximately 95 million tonnes (MtH_2) per year, of which nearly 76% is derived from natural gas and around 22% from coal. Owing to its high energy density and clean combustion characteristics, hydrogen is increasingly regarded as a future fuel for both the transportation and energy sectors (Sikander et al. 2017). Numerous studies have demonstrated the technological feasibility of hydrogen utilization, as well as the policy frameworks required to support its adoption (Baykara 2018). These studies also examine commercially available hydrogen production technologies, emerging high-efficiency methods, production costs based on energy sources, and the associated requirements for feedstock availability, raw materials, and infrastructure development (Levalley et al. 2014). The successful realization of a hydrogen-based economy depends on the seamless integration of multiple components, including efficient production technologies, long-term storage solutions, safe transportation systems, robust safety standards, and end-use applications such as mobility and power generation. Despite the fact that more than 80% of hydrogen is still produced from fossil fuels—thereby contributing to environmental pollution—significant advancements have been achieved in recent years in renewable hydrogen production technologies (Levalley et al. 2014). Although several technical, economic, and infrastructural challenges continue to hinder large-scale commercialization, sustained research and development efforts in sustainable hydrogen production have demonstrated encouraging progress toward economic feasibility and wider adoption (Abdalla et al. 2018). At present, hydrogen is predominantly utilized in hydro-treatment and hydrocracking processes within refineries, ammonia production for fertilizers, methanol manufacturing, and steel production through the direct reduction route. These established applications, coupled with emerging energy and transportation uses, position hydrogen as a key enabler of future low-carbon energy systems and an essential component of global decarbonization strategies.

1.2 Hydrogen Production from Fossil Resources

Hydrogen is increasingly recognized as a cornerstone energy carrier in the global transition toward a sustainable, low-carbon energy system, owing to its versatility, high energy density, and potential to decarbonize sectors that are difficult to electrify. However, despite its strategic importance, present-day hydrogen production is still overwhelmingly dominated by fossil fuel–based pathways. Nearly 96% of global hydrogen output is derived from carbon-intensive resources, underscoring a significant disconnect between hydrogen's clean-energy potential and its current production reality. Among these sources, natural gas is the primary contributor, accounting for approximately 48% of hydrogen generation, followed by naphtha and heavy oil fractions at around 30%, while coal-based routes contribute close to 18% of total production (Perkins 2018). This heavy reliance on fossil feedstocks results in substantial emissions of carbon dioxide and other greenhouse gases, thereby diminishing the environmental benefits often associated with hydrogen as a clean energy vector. As a consequence, conventional hydrogen production routes such as steam methane reforming, partial oxidation, and coal gasification have been extensively developed and deployed due to their technological maturity, economic viability, and established industrial infrastructure. Nevertheless, these methods are intrinsically linked to high carbon footprints and face growing scrutiny under increasingly stringent climate policies. In this context, understanding the technical foundations, operational characteristics, and inherent limitations of fossil fuel–based hydrogen production technologies is essential for evaluating their role in the current energy landscape and for identifying pathways toward cleaner alternatives. The following sections therefore examine the principal conventional methods used for hydrogen generation from fossil resources, highlighting their process mechanisms, efficiency considerations, and environmental constraints.

1.2.1 Coal Gasification

Gasification is a thermochemical process in which carbonaceous feedstocks are converted into synthesis gas (syngas) through reactions with steam, air, or oxygen under controlled conditions. In the case of coal gasification, coal undergoes partial combustion in the presence of oxygen and steam within a gasifier. This process is typically conducted at elevated pressures of approximately 5 MPa and high temperatures around 1000 °C, which facilitate the partial oxidation and breakdown of the coal structure (Wang et al. 2020). The main products of coal gasification are syngas components, including hydrogen (H_2), carbon monoxide (CO), carbon dioxide (CO_2), and residual steam. Due to the inherently low hydrogen content of coal, a significant proportion of hydrogen generated during gasification originates from the reaction between steam and the carbon-based feedstock rather than from the coal itself. It has been reported that nearly 83% of the hydrogen yield in coal gasification is derived from water, whereas gasification of heavy oil feedstocks results in approximately 69% hydrogen contribution from steam (O'Keefe et al. 2013). The fundamental reaction pathways governing these processes are represented in Eqs. (1.1) and (1.2). Subsequently, the produced syngas is subjected to the water–gas shift (WGS) reaction, a critical downstream step that enhances hydrogen yield by converting carbon monoxide into carbon dioxide while generating additional hydrogen (Siddiqui And Dincer 2019).

$$C + 1/2\,O_2 \rightarrow CO_2, \quad \Delta H'' = -394\,\mathrm{kJ\,mol^{-1}} \tag{1.1}$$

$$C + H_2\;O \rightarrow CO + H_2, \quad \Delta H'' = 131\,\mathrm{kJ\,mol^{-1}} \tag{1.2}$$

Coal is a highly heterogeneous fuel comprising a complex mixture of organic matter and inorganic minerals, and its physicochemical composition varies widely with coal rank and origin. Owing to this inherent variability, the thermodynamic characteristics of coal gasification—particularly the heat of reaction—are not fixed but differ substantially across coal types. Coal rank, which represents the degree of carbonization and overall fuel quality, plays a decisive role in gasification behavior. In general, higher-rank coals such as anthracite and bituminous coal exhibit lower reactivity due to their more ordered carbon structure, whereas lower-rank coals, including lignite and sub-bituminous coal, tend to be more reactive because of their higher volatile content and porous structure (Siddiqui And Dincer 2019). A major drawback of hydrogen production via coal gasification is the significant release of carbon dioxide, stemming from the inherently high carbon content of coal. Without mitigation, this results in a large greenhouse gas (GHG) footprint, undermining the sustainability of the process. Therefore, the deployment of efficient CO_2 capture and storage or utilization technologies is critical to lowering emissions and enhancing the environmental acceptability of coal-based hydrogen production. Carbon capture integration not only addresses climate concerns but is increasingly viewed as a prerequisite for the continued use of coal in a carbon-constrained energy landscape. Coal gasification itself is performed in a dedicated high-temperature reactor known as a gasifier, in which coal undergoes thermochemical conversion to produce synthesis gas (syngas), primarily composed of hydrogen (H_2), carbon monoxide (CO), carbon dioxide (CO_2), methane (CH_4), and minor impurities (Wang et al. 2021). The composition, calorific value, and suitability of the syngas for downstream hydrogen separation are strongly influenced by operating conditions such as temperature, pressure, residence time, and the type of oxidizing medium employed (air, oxygen, steam, or their combinations). The process is initiated by introducing air or oxygen, leading to partial combustion of the coal. This exothermic step supplies the thermal energy required to drive the subsequent endothermic gasification reactions, including char steam and char CO_2 reactions. Based on reactor configuration, flow dynamics, and operating principles, gasifiers are broadly categorized into several types. Fixed bed gasifiers, which include updraft and downdraft designs, are relatively simple and robust but offer limited scalability and syngas purity control. Fluidized bed gasifiers provide better mixing and temperature uniformity, making them suitable for a wider range of coal types. Entrained flow gasifiers operate at very high temperatures and pressures, producing a cleaner syngas with minimal tar formation, while plasma gasifiers employ extremely high temperatures generated by plasma arcs to achieve near-complete feedstock conversion (Pinsky et al. 2020). Each gasifier type presents distinct advantages and limitations in terms of efficiency, feedstock flexibility, capital cost, and environmental performance, thereby influencing its suitability for large-scale hydrogen production.

1.2.2 Steam Methane Reforming (SMR)

Steam methane reforming (SMR) is the most extensively implemented route for large-scale hydrogen production, owing to its economic competitiveness, high efficiency, and advanced state of technological maturity (Dalena et al. 2018). The process primarily relies on methane as the main feedstock; however, a range of other light hydrocarbons including ethane, propane, naphtha, and mixed natural gas streams can also be employed depending on availability and process requirements (Pen et al. 1996). Among these alternatives, natural gas remains the most economically favorable option and offers a comparatively higher H_2/CO ratio than heavier hydrocarbons or coal-derived feedstocks, making it particularly suitable for downstream hydrogen separation and purification. Fossil fuel–based reforming technologies have reached a high level of industrial maturity and are widely deployed across refineries, ammonia plants, and petrochemical complexes, collectively meeting the majority of global hydrogen demand. In recent years, steam reforming (SR) has garnered renewed attention as a practical and reliable pathway for converting primary hydrocarbon resources into hydrogen-rich gas streams, especially for applications in fuel cells and other clean energy technologies (Malik et al. 2023). Typical feedstocks for SR include natural gas, liquefied petroleum gas (LPG), and naphtha, reflecting the flexibility of the process with respect to hydrocarbon inputs. At its core, the SMR process involves the highly endothermic reaction between methane and steam at elevated temperatures, typically in the presence of a nickel-based catalyst. This reaction produces a synthesis gas rich in hydrogen and carbon monoxide, which subsequently undergoes further processing most notably the water gas shift reaction to enhance hydrogen yield and reduce carbon monoxide content. Additional purification steps, such as pressure swing adsorption, are then applied to obtain high-purity hydrogen. The fundamental reactions governing methane steam reforming and subsequent hydrogen enrichment are represented in Eqs. (1.3) and (1.4) (Malik et al. 2023).

$$CH + nH_2\ O \rightarrow nCO + (m/2 + n)H_2,\quad \Delta H^\circ = +(131 - 206.2)\,kJ\,mol^{-1} \quad (1.3)$$

$$CO + H_2\ O \rightarrow CO_2 + H_2,\quad \Delta H^{''} = -41.2\,kJ\,mol^{-1} \quad (1.4)$$

In this framework, *n* and *m* represent positive stoichiometric coefficients that indicate the number of molecules participating in the respective reactions. The steam methane reforming process is initiated by thoroughly mixing methane with steam, after which the reactant mixture is introduced into a reformer containing an appropriate catalyst. The reaction is conducted under elevated pressures, typically in the range of 1.5–3 MPa, and at high operating temperatures between 700 and 900 °C to promote efficient conversion (Van Hook 1980). During this primary reforming stage, methane reacts with steam to generate synthesis gas (syngas), consisting predominantly of hydrogen (H_2) and carbon monoxide (CO), as expressed in Eq. (1.3). Steam methane reforming is inherently and strongly endothermic, requiring a continuous and significant supply of external heat to sustain the reaction. The enthalpy change (ΔH) associated with the reforming reaction generally falls within the range of 131–206.2 kJ mol^{-1}. In industrial practice, this thermal energy demand is most commonly met through the combustion of hydrocarbons in reformer furnaces, which contributes to the overall energy intensity of the process (Soo And Gibbs 1979). Following the reforming step, the product gas mixture undergoes downstream processing to recover hydrogen, with pressure swing adsorption (PSA) being the most widely adopted separation technique due to its reliability and ability to deliver high-purity hydrogen streams. Despite its widespread industrial use and technical robustness, SMR is associated with substantial carbon dioxide emissions. On average, approximately 0.3–0.4 m^3 of CO_2 is emitted for every cubic meter of hydrogen produced, highlighting the environmental limitations of the process. To address these concerns, alternative reactor concepts have been intensively explored over the past two decades, including membrane reactors and fluidized-bed membrane reactors. Among these advanced configurations, membrane reactors have attracted particular interest because they enable in situ hydrogen separation, thereby shifting reaction equilibria, increasing hydrogen yield, and potentially improving overall process efficiency. However, challenges related to reactor scale-up, membrane stability under harsh reforming conditions, and cost-effective manufacturing remain unresolved. Consequently, continued research and development are essential to optimize reactor designs, enhance membrane durability, and facilitate the economically viable deployment of these technologies at an industrial scale (Chen et al. 2014).

1.2.3 Autothermal Reforming (ATR)

ATR integrates the fundamental operating principles of partial oxidation (POX) and steam methane reforming (SMR) to capitalize on the advantages of both processes (Abatzoglou And Fauteux-Lefebvre 2016). This hybrid approach enables the production of hydrogen-rich reformate while offering a compact reactor design and faster dynamic response compared with conventional reforming systems. In the ATR process, methane reacts with a controlled supply of pure oxygen and steam within the reformer, allowing reforming and oxidation reactions to occur simultaneously and resulting in a thermodynamically balanced process (Abatzoglou And Fauteux-Lefebvre 2016). ATR systems typically operate at temperatures around 900 °C, with carefully optimized carbon-to-steam (C/S) and oxygen-to-carbon (O_2/C) ratios to ensure efficient hydrogen production while maintaining thermal stability. Under these conditions, the heat generated from exothermic oxidation reactions is internally utilized to drive the endothermic reforming reactions, thereby reducing the need for external heat input. The overall chemical reaction governing the ATR process is presented in Eq. (1.5) (Liu et al. 2004).

$$CH + n/4O_2 + n/2H_2O \rightarrow nCO + (m/2 + n/2)H_2, \quad \Delta H^{''} \approx 0\,kJ\,mol^{-1} \qquad (1.5)$$

In partial oxidation (POX), the reforming reaction is exothermic in nature, as reflected by a negative standard enthalpy change ($\Delta H° < 0$ kJ mol^{-1}), whereas steam reforming (SR) is intrinsically endothermic and proceeds with a positive enthalpy change ($\Delta H° > 0$ kJ mol^{-1}). When these two reaction pathways are combined under thermally balanced conditions, the overall process can be operated in a near-enthalpy-neutral regime, where the net heat of reaction approaches zero ($\Delta H° \approx 0$ kJ mol^{-1}). Autothermal reforming (ATR) is based on this principle of thermal integration and, similar to POX and SR, requires little to no external energy input. In ATR, the heat released from the exothermic partial oxidation of the hydrocarbon feed is internally utilized to drive the endothermic steam reforming reactions occurring simultaneously within the same reactor. This intrinsic thermal coupling results in an overall near-zero enthalpy change ($\Delta H \approx 0$ kJ mol^{-1}), enabling stable operation without the need for externally heated furnaces. The effectiveness of ATR is strongly dependent on catalyst selection, which governs reaction rates, selectivity, and long-term stability. Nickel-based catalysts are most commonly employed due to their high catalytic activity toward hydrocarbon reforming, mechanical and thermal robustness, and relatively low cost compared with noble metal alternatives. From a performance perspective, ATR typically achieves higher hydrogen yields than POX alone, as the inclusion of steam reforming enhances hydrogen generation. However, under comparable operating conditions, ATR generally produces lower hydrogen yields than conventional steam reforming, which remains the most hydrogen-efficient reforming route (Horng et al. 2006).

1.2.4 Partial Oxidation (POX)

Partial oxidation (POX) has emerged as a practical alternative to steam reforming (SR) for hydrogen production, particularly in applications where compact system design and rapid start-up are desirable (Specchia et al. 2010). One of the key advantages of POX is its broad feedstock flexibility, as hydrogen can be produced from a wide range of carbonaceous materials, including methane, heavy fuel oils, and even coal. Because the oxidation reactions involved are strongly exothermic, POX requires little to no external heat input, which enhances its suitability for decentralized, mobile, or on-demand hydrogen generation systems (Corbo And Migliardini 2007). Prior to the reforming stage, sulfur-containing compounds are removed from the feedstock to avoid catalyst poisoning and loss of activity. Subsequently, pure oxygen is introduced to partially oxidize the hydrocarbon feed, generating a synthesis gas composed mainly of hydrogen and carbon monoxide. The resulting syngas is then processed in a manner similar to that used in SMR-based systems, typically involving downstream water gas shift (WGS) reactors to increase hydrogen content, followed by pressure swing adsorption (PSA) units to achieve high hydrogen purity. In catalytic POX processes, reforming reactions are initiated at temperatures of approximately 950 °C, with light hydrocarbons such as methane and naphtha being the most commonly used feedstocks. Non-catalytic POX, by contrast, operates at significantly higher temperatures, generally in the range of 1150–1315 °C. Although this increases energy intensity, it offers enhanced fuel flexibility, enabling the conversion of heavier hydrocarbons, including heavy oils and coal. With respect to catalyst development, cobalt-based materials have been investigated for the catalytic POX of methane. Studies indicate that while the addition of cobalt to nickel-based catalysts can effectively mitigate carbon deposition, it often results in a measurable decrease in overall catalytic activity, highlighting the trade-offs involved in catalyst design for POX systems (Freni et al. 2000).

1.2.5 Pyrolysis

Hydrocarbon pyrolysis represents an additional and increasingly attractive route for hydrogen production, distinguished by the absence of oxidizing agents and the potential to avoid direct carbon dioxide formation. Depending on key operating parameters, including heating rate, reaction temperature, feedstock particle size, and vapor residence time, pyrolysis processes are generally categorized as slow, fast, or flash pyrolysis. Slow pyrolysis is typically performed at relatively low temperatures, in the range of 300–700 °C, using very low heating rates of less than 1 °C s^{-1} and long residence times exceeding 5 min. These conditions favor extensive solid carbon formation and gradual thermal decomposition of the feedstock. Fast pyrolysis, by comparison, operates at elevated temperatures between 450 and 800 °C, with moderate heating rates on the order of 10–200 °C s^{-1} and short vapor residence times of approximately 0.5–10 s, enabling rapid conversion of hydrocarbons. Flash pyrolysis represents the most severe regime, characterized by very high temperatures of 800–1200 °C, extremely rapid heating rates exceeding 1000 °C s^{-1}, and ultra-short residence times of less than 2 s, resulting in near-instantaneous thermal cracking reactions. The thermal decomposition pathways and reaction mechanisms associated with the pyrolysis of a wide variety of hydrocarbons have been studied extensively and are comprehensively documented by Moldoveanu (2019). From a fundamental perspective, when methane is heated to around 500 °C in the absence of steam, oxygen, or other oxidizing species, it undergoes non-oxidative thermal decomposition. This process leads to the direct formation of molecular hydrogen along with solid carbon (commonly referred to as coke), without the generation of carbon oxides. The overall reaction describing methane pyrolysis is presented in Eq. (1.6) below.

$$CH_4 \rightarrow \mathrm{C(s)} + 2\mathrm{H_2} \quad (\approx 500\,^\circ C, no\ \mathrm{O_2}/\mathrm{H_2O}) \tag{1.6}$$

This hydrogen production route is strictly catalytic in nature and is characterized by the formation of carbon nanofibers and multiwalled carbon nanotubes as solid by-products (Abbas And Daud 2010). The process is most active at elevated temperatures ranging from 700 to 1000 °C (Moliner And Palacios 2010). A wide variety of catalysts have been investigated, including supported nickel, iron, cobalt, and molybdenum catalysts, as well as catalysts based on platinum-group metals. The incorporation of transition-metal promoters has been shown to facilitate the production of CO_x-free hydrogen, thereby enhancing the environmental performance of the process (Shah et al. 2022). Notably, Ni–Co–Cu–Zn catalysts have demonstrated exceptional catalytic activity, achieving methane conversions as high as 85% at 700 °C. The addition of copper significantly enhances catalyst performance, likely by improving metal dispersion and suppressing carbon deactivation (Awadallah et al. 2021). Catalyst performance is also strongly influenced by both the nature of the active phase and the morphology of the support material. For instance, increased porosity and optimized pore structure promote higher methane conversion by improving reactant diffusion and active-site accessibility. Furthermore, the catalyst preparation method plays a critical role, as variations in synthesis routes can lead to significant differences in activity and stability. In addition to metal-based catalysts, carbon-based materials constitute another effective class of catalysts for methane pyrolysis (Liu et al. 2018). Materials such as graphite, carbon black, carbon nanotubes, carbon nanofibers, and activated carbon either in their pristine form or doped with metal species are relatively inexpensive and exhibit long-term thermal stability. More advanced structured catalytic materials, including perovskites, composite systems, and zeolites, have also been explored for methane decomposition, offering advantages such as enhanced catalytic durability and tunable surface properties (Alalga et al. 2021). To overcome challenges related to catalyst deactivation caused by carbon deposition, methane pyrolysis in molten metals or molten salts has been developed. In this approach, methane cracking occurs within a molten medium, effectively preventing carbon accumulation on solid catalyst surfaces. Molten-media pyrolysis can be realized using a variety of reactor designs and offers several notable advantages over conventional solid wall reactors. The presence of a molten medium enhances heat transfer, ensures more uniform temperature distribution, and effectively prevents the adhesion and buildup of solid carbon on reactor surfaces, thereby mitigating fouling and extending reactor lifetime. These features make molten-media systems particularly attractive for sustained and stable pyrolysis operation. In parallel, the pyrolysis of crude oil and its different fractions have progressed to industrial-scale deployment, reflecting the technological maturity of this approach. Oil pyrolysis is commonly carried out in the presence of supported metal catalysts most notably nickel, molybdenum, or rhodium at moderate to high temperatures, typically in the range of 420–600 °C. Under these conditions, large hydrocarbon molecules undergo thermal and catalytic cracking reactions. As described by Eqs. (1.7) and (1.8) below, the process not only upgrades the feedstock by lowering its average molecular weight and improving product quality, but can also drive extensive, and in some cases complete, decomposition of the carbon backbone. This results in the liberation of molecular hydrogen as a valuable product, alongside lighter hydrocarbons or solid carbon, depending on operating conditions and catalyst selection (Patzschke et al. 2021).

$$CH \rightarrow H_2 + \text{olefins} - Q_1 \tag{1.7}$$

$$CH \rightarrow (m/2)H_2 + nC - Q_2 \tag{1.8}$$

In this framework, Q_1 and Q_2 denote the thermal effects associated with the respective reactions and are determined based on the chemical composition of the hydrocarbon feedstock undergoing conversion. Equation (1.9) presents a simplified expression describing the complete transformation of hydrocarbon molecules, including olefinic species, into gaseous products. A diverse range of feedstocks can be processed through pyrolysis for syngas generation, encompassing heavy oil fractions, coal, wastewater-derived organic matter, biomass, and cellulose-rich materials. This broad feedstock flexibility highlights the adaptability of pyrolysis-based routes for hydrogen and syngas production. Depending on the desired product distribution and process objectives, pyrolysis can be conducted either catalytically or non-catalytically. Catalytic pyrolysis commonly employs materials such as zeolites, supported metal catalysts most frequently cobalt, zinc, gallium, or nickel—and char-derived catalysts, which can significantly influence reaction pathways, enhance gas yields, and tailor syngas composition. Beyond syngas and hydrogen, pyrolysis processes are capable of generating a range of value-added by-products, including liquid fuels, solid char, and advanced carbon materials such as carbon nanostructures. The co-production of these materials improves overall resource efficiency and enhances the economic and technological versatility of the process (Devi et al. 2020). In recent years, microwave-assisted heating has gained attention as an alternative energy input for hydrocarbon pyrolysis. This approach offers several advantages, including rapid and volumetric heating, improved energy efficiency, accelerated reaction kinetics, and relative ease of integration into existing systems.

For example, microwave-assisted pyrolysis of waste automotive engine oil has been reported to yield syngas contents of up to 35 vol%, with hydrogen concentrations reaching approximately 19 vol%, with maximum gaseous product formation observed at temperatures around 700 °C. Similarly, continuous microwave-assisted pyrolysis of sewage sludge has demonstrated biogas yields of approximately 51.7 wt%, with syngas concentrations as high as 83.6 vol% (Lin et al. 2022).

1.3 Hydrogen Production from Renewable Resources

The majority of hydrogen produced today is derived from fossil fuels, and given the significant environmental impacts associated with these methods, such production pathways are inherently unsustainable. In addition, the rapid depletion of fossil fuel reserves further underscores the need for alternative approaches (Ji and Wang 2021). Renewable energy sources currently contribute only a small fraction of global hydrogen production. Consequently, recent research has increasingly focused on developing environmentally friendly and low-emission hydrogen production technologies utilizing renewable resources (Kim et al. 2018). The ongoing advancement of renewable energy technologies, including biomass, solar, and wind energy, has facilitated the development of more efficient and sustainable hydrogen production methods (Acar And Dincer 2019; Mir 2025b). In this section, the primary methods for hydrogen generation from renewable sources, such as biomass conversion, solar-driven processes, wind energy integration, and water electrolysis, will be examined. Additionally, emerging approaches for renewable hydrogen production, including nuclear-assisted methods and algae-based systems, will also be discussed, highlighting their potential to contribute to a sustainable hydrogen economy. Figure 1.1 depicts the regional proportions of total renewable energy capacity devoted to hydrogen production under the main scenario during 2021–2027, providing insight into the evolving role of hydrogen within regional renewable energy portfolios (IEA 2022).

1.3.1 Biomass to Hydrogen

The global increase in anthropogenic CO_2 emissions presents a critical environmental challenge. To mitigate this, it is essential to expand the deployment of renewable energy sources, including alternative fuels and biomass, within current power generation systems. Such measures can reduce dependence on fossil fuels and minimize the release of CO_2 and other harmful emissions. Beyond electricity generation, renewable resources can also serve as feedstocks for producing value-added energy carriers, with hydrogen emerging as a particularly promising option. Hydrogen is not only widely applicable in industrial and power sectors but also has growing utility in transportation and as a fuel for mobility applications (Lepage et al. 2021). Currently, over 85% of global hydrogen is produced from fossil fuel sources, with

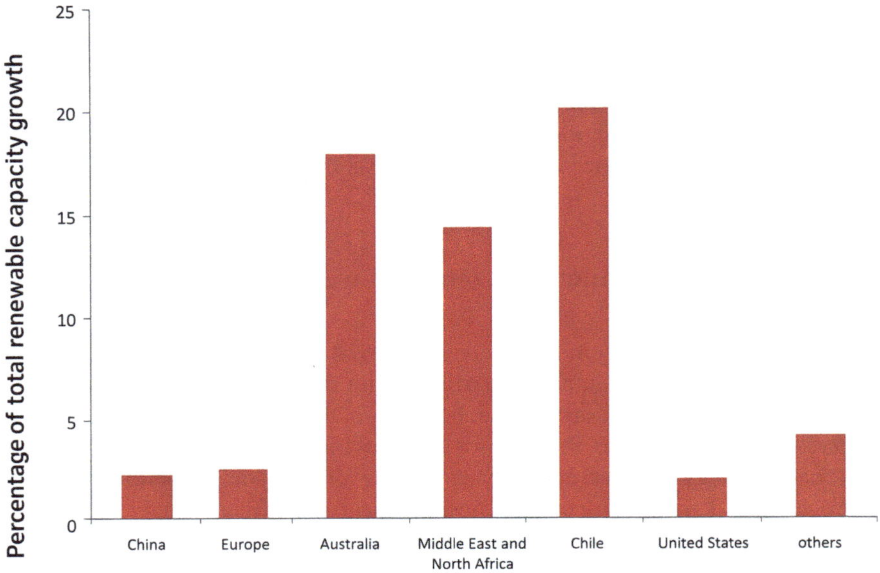

Fig. 1.1 Regional share of renewable energy capacity dedicated to hydrogen production in the main scenario, 2021–2027

steam methane reforming (SMR) representing the most cost-effective method for industrial-scale production. In contrast, less than 5% of hydrogen is derived from renewable sources, primarily via water electrolysis, and no industrial-scale hydrogen production from algal biomass has been realized to date (Rodriguez Correa And Kruse 2018). Among renewable options, biomass was the earliest energy source used by humans and continues to serve as a viable alternative to fossil fuels. Biomass encompasses a wide array of feedstocks, including agricultural residues, wood, plant matter, and animal wastes. Research has largely focused on lignocellulosic biomass, such as wood, agricultural residues, and energy crops, using thermochemical conversion techniques like gasification and pyrolysis (Balat And Kirtay 2010). Hydrogen production from biomass can be achieved through several pathways. Biological processes, such as anaerobic digestion, generate biogas through microbial conversion of organic matter. Fermentation processes similarly produce hydrogen along with alcohols, acids, and other gases. Thermochemical approaches, particularly biomass gasification, are analogous to coal gasification and convert feedstocks into a mixture of H_2, CO, CO_2, and methane (Sun et al. 2019). Among these methods, thermochemical conversion is the most mature and holds the greatest potential for large-scale hydrogen production. This category encompasses gasification, liquefaction, and pyrolysis, with gasification currently representing the most economical route. Presently, the efficiency of biomass gasification for hydrogen production is approximately 50%, with scope for improvement (Peng et al. 2017). Biomass steam gasification is an endothermic process conducted at high temperatures between 700 and 1200 °C. The primary products, including H_2 and

CO_2, are further processed through pressure swing adsorption (PSA) units to yield high-purity hydrogen. A variety of feedstocks can be utilized, such as sugarcane, sorghum, algal biomass, oil palm residues, and rice husk. The gasification mixture typically contains hydrogen, carbon monoxide, carbon dioxide, steam, tar, and char (Zhang et al. 2019; Mir 2025b). Although biomass gasification generally produces lower environmental impacts compared with fossil fuel–based hydrogen production, challenges remain, including the complex composition of biomass, tar formation that can cause equipment clogging and catalyst deactivation, and potential process inefficiencies (Singh et al. 2022). To address these limitations, high-temperature plasma-assisted gasification units have been proposed, where temperatures exceeding 2000 °C enable complete tar decomposition. Supercritical water gasification (SWG) represents another promising approach, utilizing water above its critical temperature and pressure as both the reaction medium and the reactant (Saleem et al. 2020). Under supercritical conditions, biomass undergoes complex reactions, with large polymeric molecules such as lignin and cellulose breaking down into smaller intermediates, including alcohols and sugars. These intermediates are subsequently catalytically converted into synthesis gas, with hydrogen recovered through the water–gas shift (WGS) reaction. The use of alkali salt catalysts enhances the WGS reaction and promotes the formation of various hydrogen-rich intermediates. The overall reaction for hydrogen production via SWG is given in Eq. (1.9) (Nanda et al. 2017).

$$CH_xO_\gamma + (2 - y)H_2O \rightarrow CO_2 + (2 - y + 0.5x)H_2 \tag{1.9}$$

During supercritical water gasification (SWG), the formation of CO_2 can be effectively managed by controlling the reaction temperature and pressure, allowing selective separation of carbon dioxide from the product stream. Currently, both steam gasification and solar-assisted gasification technologies have demonstrated feasibility for enhancing hydrogen yields, achieving efficiencies of up to approximately 50%. Biological hydrogen production through anaerobic digestion is also widely employed; however, its application is limited to specific organic feedstocks, such as sewage sludge, agricultural residues, and household waste. Fermentation-based methods, in contrast, are capable of processing nonedible, cellulosic portions of certain plants, expanding the range of usable biomass.

1.3.2 Solar to Hydrogen

Among renewable energy sources, solar energy is considered to have the highest potential to meet global energy demands in the future. In recent years, fluctuations in oil and natural gas prices have coincided with a significant reduction in the cost of renewable energy technologies (Kabir et al. 2018). The ability of solar energy to generate low-cost electricity has enabled the development of solar-driven hydrogen production, offering a clean and sustainable alternative fuel. While solar energy storage addresses intermittency challenges, direct conversion of solar energy into hydrogen is considered a more reliable and efficient approach (Zhang et al. 2022). Solar electricity generation can be broadly categorized into photovoltaic (PV) and solar thermal technologies. PV systems convert solar radiation directly into electricity via the photoelectric effect, while solar thermal systems use arrays of mirrors to concentrate sunlight and produce heat, which can then drive steam turbines for electricity generation (Zhang et al. 2022). Hydrogen can be produced from solar-generated electricity through water electrolysis, with the main commercial technologies being alkaline water electrolysis (AWE), proton exchange membrane electrolysis (PEM), and solid oxide electrolysis cells (SOECs). Among these, AEL and PEM are currently the most attractive, particularly PEM systems, which can be directly coupled with variable renewable energy sources. In a PEM electrolyzer, pure water is split into hydrogen and oxygen by applying a voltage across an anode and a cathode separated into two chambers, with hydrogen generated at the cathode and oxygen at the anode (Carmo et al. 2013). Thermochemical water-splitting processes are also being explored, with the UT-3 and sulfur–iodine (S–I) cycles representing the most efficient methods. In the S–I cycle, water splitting occurs through three closed-loop reactions, generating sulfuric acid and hydroiodic acid via the Bunsen reaction. The acids are subsequently purified, concentrated, and thermally decomposed, yielding oxygen from sulfuric acid and hydrogen from hydroiodic acid (Zhang et al. 2022). Other approaches, such as solar photonic–based hydrogen production, remain in the development stage. The efficiency of solar-to-hydrogen conversion is a critical performance parameter. Using conventional PV systems coupled with water electrolysis, efficiencies of up to 12% are achievable, whereas concentrated solar photovoltaic (CSPV) systems can reach efficiencies of approximately 25% (Acar et al. 2019). Concentrated Solar Power (CSP) technologies, employing parabolic dishes or troughs, offer the potential to maximize hydrogen production at relatively low cost. Early solar-to-hydrogen systems typically involved PV-generated electricity driving low-temperature water electrolysis. Although these first-generation PV-electrolysis systems suffer from limited efficiency, they still offer substantial environmental benefits (Wang et al. 2019). Recent research focuses on replacing first-generation PV cells with third-generation multi-junction solar cells (MJSCs) to harvest more electricity, often augmented by reflective mirrors to increase solar irradiation and process efficiency. Two-step water electrolysis is another commercially applied method, yielding clean hydrogen free from carbon emissions (Hosseini And Wahid 2020). While direct solar thermolysis of water requires extremely high temperatures (>2900 K), limiting material options, thermochemical solar-to-hydrogen processes can operate at moderate temperatures through redox cycles involving metals and metal oxides in solar reactors. Overall, solar-based hydrogen production, including electrolysis (AEL, PEM) and thermochemical methods, represents a clean, decarbonized pathway for hydrogen generation (Shiva Kumar And Himabindu 2019). Table 1.1 presents the regional distribution of electrolyzer capacity for hydrogen production by commissioning year over the period 2010–2020. Table 1.1 highlights temporal and geographical trends in the

Table 1.1 Regional distribution of electrolyzer capacity for hydrogen production by commissioning year, 2010–2020

Region	2010	2011	2012	2013	2014	2015	2016	2017	2018	2019	2020
China	–	–	–	–	–	–	–	–	–	4 MWe	46 MWe
North America	–	–	–	–	–	–	–	1 MWe	2 MWe	3 MWe	21 MWe
Europe	–	6 MWe	1 MWe	14 MWe	3 MWe	11 MWe	4 MWe	6 MWe	9 MWe	16 MWe	55 MWe
Japan	–	–	–	–	–	–	–	–	–	–	10 MWe
Rest of world	1 MWe	12 MWe	6 MWe	2 MWe	6 MWe	2 MWe	2 MWe	3 MWe	3 MWe	2 MWe	11 MWe

Mwe Megawatt electrical

deployment of electrolyzer technologies, reflecting the pace and scale of hydrogen infrastructure development across different regions.

1.3.3 Wind to Hydrogen

The declining costs and growing societal demand for renewable energy are driving the expansion of installed capacities worldwide. With rapid technological advancements, many countries are actively promoting the use of renewable energy to achieve sustainable and low-carbon electricity generation. In particular, wind energy has received considerable attention as a reliable and clean source of electricity, and it is increasingly considered a key component of hybrid power systems, where multiple generation sources work together to ensure stable energy supply (Almutairi et al. 2021). Despite its advantages, wind energy is inherently variable and can pose challenges to power producers and investors due to fluctuations in generation. To prevent curtailment of excess electricity, energy storage solutions are often integrated with wind power systems. Common approaches include electrochemical storage (e.g., lithium-ion batteries), energy conversion to other carriers such as gas, and energy conversion to hydrogen (He et al. 2016; Xiao et al. 2020). Although significant research efforts continue to focus on enhancing battery storage capacity, hydrogen production using wind energy has gained increasing attention as a highly attractive alternative due to its flexibility and wide range of end-use applications, including fuel cell vehicles, power generation, and industrial energy supply. Among the available hydrogen production technologies, water electrolysis is a well-established and technologically mature approach, with reported conversion efficiencies exceeding 75% under optimized operating conditions (An et al. 2014). In this pathway, renewable electricity generated by wind turbines is utilized to split water molecules into hydrogen and oxygen within an electrolyzer, enabling the production of green hydrogen without direct carbon emissions. Hybrid renewable energy systems that integrate wind and photovoltaic (PV) sources have been extensively investigated to enhance system reliability and energy availability. Studies have shown that, within such hybrid configurations, wind power often makes the most significant contribution to overall hydrogen production, owing to its higher capacity factors and greater energy output compared with solar PV in many regions (Al-Sharafi et al. 2017). Since hydrogen cannot be directly harvested from renewable resources, these integrated systems play a critical role in ensuring continuous and efficient hydrogen generation while simultaneously reducing greenhouse gas emissions associated with conventional fossil fuel–based hydrogen production. A typical wind-to-hydrogen configuration comprises a wind farm, an AC/DC converter (rectifier), a power management and control unit, and an alkaline electrolyzer (Li et al. 2019). Electricity generated by the wind turbines is first conditioned and converted before being supplied to the electrolyzer, where water is electrochemically split at the anode and cathode to produce oxygen and hydrogen, respectively. The overall hydrogen output is strongly influenced by the energy consumption of the electrolyzer typically in the range of 5–6 kWh per Nm^3 of hydrogen as well as by the efficiency of the rectifier, which generally lies between 85% and 95%. System performance can be further enhanced through optimization of wind turbine operation, which depends on factors such as wind speed distribution, turbine size and design characteristics, and the geographical location of the installation (Mostafaeipour et al. 2019). Among the various renewable hydrogen production pathways, wind-based hydrogen is widely regarded as one of the lowest in terms of greenhouse gas emissions across the lifecycle. Techno-economic assessments support its growing viability; for example, in Algeria, the cost of hydrogen production from wind energy has been estimated at approximately USD 1.214 per kg of H_2, with a payback period of about 3.1 years and an annual profit of USD 2224 reported for the Manjil wind plant (Rezaei et al. 2018). Despite these advantages, several challenges remain, notably the inherent intermittency of wind resources, as well as the technical and economic constraints associated with large-scale hydrogen storage and transportation. Addressing these issues will be critical for the widespread adoption of wind-to-hydrogen systems.

1.4 Waste-to-Hydrogen Technologies

To achieve waste-free urban environments, it is essential to focus on waste-to-energy solutions. This approach involves not only modernizing municipal solid waste (MSW) management systems but also implementing incentive policies for industries capable of reducing greenhouse gas (GHG) emissions. In many developing countries, households contribute the majority of MSW (55–80%), whereas the commercial sector accounts for a smaller share (10–30%). Nonresidential MSW is highly heterogeneous, including plastics, paper, wood, leather, fabrics, food waste, yard waste, demolition debris, and other materials (Hoang et al. 2022). Global MSW generation is estimated at approximately two billion tons annually and is projected to reach 3.4 billion tons by 2050, outpacing population growth. In low- and middle-income countries, more than 50% of MSW consists of food and green waste. Currently, around 37% of MSW worldwide is disposed of via landfills (Das et al. 2019). Improper disposal, such as uncontrolled landfilling or open burning, leads to both short- and long-term environmental pollution. Hydrogen production from MSW gasification has emerged as a promising solution to address waste management challenges while reducing reliance on fossil fuels (Mondal 2022). Municipal solid waste can serve as a feedstock for hydrogen generation through anaerobic digestion or thermochemical processes such as steam reforming (Mir 2025b). Gasification, in particular, converts MSW into energy-rich syngas, comprising hydrogen, carbon monoxide, methane, and other hydrocarbons. Compared to incineration, gasification is preferred due to its adaptability, higher efficiency, and lower environmental impact, especially for smaller municipalities (Zhou et al. 2024). In a typical gasification process, MSW is fed from the top of the gasifier, while the oxidizing agent enters from the bottom. Syngas is collected at the top, and residue or slag is discharged from the bottom (Chanthakett et al. 2021). Among gasification pathways, indirect gasification using steam as the gasifying agent has demonstrated the highest hydrogen yields, whereas direct gasification shows lower performance; however, steam injection during direct gasification can enhance hydrogen production. Gasification parameters, including temperature, pressure, and feed composition, can be adjusted to optimize syngas composition. The hydrogen content in syngas strongly depends on MSW composition. Higher plastic content increases H_2 concentrations, while biomass-rich waste contributes more to CO and CO_2 production (Erdem et al. 2024). The gaseous products from MSW gasification—H_2, CO, CH_4, and C_2–C_4 hydrocarbons—are valuable due to their high calorific value and industrial applicability. Volatile compounds generated also include NO_x, SO_x, tars, and C_1–C_4 aromatics, along with particulate and inorganic residues (Erdem et al. 2024). Several studies have explored MSW gasification for hydrogen production. Behzadi et al. (2018) reported exergy efficiencies of 19.02% and 17.98% for digestion and gasification processes, with respective product costs of $27.68/GJ and $28.31/GJ. Ozturk and Dincer (2021) demonstrated hydrogen production from MSW at 3.014 kg/s using a gasifier, cyclone, catalytic reactor, and pressure swing adsorption, yielding H_2 at 18.5 vol%. Qi et al. (2023) combined carbon capture with chemical-looping gasification, achieving a hydrogen selectivity of 51.44% at 750 °C, which increased to 56.29% when the solid waste ratio was adjusted from 0.1 to 2.1. Irfan et al. (2019) utilized waste marble powder (primarily $CaCO_3$) as a catalyst and CO_2 sorbent, resulting in enhanced hydrogen production. Hu et al. (2015) found that at 750 °C, with a CaO-to-carbon ratio of 0.7 and MSW moisture content of 40%, the hydrogen mole fraction peaked at 49.4% in a fixed-bed reactor. He et al. (2009) reported that steam-assisted MSW gasification with calcined dolomite as a sorbent produced hydrogen at 42.98 mol H_2/kg MSW and 53.22% mole fraction. Zheng et al. (2018) observed that CO_2 recycling in a drop tube reactor increased H_2 production to 18.82 mol/kg MSW, with CO_2/steam ratios between 0.5 and 2.5 improving hydrogen yields. In gasification processes, organic fractions of MSW can undergo steam reforming similar to industrial methane reforming, followed by the water–gas shift reaction, converting CO and H_2O into additional H_2 and CO_2. Key operational parameters, including temperature, steam flow rate, and sample-to-catalyst ratios, significantly influence hydrogen yield. Due to the endothermic nature of steam reforming, higher temperatures favor H_2 production, while increased steam flow enhances hydrogen generation during catalytic reforming (Nugroho et al. 2025; Rahman et al. 2025). Although the variability in MSW composition presents challenges, it also offers opportunities for optimizing hydrogen production.

1.5 Conclusion

Hydrogen has emerged as a highly promising low-emission energy carrier, particularly when produced from renewable or so-called "green" sources. This chapter has provided a comprehensive assessment of hydrogen production pathways derived from both conventional fossil fuels and renewable resources, while also critically examining the technical, economic, and environmental barriers that currently limit large-scale deployment. Although hydrogen production from fossil feedstocks such as methane and coal is commercially mature and widely implemented, the substantial environmental impacts associated with these routes have increasingly redirected research and industrial focus toward the development of green hydrogen from renewable feedstocks, alongside efficient storage and transportation solutions. Despite the clear benefits of renewable hydrogen, including carbon-free production and the potential for integration into sustainable energy supply networks, its widespread commercialization remains constrained by several challenges. These include difficulties associated with hydrogen storage and transport, the absence of a clearly established business model and value chain, high production costs, investment and market risks, the lack of harmonized international standards, and safety concerns arising from hydrogen's high flammability. Addressing these issues is essential for enabling the transition from pilot-scale demonstrations to fully commercial hydrogen systems. Enhancing process efficiency is expected to remain a central research priority in the advancement of clean hydrogen technologies. Comparative studies indicate that biomass-based hydrogen production can deliver hydrogen yields comparable to those of water electrolysis, while potentially offering lower operating costs and improved energy efficiency. In contrast to fossil fuel–based pathways, sustainable hydrogen production from renewable resources is fundamental to the realization of a future hydrogen economy. However, achieving this vision requires coordinated progress in scientific understanding, technological innovation, social acceptance, and economic feasibility. Further improvements in renewable hydrogen production efficiency may be achieved through the application of advanced process intensification techniques, such as ultrasonic irradiation or pulsating electric fields. In the context of water electrolysis, the economic viability of large-scale hydrogen production could be enhanced by the development and deployment of low-cost polymeric membrane materials, including polyether ketones, polyimides, and polyethylene-based membranes. Future research should also focus on improving stakeholder awareness and technical understanding of biohydrogen production pathways to support scale-up and commercialization. In particular, biomass-based hydrogen generation methods—such as supercritical water gasification—continue to face significant challenges related to tar formation, high energy demands, and reactor corrosion. Addressing these issues will require detailed investigations into reaction mechanisms, thermodynamic behavior, and optimal operating conditions to enable the development of efficient, sustainable, and commercially viable hydrogen production technologies.

References

Abatzoglou N, Fauteux-Lefebvre C (2016) Review of catalytic syngas production through steam or dry reforming and partial oxidation of studied liquid compounds. Wiley Interdiscip Rev Energy Environ 5(2):169–187

Abbas HF, Daud WMAW (2010) Hydrogen production by thermocatalytic decomposition of methane using a fixed bed activated carbon in a pilot scale unit: apparent kinetic, deactivation and diffusional limitation studies. Int J Hydrogen Energy 35:12268–12276. https://doi.org/10.1016/j.ijhydene.2010.08.036

Abdalla AM, Hossain S, Nisfindy OB, Azad AT, Dawood M, Azad AK (2018) Hydrogen production, storage, transportation and key challenges with applications: a review. Energy Convers Manage 165:602e27. https://doi.org/10.1016/j.enconman.2018.03.088

Abdin Z, Zafaranloo A, Rafiee A, Mérida W, Lipiński W, Khalilpour KR (2020) Hydrogen as an energy vector. Renew Sust Energ Rev 120:109620

Acar C, Dincer I (2019) Review and evaluation of hydrogen production options for better environment. J Clean Prod 218:835e49. https://doi.org/10.1016/j.jclepro.2019.02.046

Acar C, Bicer Y, Demir ME, Dincer I (2019) Transition to a new era with light-based hydrogen production for a carbon-free society: an overview. Int J Hydrogen Energy 44:25347e64. https://doi.org/10.1016/j.ijhydene.2019.08.010

Alalga L, Benamar A, Trari M (2021) Hydrogen production via methane decomposition over nickel supported on synthesized ZSM-5/MCM-41 zeolite composite material. Int J Hydrog Energy 46:28501–28512. https://doi.org/10.1016/j.ijhydene.2021.06

Almutairi K, Hosseini Dehshiri SS, Hosseini Dehshiri SJ, Mostafaeipour A, Jahangiri M, Techato K (2021) Technical, economic, carbon footprint assessment, and prioritizing stations for hydrogen production using wind energy: a case study. Energy Strategy Rev 36. https://doi.org/10.1016/j.esr.2021.100684

Al-Sharafi A, Sahin AZ, Ayar T, Yilbas BS (2017) Techno-economic analysis and optimization of solar and wind energy systems for power generation and hydrogen production in Saudi Arabia. Renew Sust Energ Rev 69:33e49. https://doi.org/10.1016/j.rser.2016.11.157

An L, Zhao TS, Chai ZH, Tan P, Zeng L (2014) Mathematical modeling of an anion-exchange membrane water electrolyzer for hydrogen production. Int J Hydrogen Energy 39:19869e76. https://doi.org/10.1016/j.ijhydene.2014.10.025

Awadallah AE, Deyab MA, Ahmed HA (2021) Mo/MgO as an efficient catalyst for methane decomposition into COx-free hydrogen and multi-walled carbon nanotubes. J Environ Chem Eng 9:106023. https://doi.org/10.1016/j.jece.2021.106

Balat H, Kirtay E (2010) Hydrogen from biomass – present scenario and future prospects. Int J Hydrog Energy 35:7416e26. https://doi.org/10.1016/j.ijhydene.2010.04.137

Baykara SZ (2018) Hydrogen: a brief overview on its sources, production and environmental impact. Int J Hydrogen Energy 43:10605e14. https://doi.org/10.1016/j.ijhydene.2018.02.022

Behzadi A, Houshfar E, Gholamian E, Ashjaee M, Habibollahzade A (2018) Multi-criteria optimization and comparative performance analysis of a power plant fed by municipal solid waste using a gasifier or digester. Energy Convers Manag 171:863–878. https://doi.org/10.1016/j.enconman.2018.06.014

Brockway PE, Owen A, Brand-Correa LI, Hardt L (2019) Estimation of global final-stage energy-return-on-investment for fossil fuels with comparison to renewable energy sources. Nat Energy 4:612e21. https://doi.org/10.1038/s41560-019-0425-z

Carmo M, Fritz DL, Mergel J, Stolten D (2013) A comprehensive review on PEM water electrolysis. Int J Hydrogen Energy 38:4901e34. https://doi.org/10.1016/j.ijhydene.2013.01.151

Chanthakett A, Arif MT, Khan MMK, Oo AMT (2021) Hydrogen production from municipal solid waste (MSW) for cleaner environment. Bioenergy Resour Technol:219–247. https://doi.org/10.1016/B978-0-12-822525-7.00017-2. Elsevier

Chen Y, Mahecha-Botero A, Lim CJ, Grace JR, Zhang J, Zhao Y, Zheng C (2014) Hydrogen production in a sorption-enhanced fluidized-bed membrane reactor: operating parameter investigation. Ind Eng Chem Res 53(14):6230–6242

Corbo P, Migliardini F (2007) Hydrogen production by catalytic partial oxidation of methane and propane on Ni and Pt catalysts. Int J Hydrog Energy 32(1):55–66

Cozzi L, Gould T, Bouckart S, Crow D, Kim TY, McGlade C, Olejarnik P, Wanner B, Wetzel D (2020) World energy outlook 2020. Energy 2019:30

Dalena F, Senatore A, Basile M, Knani S, Basile A, Iulianelli A (2018) Advances in methanol production and utilization, with particular emphasis toward hydrogen generation via membrane reactor technology. Membranes 8(4):98

Das S, Lee S-H, Kumar P, Kim K-H, Lee SS, Bhattacharya SS (2019) Solid waste management: scope and the challenge of sustainability. J Clean Prod 228:658–678

Devi M, Rawat S, Sharma S (2020) A comprehensive review of the pyrolysis process: from carbon nanomaterial synthesis to waste treatment. Oxford Open Mater Sci 1:1–30. https://doi.org/10.1093/oxfmat/itab014

Erdem K, Gündüz Han D, Midilli A (2024) A parametric study on hydrogen production by fluidized bed co-gasification of biomass and waste plastics. Int J Hydrog Energy 52:1434–1444. https://doi.org/10.1016/j.ijhydene.2023.10.115

Freni SCSC, Calogero G, Cavallaro S (2000) Hydrogen production from methane through catalytic partial oxidation reactions. J Power Sources 87(1–2):28–38

He M, Xiao B, Liu S, Guo X, Luo S, Xu Z et al (2009) Hydrogen-rich gas from catalytic steam gasification of municipal solid waste (MSW): influence of steam to MSW ratios and weight hourly space velocity on gas production and composition. Int J Hydrogen Energy 34:2174–2183. https://doi.org/10.1016/j.ijhydene.2008.11.115

He G, Chen Q, Kang C, Pinson P, Xia Q (2016) Optimal bidding strategy of battery storage in power markets considering performance-based regulation and battery cycle life. IEEE Trans Smart Grid 7:2359e67. https://doi.org/10.1109/TSG.2015.2424314

Hoang AT, Varbanov PS, Niˇzeti'c S, Sirohi R, Pandey A, Luque R et al (2022) Perspective review on municipal solid waste-to-energy route: characteristics, management strategy, and role in circular economy. J Clean Prod 359:131897. https://doi.org/10.1016/j.jclepro.2022.131897

Horng RF, Chou HM, Lee CH, Tsai HT (2006) Characteristics of hydrogen produced by partial oxidation and auto-thermal reforming in a small methanol reformer. J Power Sources 161(2):1225–1233

Hosseini SE, Wahid MA (2020) Hydrogen from solar energy, a clean energy carrier from a sustainable source of energy. Int J Energy Res 44:4110e31. https://doi.org/10.1002/er.4930

Hu M, Guo D, Ma C, Hu Z, Zhang B, Xiao B et al (2015) Hydrogen-rich gas production by the gasification of wet MSW (municipal solid waste) coupled with carbon dioxide capture. Energy 90:857–863. https://doi.org/10.1016/j.energy.2015.07.122

IEA (2022) Total renewable capacity dedicated to hydrogen production in the main case by region, *2021–2027*, IEA, Paris. https://www.iea.org/data-and-statistics/charts/total-renewable-capacity-dedicated-to-hydrogen-production-in-the-main-case-by-region-2021-2027

Irfan M, Li A, Zhang L, Wang M, Chen C, Khushk S (2019) Production of hydrogen enriched syngas from municipal solid waste gasification with waste marble powder as a catalyst. Int J Hydrogen Energy 44:8051–8061. https://doi.org/10.1016/j.ijhydene.2019.02.048

Ji M, Wang J (2021) Review and comparison of various hydrogen production methods based on costs and life cycle impact assessment indicators. Int J Hydrog Energy 46:38612e35. https://doi.org/10.1016/j.ijhydene.2021.09.142

Kabir E, Kumar P, Kumar S, Adelodun AA, Kim KH (2018) Solar energy: potential and future prospects. Renew Sust Energ Rev 82:894e900. https://doi.org/10.1016/j.rser.2017.09.094

Kim J, Jun A, Gwon O, Yoo S, Liu M, Shin J et al (2018) Hybrid-solid oxide electrolysis cell: a new strategy for efficient hydrogen production. Nano Energy 44:121e6. https://doi.org/10.1016/j.nanoen.2017.11.074

Lepage T, Kammoun M, Schmetz Q, Richel A (2021) Biomass-tohydrogen: a review of main routes production, processes evaluation and techno-economical assessment. Biomass Bioenergy 144:105920. https://doi.org/10.1016/j.biombioe.2020.105920

Levalley TL, Richard AR, Fan M (2014) The progress in water gas shift and steam reforming hydrogen production technologies- a review. Int J Hydrogen Energy 39:16983e7000. https://doi.org/10.1016/j.ijhydene.2014.08.041

Li Z, Guo P, Han R, Sun H (2019) Current status and development trend of wind power generation-based hydrogen production technology. Energy Explor Exploit 37:5e25. https://doi.org/10.1177/0144598718787294

Lin J, Cui C, Sun S, Ma R, Yang W, Chen Y (2022) Synergistic optimization of syngas quality and heavy metal immobilization during continuous microwave pyrolysis of sludge: competitive relationships, reaction mechanisms, and energy efficiency assessment. J Hazard Mater 438:129451. https://doi.org/10.1016/j.jhazmat.2022.12945

Liu DJ, Kaun TD, Liao HK, Ahmed S (2004) Characterization of kilowatt-scale autothermal reformer for production of hydrogen from heavy hydrocarbons. Int J Hydrog Energy 29(10):1035–1046

Liu F, Yang L, Song C (2018) Chemical looping hydrogen production using activated carbon and carbon black as multi-function carriers. Int J Hydrog Energy 43:5501–5511. https://doi.org/10.1016/j.ijhydene.2018.01.098

Malik FR, Yuan HB, Moran JC, Tippayawong N (2023) Overview of hydrogen production technologies for fuel cell utilization. Eng Sci Technol Int J 43:101452

Mir MY (2025a) Sustainable fuels and Decarbonization: challenges and opportunities. In: Sustainable fuels production, pp 1–19. https://doi.org/10.1007/978-3-031-97337-6_1

Mir MY (2025b) Comprehensive analysis of zero-carbon and low-carbon sustainable fuels and their combustion emissions. In: Sustainable fuels production. Springer Nature Switzerland, Cham, pp 21–42

Mir MY (2025c) Comparative perspectives of biofuel generations: new insights into challenges and opportunities. Sustainable fuels production:61–78. https://doi.org/10.1007/978-3-031-97337-6_4

Moldoveanu SC (2019) Pyrolysis of hydrocarbons. https://doi.org/10.1016/b978-0-444-6400

Moliner R, Palacios JM (2010) Parametric study of the decomposition of methane using a NiCu/Al2O3 catalyst in a fluidized bed reactor. 35:9801–9809. https://doi.org/10.1016/j.ijhydene

Mondal P (2022) From municipal solid waste (MSW) to hydrogen: performance optimization of a fixed bed gasifier using box-Benkhen method. Int J Hydrog Energy 47:20064–20075. https://doi.org/10.1016/j.ijhydene.2022.04.150

Mostafaeipour A, Qolipour M, Goudarzi H (2019) Feasibility of using wind turbines for renewable hydrogen production in Firuzkuh, Iran. Front Energy 13:494e505. https://doi.org/10.1007/s11708-018-0534-6

Nanda S, Rana R, Zheng Y, Kozinski JA, Dalai AK (2017) Insights on pathways for hydrogen generation from ethanol. Sustain Energy Fuels 1:1232e45. https://doi.org/10.1039/C7SE00212B

Nugroho RAA, Hsu HW, Wang WC, Utomo VL, Saputro H, Sittijunda S, Kuo JK, Surjasatyo A (2025) Energy/exergy, economic, and environment (3E) analysis of the hydrogen energy production process. Int J Hydrog Energy 103:467–479

O'Keefe JMK, Bechtel A, Christanis K, Dai S, DiMichele WA, Eble CF et al (2013) On the fundamental difference between coal rank and coal type. Int J Coal Geol 118:58e87. https://doi.org/10.1016/j.coal.2013.08.007
Ozturk M, Dincer I (2021) An integrated system for clean hydrogen production from municipal solid wastes. Int J Hydrogen Energy 46:6251–6261. https://doi.org/10.1016/j.ijhydene.2020.11.145
Patzschke CF, Parkinson B, Willis JJ, Nandi P, Love AM, Raman S et al (2021) Co-Mn catalysts for H2 production via methane pyrolysis in molten salts. Chem Eng J 414:128730. https://doi.org/10.1016/j.cej.2021.12873
Pen MA, Gomez JP, Fierro JG (1996) New catalytic routes for syngas and hydrogen production. Appl Catal A Gen 144(1–2):7–57
Peng WX, Wang LS, Mirzaee M, Ahmadi H, Esfahani MJ, Fremaux S (2017) Hydrogen and syngas production by catalytic biomass gasification. Energy Convers Manage 135:270e3. https://doi.org/10.1016/j.enconman.2016.12.056
Perkins G (2018) Underground coal gasification e Part I: field demonstrations and process performance. Prog Energy Combust Sci 67:158e87. https://doi.org/10.1016/j.pecs.2018.02.004
Pinsky R, Sabharwall P, Hartvigsen J, O'Brien J (2020) Comparative review of hydrogen production technologies for nuclear hybrid energy systems. Prog Nucl Energy 123. https://doi.org/10.1016/j.pnucene.2020.103317
Qi J, Liu J, Chen G, Yao J, Yan B, Yi W et al (2023) Hydrogen production from municipal solid waste via chemical looping gasification using CuFe2O4 spinel as oxygen carrier: an Aspen plus modeling. Energy Convers Manage 294:117562. https://doi.org/10.1016/j.enconman.2023.117562
Rahman MAA, Robin HM, Hossain MS, Aziz M, Mourshed M (2025) Transforming waste into clean energy: the future of hydrogen energy generation in Bangladesh. Int J Hydrog Energy 136:275–293
Rezaei M, Salimi M, Momeni M, Mostafaeipour A (2018) Investigation of the socio-economic feasibility of installing wind turbines to produce hydrogen: case study. Int J Hydrogen Energy 43:23135e47. https://doi.org/10.1016/j.ijhydene.2018.10.184
Rodriguez Correa C, Kruse A (2018) Supercritical water gasification of biomass for hydrogen production e review. J Supercrit Fluids 133:573e90. https://doi.org/10.1016/j.supflu.2017.09.019
Ruocco C, Palma V, Ricca A (2019) Kinetics of oxidative steam reforming of ethanol over bimetallic catalysts supported on CeO2–SiO2: a comparative study. Top Catal 62(5):467–478
Saleem F, Harris J, Zhang K, Harvey A (2020) Non-thermal plasma as a promising route for the removal of tar from the product gas of biomass gasification e a critical review. Chem Eng J 382. https://doi.org/10.1016/j.cej.2019.122761
Shah M, Al Mesfer MK, Danish M (2022) Facile synthesis of Co-Rh bimetallic catalysts for methane decomposition: effect of support morphology. Fuel 330:125596. https://doi.org/10.1016/j.fuel.2022.125596
Sharma S, Basu S, Shetti NP, Aminabhavi TM (2020) Waste-to-energy nexus for circular economy and environmental protection: recent trends in hydrogen energy. Sci Total Environ 713:136633. https://doi.org/10.1016/j.scitotenv.2020.136633
Shiva Kumar S, Himabindu V (2019) Hydrogen production by PEM water electrolysis e a review. Mater Sci Energy Technol 2:442e54. https://doi.org/10.1016/j.mset.2019.03.002
Siddiqui O, Dincer I (2019) A well to pump life cycle environmental impact assessment of some hydrogen production routes. Int J Hydrog Energy 44:5773e86. https://doi.org/10.1016/j.ijhydene.2019.01.118
Sikander U, Sufian S, Salam MA (2017) A review of hydrotalcite based catalysts for hydrogen production systems. Int J Hydrog Energy 42(31):19851–19868
Singh S, Kumar Bhaumik S, Dong L, Li CZ, Vuthaluru H (2022) An integrated two-step process of reforming and adsorption using biochar for enhanced tar removal in syngas cleaning. Fuel 307. https://doi.org/10.1016/j.fuel.2021.121935
Singla S, Sharma S, Basu S, Shetti NP, Reddy KR (2020) Graphene/graphitic carbon nitride-based ternary nanohybrids: synthesis methods, properties, and applications for photocatalytic hydrogen production. FlatChem 24:100200

Soo SL, Gibbs RT (1979) A steam process for coal gasification. Energy 4(2):357–364

Specchia S, Vella LD, Lorenzut B, Montini T, Specchia V, Fornasiero P (2010) Effect of the catalyst load on syngas production in short contact time catalytic partial oxidation reactors. Ind Eng Chem Res 49(3):1010–1017

Sun X, Atiyeh HK, Huhnke RL, Tanner RS (2019) Syngas fermentation process development for production of biofuels and chemicals: a review. Bioresour Technol Rep 7. https://doi.org/10.1016/j.biteb.2019.100279

Thorbecke W (2019) How oil prices affect East and Southeast Asian economies: evidence from financial markets and implications for energy security. Energy Policy 128:628e38. https://doi.org/10.1016/j.enpol.2019.01.044

Van Hook JP (1980) Methane-steam reforming. Catalysis Rev Sci Eng 21(1):1–51

Wang M, Wang G, Sun Z, Zhang Y, Xu D (2019) Review of renewable energy-based hydrogen production processes for sustainable energy innovation. Glob Energy Interconnect 2:436e43. https://doi.org/10.1016/j.gloei.2019.11.019

Wang Y, Wang Y, Song S, Wang T, Li D, Tan H (2020) Effects of coal types and combustion conditions on carbonaceous aerosols in flue gas and their light absorption properties. Fuel 277. https://doi.org/10.1016/j.fuel.2020.118148

Wang N, Akimoto K, Nemet GF (2021) Whatwentwrong? Learning from three decades of carbon capture, utilization and sequestration (CCUS) pilot and demonstration projects. Energy Policy 158. https://doi.org/10.1016/j.enpol.2021.112546

Xiao P, Hu W, Xu X, Liu W, Huang Q, Chen Z (2020) Optimal operation of a wind-electrolytic hydrogen storage system in the electricity/hydrogen markets. Int J Hydrog Energy 45:24412e23. https://doi.org/10.1016/j.ijhydene.2020.06.302

Zhang Y, Xu P, Liang S, Liu B, Shuai Y, Li B (2019) Exergy analysis of hydrogen production from steam gasification of biomass: a review. Int J Hydrog Energy 44:14290e302. https://doi.org/10.1016/j.ijhydene.2019.02.064

Zhang J, Ling B, He Y, Zhu Y, Wang Z (2022) Life cycle assessment of three types of hydrogen production methods using solar energy. Int J Hydrog Energy 47:14158e68. https://doi.org/10.1016/j.ijhydene.2022.02.150

Zheng X, Ying Z, Wang B, Chen C (2018) Hydrogen and syngas production from municipal solid waste (MSW) gasification via reusing CO2. Appl Therm Eng 144:242–247. https://doi.org/10.1016/j.applthermaleng.2018.08.058

Zhou H, Sun S, Xu Y, Zhang Y, Yi S, Wu C (2024) Upcycling municipal solid waste to sustainable hydrogen via two-stage gasification-reforming. J Energy Chem 96:611–624. https://doi.org/10.1016/j.jechem.2024.05.018

Chapter 2
Harnessing Microalgae for Sustainable Hydrogen Production: Innovations, Obstacles, and Future Opportunities

Abstract The extensive use of fossil fuels in industrial and energy sectors has led to the rapid depletion of conventional energy reserves, prompting intensified efforts to transition toward renewable energy sources through technological innovation. Among various alternatives, microalgae have emerged as a promising feedstock for the production of biofuels, including biodiesel, bioethanol, and hydrogen. Microalgae are particularly attractive for hydrogen production owing to their fast growth rates, adaptability to diverse environmental conditions, ability to avoid competition with food crops, and high efficiency in capturing and utilizing atmospheric carbon dioxide. Consequently, microalgae-based biohydrogen production has gained increasing attention as a clean, renewable energy option capable of supporting carbon neutrality and environmental sustainability. In this context, this chapter discusses recent advancements in microalgae-driven biohydrogen production, covering sustainable hydrogen generation mechanisms, key factors influencing biohydrogen yields, bioreactor configurations, and emerging strategies aimed at improving production efficiency. Additionally, existing bottlenecks and future perspectives are discussed to identify pathways for overcoming technical and economic challenges. This chapter compiles recent progress and emerging knowledge in the field, with the objective of promoting biohydrogen as a viable alternative to conventional hydrocarbon-based biofuels and accelerating the transition toward a low-carbon and sustainable energy future.

Keywords Hydrogen production · Microalgae · Photobioreactor · Dark fermentation · Detoxification

2.1 Introduction

Fossil fuels remain the dominant sources for meeting global energy demands. In 2018 alone, worldwide energy consumption reached approximately 14,000 Mt (IEA 2020), and this demand is projected to rise by nearly 50% by 2050 due to rapid urbanization and the expanding global demand for essential goods and services

D. M. Y. Mir et al., *Low-Carbon Hydrogen*, Clean Energy Production Technologies,
https://doi.org/10.1007/978-981-95-8142-9_2

(Jing et al. 2020). However, fossil fuel reserves are limited, and their extensive use is closely associated with significant greenhouse gas emissions, contributing to climate change and environmental degradation (Marchi et al. 2019). Consequently, growing research efforts have focused on identifying alternative energy sources that are sustainable, environmentally benign, and nontoxic to replace fossil-fuel-based energy systems. Hydrogen has emerged as a promising alternative fuel owing to its abundance and versatility. It can be produced from various sources, including biomass, fossil fuels, and water (Cheng et al. 2019). Hydrogen possesses several advantageous properties: it is colorless, odorless, nontoxic, and highly flammable, with a high combustion energy of approximately 120 MJ/kg (Boodhun et al. 2017). Notably, its combustion produces only water as a by-product, making it an environmentally clean energy carrier. These characteristics position hydrogen as a highly attractive option for future biofuel and clean energy applications. Nevertheless, conventional hydrogen production technologies such as steam methane reforming and water electrolysis are energy-intensive and often depend heavily on fossil fuels, thereby limiting their sustainability and economic feasibility (Ahmed et al. 2023). In recent years, biohydrogen has gained considerable attention as a sustainable and eco-friendly alternative. Biohydrogen can be generated from biomass through biological processes involving microorganisms such as bacteria, cyanobacteria, and microalgae (Mona et al. 2020; Deviram et al. 2020). Among these, microalgae have emerged as particularly promising candidates due to their role as primary producers and their ability to thrive in diverse environments, including freshwater, marine ecosystems, and terrestrial surfaces such as tree bark. Microalgae exhibit diverse metabolic modes: photoautotrophic species utilize sunlight to convert carbon dioxide into organic matter, whereas heterotrophic species rely on organic carbon sources and can grow in the absence of light (Ardo et al. 2022). Microalgal growth depends on the availability of essential nutrients, oxygen, and trace elements, all of which can be optimized to enhance biomass productivity (Woon et al. 2023). Their photosynthetic efficiency is reported to be 10–50 times higher than that of conventional terrestrial plants, and their short life cycle ensures continuous biomass production throughout the year (dos Santos et al. 2021). Additionally, microalgae possess relatively simple cellular structures, requiring only mild pretreatment processes, which reduces energy input and minimizes the formation of inhibitory by-products during fermentation (Sundaram et al. 2023). Microalgae are also highly adaptable to fluctuating environmental conditions, including variations in nutrient availability, light intensity, and extreme pH levels, making them suitable for large-scale cultivation (Rastogi et al. 2020). They serve as a rich source of valuable metabolites such as proteins, carbohydrates, and lipids. Following metabolite extraction, the residual biomass can be further utilized for energy generation through processes such as anaerobic digestion and thermochemical conversion, enabling the production of biogas, biohydrogen, biodiesel, and bioethanol from the entire biomass (Jayaseelan et al. 2021). Furthermore, under specific physiological conditions such as oxygen-depleted environments combined with high light intensity microalgae can directly produce hydrogen through various biochemical pathways. Exploiting these mechanisms offers significant potential to enhance hydrogen yields and overall process efficiency (Woon et al. 2023). As a result, microalgae-based biohydrogen production has attracted growing interest as a clean, renewable, and sustainable energy solution capable of contributing to carbon neutrality. In this context, this chapter focuses on recent advances in microalgae-based biohydrogen production, including underlying production mechanisms, key factors influencing hydrogen yield, bioreactor configurations, and innovative strategies to improve process efficiency. Additionally, current technological bottlenecks and future perspectives are discussed to highlight pathways for overcoming existing challenges and advancing the commercialization of microalgal biohydrogen systems.

2.2 Mechanisms of Sustainable Hydrogen Production

In recent years, considerable scientific and technological attention has been directed toward the sustainable production of hydrogen. The practical application of biological systems capable of converting solar energy into alternative energy carriers, such as hydrogen fuel, has proven to be highly viable (Sahrin et al. 2022). This approach relies on microorganisms that possess inherent metabolic pathways enabling hydrogen generation as a by-product of their biochemical processes (Woon et al. 2023). Biological hydrogen production offers notable advantages, including potential cost-effectiveness and a reduced reliance on hazardous chemicals throughout the entire production, utilization, and disposal cycle. Hydrogen can be generated through several biological routes, with biophotochemical and fermentative processes being the most widely investigated. Among these methods, microalgae are capable of producing hydrogen and oxygen through the biophotolysis process. This mechanism is facilitated by chlorophyll-based photosynthetic reaction centers in conjunction with key enzymes such as hydrogenase and nitrogenase. Under anaerobic conditions, hydrogen can be produced through the direct photolysis of water, as represented in Eq. (2.1).

$$2H_2O + \text{Light energy} \rightarrow 2H_2 + O_2 \tag{2.1}$$

Biophotolysis can proceed via two principal pathways: direct biophotolysis, in which water is directly split into hydrogen and oxygen using light energy, and indirect biophotolysis, where photosynthetic activity initially generates organic intermediates that are subsequently converted into hydrogen under suitable conditions. These mechanisms highlight the versatility of microalgae as biological platforms for sustainable hydrogen production. The biological pathways and recent technological advances in microalgal hydrogen production are systematically summarized in Table 2.1.

Table 2.1 Mechanistic pathways and advantages of microalgal biohydrogen production processes

Mechanism	Description	Advantages	References
Direct biophotolysis	Microalgae (e.g., *Chlamydomonas reinhardtii*) use sunlight to split water into H_2 and O_2 via photosystem II and hydrogenase enzymes	High hydrogen purity; CO_2-neutral process; direct use of solar energy	Benemann (2000)
Indirect biophotolysis	Two-stage process where O_2-evolving photosynthesis and anaerobic hydrogenase activity occur sequentially	Avoids enzyme inhibition by O_2; more sustained H_2 yield	Kumari et al. (2017)
Genetic engineering of microalgae	Modification of hydrogenase and photosynthetic pathways to enhance H_2 production and O_2 tolerance	Boosted productivity; improved enzyme stability; potential industrial scalability	Khetkorn et al. (2017)
Hybrid photo-fermentation	Combines photosynthetic and fermentative systems in engineered microalgal consortia	Efficient substrate use; reduced waste; adaptable to wastewater	Singh and Das (2020)
Photoreduction (fermentative stage)	Under dark or low-oxygen conditions, stored carbohydrates are fermented to generate H_2	Enables continuous production even without light; compatible with waste feedstocks	Mandotra et al. (2023)
Bioelectrocatalytic hydrogen production (MEC)	Microalgae act as bioanodes generating electrons that reduce protons at cathodes to form H_2	Coupled wastewater treatment; high energy recovery efficiency	Nguyen et al. (2024)
Recent advances in sustainable hydrogen production	Comprehensive analysis of mechanisms, challenges, and future perspectives for hydrogen production from microalgae, focusing on biophotolysis, hybrid systems, and bioelectrochemical routes	Integrates multidisciplinary insights; highlights optimization strategies and engineering solutions for enhanced yield and scalability	Jiao et al. (2024)
Bio-photoelectrochemical (BPEC) systems	Integrate microalgae with electrodes to enhance H_2 generation through electron transfer	Higher efficiency; reduced photoinhibition; potential for scale-up	Ramaraj et al. (2025)

2.2.1 *Direct Biophotolysis*

The mechanism of biophotolysis closely resembles that of oxygenic photosynthesis. Under anaerobic conditions, green microalgae are uniquely capable of performing biophotolysis and generating biohydrogen (Chong et al. 2022). When exposed to light, photosystem II (PSII) plays a central role in driving the photolysis of water. In the first step, PSII catalyzes the splitting of water molecules into molecular oxygen, protons, and electrons. These electrons are subsequently transferred through the photosynthetic electron transport chain. In the second step, hydrogenase enzymes utilize the generated electrons and protons to reduce protons into molecular hydrogen (Bahadar et al. 2022). The overall biochemical reactions governing these processes are expressed in Eqs. (2.2) and (2.3), which describe water oxidation and hydrogen evolution, respectively.

Photosynthesis reaction:

$$2H_2O + \text{Sunlight} \rightarrow 4H^+ + 4e^- + O_2 \tag{2.2}$$

Hydrogen production reaction:

$$4H^+ + 4e^- \rightarrow 2H_2 \tag{2.3}$$

In this process, electrons are generated through the oxidation of water mediated by photosystem II (PSII). Incident solar photons excite these electrons within PSII, after which they are transferred through the photosynthetic electron transport chain. Electron transfer is facilitated by photosystem I (PSI) and ferredoxin (Fd), ultimately leading to the reduction of $NADP^+$ to NADPH via plastoquinol intermediates. To sustain electron flow, water molecules are continuously oxidized, resulting in the release of oxygen, protons, and electrons. The electrons derived from water oxidation are utilized by PSII to drive the photosynthetic reactions. Concurrently, the accumulation of protons generated during water splitting establishes a proton gradient across the thylakoid membrane, which in turn drives ATP synthesis through ATP synthase. Within the chloroplasts of microalgae, protons act as terminal electron acceptors during hydrogenase-mediated reactions, enabling the simultaneous evolution of hydrogen and oxygen gases (Woon et al. 2023). Despite its conceptual simplicity, the biophotolysis process exhibits relatively low efficiency in biohydrogen generation. This limitation arises from both physiological constraints on microalgal growth and the inherently low energy conversion efficiency of photosynthesis. Consequently, photon conversion efficiency plays a pivotal role in determining hydrogen yield. Although direct biophotolysis has been reported to convert up to 80% of incident photons, the maximum practical efficiency achieved by green microalgae is approximately 12.5% (Nagarajan et al. 2017). Light absorption efficiency is another critical factor influencing biohydrogen production. Increasing light intensity can enhance microalgal biomass accumulation and growth rates. Under natural low-light conditions, microalgae adapt by developing large chlorophyll antenna complexes that maximize light harvesting. However, under high light intensities, photosynthetic efficiency declines due to an imbalance between photon absorption and utilization. In such conditions, photons absorbed by the upper layers of chlorophyll antennae exceed the photosynthetic processing capacity, resulting in energy dissipation through heat and fluorescence, a phenomenon known as photoinhibition (Mona et al. 2020).

2.2.2 *Indirect Biophotolysis*

Indirect biophotolysis has been developed as an alternative strategy to overcome the limitations associated with direct biophotolysis. In this approach, microalgae generate hydrogen under anoxic conditions by metabolizing internally stored carbohydrates, such as glycogen and starch, through fermentative or respiratory pathways (Anwar et al. 2019). In the first stage, microalgae harness solar energy to synthesize organic compounds, including carbohydrates and lipids, while simultaneously releasing oxygen as a by-product. This phase leads to the accumulation of endogenous carbon reserves within the cells (Shuba and Kifle 2018). During the second stage, these stored carbohydrates are metabolized, generating a substantial flux of electrons. The electrons are subsequently transferred to the plastoquinone pool and photosystem II (PSII). As oxygen becomes depleted, the photosynthetic electron transport chain is disrupted, creating anaerobic conditions that activate oxygen-sensitive hydrogenase enzymes. Under these conditions, electrons are directed via ferredoxin (Fd) and photosystem I (PSI) toward hydrogenase-mediated hydrogen evolution (Sivaramakrishnan et al. 2021). The feasibility of indirect biophotolysis has been demonstrated using glucose as a substrate and cyanobacteria such as *Anabaena* sp. (UTEX 1448), where maximum hydrogen yields of 9.73 mmol L^{-1} and production rates of 67.07 μmol L^{-1} h^{-1} were reported (Vargas et al. 2018). A key advantage of indirect biophotolysis is the spatial and temporal separation of oxygen and hydrogen evolution, which effectively mitigates hydrogenase oxygen sensitivity. Moreover, this approach generally enables higher hydrogen yields compared to direct biophotolysis (Sharma and Arya 2017). Despite these advantages, the overall hydrogen output of indirect biophotolysis remains limited. This constraint is primarily attributed to hydrogen consumption by uptake hydrogenases, which reduces net hydrogen recovery (Akhlaghi and Najafpour-Darzi 2020). The overall biochemical reactions governing indirect biophotolysis are summarized in Eqs. (2.4) and (2.5).

Aerobic reaction:

$$6H_2O + 6CO_2 + \text{Solar energy} \rightarrow C_6H_{12}O_6 + 6O_2 \tag{2.4}$$

Anaerobic reaction:

$$C_6H_{12}O_6 + 6H_2O + \text{Solar energy} \rightarrow 6CO_2 + 12H_2 \tag{2.5}$$

2.2.3 *Dark Fermentation*

Dark fermentation is among the most extensively investigated biological pathways for biohydrogen production (Srivastava et al. 2020). In this anaerobic process, microorganisms metabolize readily degradable substrates, such as carbohydrates and proteins, to generate molecular hydrogen. Dark fermentation functions as an intrinsic metabolic mechanism for the disposal of excess reducing equivalents and is typically accompanied by the formation of organic acids and/or alcohols as metabolic by-products (Sambusiti et al. 2015). Hydrogen generation during dark fermentation proceeds through two principal metabolic pathways, which govern electron flow and product distribution (Bhatia et al. 2021). The efficiency and hydrogen yield of this process are highly dependent on both the microbial inoculum and the characteristics of the substrate, as these factors directly influence metabolic flux distribution and the overall fermentation performance (Srivastava et al. 2020). The generalized biochemical reaction describing dark fermentative hydrogen production is presented in Eq. (2.6).

$$C_6H_{12}O_6 + 6H_2O \rightarrow 6CO_2 + 12H_2 \tag{2.6}$$

Biohydrogen production via dark fermentation is governed by a range of interdependent factors, including the selection of anaerobic microbial strains, the kinetics of substrate hydrolysis, and the efficiency with which microorganisms assimilate and metabolize available substrates (Ardo et al. 2022). Achieving optimal process performance therefore necessitates precise control of critical operational parameters, such as temperature, pH, and organic loading rate, which collectively determine microbial activity, metabolic flux distribution, and hydrogen yield stability. Such regulation is especially important for ensuring reproducible performance and sustained hydrogen production. For large-scale and industrial applications involving high substrate loading and continuous operation, bioreactor configuration and operational strategy become decisive factors. Continuous-mode bioreactor systems are generally considered more robust and reliable than batch processes, as they provide enhanced process stability, improved biomass retention, and consistent hydrogen generation over extended operational periods. In addition to hydrogen, dark fermentation pathways facilitate the concurrent production of value-added metabolites, including acetic, lactic, and butyric acids. These organic acids can be further exploited as substrates for downstream fermentation or upgraded through separation and purification for commercial use, thereby enhancing the overall economic feasibility and process integration potential of biohydrogen production systems (Bhatia et al. 2021).

2.2.4 *Photo-fermentation*

Photo-fermentation is a light-driven biological process in which organic compounds are converted into biohydrogen and carbon dioxide. Unlike dark fermentation, this pathway utilizes solar energy as the primary energy input rather than relying solely on glucose or other reduced substrates. Photo-fermentation is predominantly carried out by purple non-sulfur photosynthetic bacteria, which exploit light energy to metabolize carbon-containing compounds such as organic acids into hydrogen and carbon dioxide through the activity of nitrogenase enzymes (Iqbal et al. 2022). The catalytic reactions mediated by nitrogenase in the presence and absence of molecular nitrogen are represented by Eqs. (2.7) and (2.8), respectively. Owing to their ability to efficiently convert a wide range of organic substrates into hydrogen, purple non-sulfur bacteria are regarded as among the most effective biological hydrogen producers, exhibiting relatively high substrate-to-hydrogen conversion efficiencies (Adessi et al. 2017). The overall biochemical transformation of organic acids into hydrogen during photo-fermentation can be described by Eq. (2.9).

$$N_2 + 8H^+ + 8e^- + 16ATP \rightarrow 2NH_3 + H_2 + 16ADP + 16P_i \tag{2.7}$$

$$8H^+ + 8e^- + 16ATP \rightarrow 4H_2 + 16ADP + 16P_i \tag{2.8}$$

$$CH_3COOH\left(\text{acetic acid}\right) + H_2O + \text{Sunlight} \rightarrow 2CO_2 + 4H_2 \tag{2.9}$$

Microalgae represent a highly promising source of organic substrates for photo-fermentation owing to their rapid growth rates, efficient carbon dioxide fixation, and capacity to synthesize a diverse range of organic compounds (Melitos et al. 2021). In photo-fermentative systems, specialized microorganisms utilize volatile fatty acids (VFAs) and other organic acids generated during dark fermentation as natural substrates. Under light conditions, these compounds are metabolized to produce biohydrogen and carbon dioxide, with ATP and NADPH serving as essential energy and reducing equivalents (Mishra et al. 2019). Photo-fermentation offers the potential for comparatively higher biohydrogen yields, with a theoretical maximum of 8 mol H_2 per mol of glucose (Das and Basak 2021). Unlike certain photobiological pathways, this process is not adversely affected by oxygen-mediated inhibition of the nitrogenase enzyme. This tolerance arises from the absence of oxygen-sensitive photosystem II (PSII) in photo-fermentative microorganisms, allowing sustained hydrogen production even in the presence of residual oxygen (Kapdan et al. 2009). Furthermore, these organisms can transition between photosynthetic growth and photo-fermentative hydrogen production under conditions characterized by excess carbon-rich substrates and nitrogen limitation. Despite these advantages, the relatively long operational times required for photo-fermentation present a notable limitation. To maximize hydrogen productivity, several operational parameters must be carefully optimized, including light intensity, nutrient composition, and pH. In addition, the choice of microalgal species and their cultivation conditions significantly influence process performance. Certain species, such as *Chlamydomonas reinhardtii*, have demonstrated high biohydrogen production under specific physiological conditions, whereas other strains may require metabolic or genetic modifications to enhance hydrogen yields. Although photo-fermentation of microalgae holds considerable promise as a sustainable biohydrogen production pathway, several challenges hinder its large-scale implementation. These include relatively low hydrogen yields under practical conditions, limitations in reactor design, economic constraints, and the need for advanced separation and recovery technologies.

Consequently, further research and technological innovation are essential to optimize process efficiency, reduce costs, and enable the commercialization of microalgae-based photo-fermentative hydrogen production systems.

2.3 Factors Affecting Biohydrogen Production by Microalgae

A diverse array of microalgal taxa have been reported to possess hydrogen-producing capabilities, including *Euastrum*, *Cosmarium*, *Chlorella* spp., *Ulothrix*, *Selenastrum*, *Chlamydomonas* spp., *Closterium*, *Micractinium* spp., *Pandorina*, *Stigeoclonium*, and *Staurastrum* (Wang et al. 2020). The efficiency of hydrogen evolution in microalgal systems is highly sensitive to environmental conditions, as hydrogen production is tightly regulated by cellular metabolism and hydrogenase activity. In particular, parameters such as temperature, pH, light intensity, and nutrient availability must be carefully optimized and maintained, as they collectively influence photosynthetic performance, redox balance, and enzymatic functionality (Ardo et al. 2022).

2.3.1 Temperature

Temperature is a fundamental physical parameter that governs both microalgal growth and hydrogen production efficiency in cultivation systems (Ardo et al. 2022). Biohydrogen generation in microalgae is mediated by multiple hydrogenase enzymes, and targeted modulation of associated metabolic pathways can substantially improve hydrogen yields (Woon et al. 2023). Temperature exerts a direct influence on these processes by regulating enzyme kinetics, reaction rates, and overall metabolic efficiency (Javed et al. 2022). Each enzymatic system operates within a specific temperature range, beyond which catalytic activity declines. Most microalgal species exhibit optimal growth between 15 and 35 °C; however, certain hydrogen-producing strains, such as *Chaetoceros* and *Anacystis nidulans*, demonstrate enhanced tolerance to elevated temperatures approaching 40 °C (Khan et al. 2018). Exposure to temperatures above the optimal threshold can result in the denaturation of key metabolic enzymes, leading to impaired cellular function and reduced biohydrogen productivity. Conversely, suboptimal low temperatures—particularly under carbon-limited conditions—can suppress photosynthetic activity and disrupt energy metabolism. Prolonged thermal stress also compromises photosynthetic machinery by inactivating essential proteins, thereby diminishing photosynthetic efficiency and slowing biomass accumulation. This effect is largely attributed to the thermal sensitivity of ribulose-1,5-bisphosphate carboxylase/oxygenase (Rubisco), a critical enzyme in carbon fixation, whose inactivation under heat stress significantly constrains metabolic flux and growth (Ayub et al. 2024).

2.3.2 pH

The pH of the cultivation medium is a key parameter influencing microalgal growth, metabolic activity, and ultimately biohydrogen production (Sathinathan et al. 2023). Optimal pH requirements are species-specific and vary between marine and freshwater microalgae, largely due to differences in nutrient utilization, particularly nitrate availability. Maintaining a stable pH within the optimal range is therefore essential for maximizing biohydrogen yields. In general, biohydrogen production is most efficient within a pH range of approximately 4.5–8.0 (Woon et al. 2023), while the optimal pH for the growth and hydrogen production of most microalgal strains typically lies between 6.5 and 8.5 (Arimbrathodi et al. 2023). For instance, *Tetraspora* sp. CU2551 exhibited maximum biohydrogen productivity of 0.22 mL mg^{-1} Chl *a* h^{-1} at an optimal pH of 7.5 (Maneeruttanarungroj and Phunpruch 2017). Similarly, other studies have reported that biohydrogen production is strongly pH-dependent, with optimal performance commonly observed in the pH range of 6.5–7.5 (Tang et al. 2022). Microalgal cultivation generally tolerates a broader pH window, ranging from 5 to 9. Within this range, moderately higher pH values have been shown to reduce the lag phase and enhance hydrogen production rates (Chowdury et al. 2020). When glucose is used as a substrate, biohydrogen generation is maximized at around pH 6, resulting in higher hydrogen yields (Fangkum and Reungsang 2011). In contrast, under more acidic conditions, hydrogen production proceeds via the pyruvate–ferredoxin oxidoreductase pathway, during which Fe-hydrogenase activity is inhibited. As a result, microalgae shift toward a more energetically favorable metabolic route that converts acetyl-CoA into acetate and ATP rather than hydrogen. Additionally, metabolic processes such as carbon fixation and nitrate assimilation can induce fluctuations in culture pH. The accumulation of volatile fatty acids (VFAs) may lower the pH to values below 4.5, potentially triggering a metabolic shift toward solventogenesis. This shift reduces hydrogen evolution and negatively impacts overall biohydrogen productivity (Srivastava et al. 2020).

2.3.3 Light

Light plays a pivotal role in regulating microalgal growth as well as the accumulation of energy reserves such as starch and lipids (Maltsev et al. 2021). Under illuminated conditions, microalgae convert light energy into chemical energy through photosynthesis, leading to the synthesis and storage of glucose and lipids (Cheirsilp and Torpee 2012). Manipulation of light exposure, including intensity and duration, can therefore be strategically employed to optimize hydrogen production during the subsequent anaerobic phase. Several studies have reported that sudden or intermittent illumination can stimulate hydrogen evolution, resulting in enhanced biohydrogen yields (Anwar et al. 2019). Light availability also significantly influences the lag phase preceding hydrogen production. Exposure to light has been shown to shorten the lag phase while simultaneously increasing hydrogen yield, thereby improving overall process efficiency (Oncel and Sukan 2011). Moreover, the pattern and timing of light supplementation serve as key indicators of hydrogen production rates. However, excessive light intensity can have detrimental effects; for instance, *Chlamydomonas reinhardtii* exhibited photoinhibition when exposed to high irradiance levels of approximately 200 μmol m^{-2} s^{-1} for durations of 15–30 min (Dalena et al. 2017). A correlation between light intensity, sulfur metabolism, and hydrogen production has also been observed. As light intensity increases, the rate of sulfur consumption rises, reaching a maximum at around 200 μmol m^{-2} s^{-1}. Consequently, hydrogen production has been linked to the residual sulfate concentration in sulfur-depleted culture media (Kim and Kim 2012). Furthermore, the impact of alternating photoperiods on hydrogen production under sulfur-deprived conditions has been investigated in *C. reinhardtii*. When subjected to alternating light–dark cycles of 2, 3, and 4 h, hydrogen concentrations of 47%, 86%, and 87%, respectively, were achieved (Vijayaraghavan et al. 2009). These findings collectively underscore the necessity of controlled light exposure for efficient hydrogen generation in microalgal systems.

2.3.4 Nutrients

Adequate nutrient supplementation is essential for promoting microalgal growth and enhancing biohydrogen production (Khetkorn et al. 2017). Nutrient requirements vary depending on the metabolic mode of the microalgae, which may include autotrophic, heterotrophic, or mixotrophic growth. For example, *Chlorella* spp. are capable of producing hydrogen under heterotrophic conditions by utilizing both endogenous reserves and externally supplied substrates (Morales-Sánchez et al. 2017). External carbon sources not only support microalgal proliferation but also play a critical role in stimulating hydrogen production. Among internal carbon reserves, starch synthesized during the photosynthetic phase serves as a major electron donor during hydrogen evolution. In addition, microalgae can metabolize a wide range of external carbon substrates including glucose, sucrose, fructose, and acetate for biohydrogen generation (Woon et al. 2023). The nature and concentration of the carbon source strongly influence fermentation pathways, metabolite profiles, and hydrogen yields. Carbohydrates, in particular, are central to fermentative hydrogen production. The initial substrate concentration is a key determinant of process efficiency; for instance, an optimal glucose concentration of 30 mM was shown to reduce the lag phase of hydrogen production by approximately 10 h (Ahmed et al. 2022). Several studies have demonstrated the feasibility of using low-cost substrates for hydrogen production. *Chlorella vulgaris* MSU 01 achieved a hydrogen yield of 220 mL L^{-1} when cultivated with corn stalk at a concentration of 4 g L^{-1} (Amutha and Murugesan 2011). Similarly, fructose and sucrose were identified as the most effective substrates for hydrogen production by *Anabaena* sp. CH3, yielding 1.6 and 1.0 mmol of hydrogen, respectively (Chen et al. 2008). In addition to carbon sources, the balanced supplementation of macronutrients and trace elements, including nitrogen, phosphorus, and essential minerals, is crucial for robust microalgal growth and hydrogen productivity (Goswami et al. 2021). Nitrogen is vital for protein synthesis and cellular growth, while phosphorus plays a central role in energy transfer and metabolic regulation (Su 2021). Trace elements such as Mg^{2+}, Na^{+}, Zn^{2+}, and Fe^{3+} are required in small quantities but are indispensable for enzymatic function and metabolic control. In particular, iron (Fe^{3+}) is closely associated with hydrogenase activity, making it a key determinant of biohydrogen production efficiency (Hawkes et al. 2002). Sulfur availability also exerts a significant influence on hydrogen production. Under sulfur-deprived conditions, protein synthesis is inhibited and photosystem II (PSII) activity is partially suppressed. This reduction in photosynthetic oxygen evolution alleviates oxygen-induced inhibition of hydrogenase, thereby favoring hydrogen production (Veeravalli et al. 2019). Consequently, precise control of nutrient composition—especially carbon sources, trace elements, and sulfur—is essential for optimizing microalgal cultivation and maximizing biohydrogen yields.

2.4 Photobioreactor and Hydrogen Production

Economically viable hydrogen production requires achieving high hydrogen productivity while maintaining low capital and operating costs. Large-scale biohydrogen generation using microalgae therefore necessitates the development of specialized bioreactor systems capable of sustaining high performance under controlled conditions (Ruiz-Marin et al. 2020). Bioreactors enable precise regulation of critical operational parameters such as light intensity, temperature, and pH, all of which are essential for maximizing hydrogen production rates. Among these parameters, hydraulic retention time (HRT) has been identified as a key factor influencing reactor productivity; shorter HRTs can enhance hydrogen yields, provided that biomass washout is effectively prevented. Because light availability is fundamental for microalgal hydrogen production, most reactor configurations are transparent and are classified as photobioreactors (PBRs). Based on their operational strategies and design features, PBRs can be further categorized into several types. While most PBRs rely on solar irradiation as the primary light source, some systems are equipped with artificial lighting to allow greater control over illumination conditions (Ahmed et al. 2022). However, PBRs are generally applicable to a limited range of microalgal strains and present challenges related to area-to-volume ratio, temperature control, mixing efficiency, and gas exchange (Wang et al. 2018a, b). To enhance mixing and establish anaerobic conditions, inert gases such as argon have been used, although argon has been reported to negatively affect the growth of certain species, including *Rhodopseudomonas*. Membrane bioreactors (MBRs) have also demonstrated potential for hydrogen production due to their ability to regulate biomass concentration. Nevertheless, their large-scale application is constrained by high operational costs and membrane fouling. Fouling typically arises from sludge accumulation, microbial flocs, and deposits of dissolved organic matter, which reduce membrane efficiency and lifespan (Wang et al. 2018a, b).

In recent years, increasing attention has been directed toward multistage reactor (MSR) systems, as they offer enhanced hydrogen production capacity compared to single-stage configurations. MSRs integrate multiple reactor units, each optimized for specific biochemical reactions, thereby improving overall process efficiency (El-Dalatony et al. 2020). Typical MSR configurations may include stages for light filtration using cyanobacteria, photo-fermentative hydrogen production utilizing infrared light, dark fermentation for hydrogen and organic acid generation, and additional reactors dedicated to organic acid production (Razu et al. 2019). Despite their advantages, MSRs are inherently complex systems that require sophisticated design, strict process control, and advanced hydrogen recovery methods. Issues such as strain compatibility, contamination risks, and elevated construction costs limit their widespread industrial adoption. Life cycle sustainability assessments further highlight the challenges associated with biological hydrogen production. Barghash et al. (2024) conducted a comparative life cycle assessment of hydrogen production in Oman using activated sludge methods (ASM), fluidized-bed bioreactors (FBB), dark fermentation (DF), and electrolysis. Their findings indicated that hydrogen yields from DF were lower than those obtained through electrolysis, primarily due to DF's dependence on sludge generation, whereas electrolysis relied on treated water. However, the study also demonstrated reduced CO_2 emissions when solar energy was integrated into ASM-based systems, emphasizing the importance of renewable energy coupling. For large-scale microalgal cultivation, open raceway ponds are often employed due to their low cost; however, they are unsuitable for effective hydrogen gas collection. This limitation has driven the development of closed photobioreactor systems to enable commercial scale hydrogen production (Geada et al. 2017). Adequate illumination typically provided by sunlight is essential for PBR operation, although artificial lighting can be incorporated to achieve controlled and uniform light distribution. PBR design must account for several limiting factors, including light penetration depth, surface area-to-volume ratio, mixing efficiency, temperature regulation, and gas exchange. The optimal configuration depends largely on the physiological characteristics of the selected microalgal strain and the intended operational conditions. The most commonly employed PBR designs for hydrogen production include vertical column reactors, tubular reactors, flat-panel reactors, and rhomboidal photobioreactors. Vertical column reactors consist of transparent columns enclosed by water jackets that regulate temperature and control light exposure. Tubular PBRs feature long transparent tubes through which algal cultures circulate, offering high surface-area-to-volume ratios, flexible orientation for light capture, and scalability for continuous operation (Peter et al. 2022). However, their installation and maintenance costs are relatively high. Flat-panel photobioreactors are constructed using stainless-steel frames and polycarbonate panels arranged into adjacent compartments. These systems are equipped with advanced control modules that continuously monitor parameters such as pH, temperature, optical density, and dissolved hydrogen and oxygen concentrations.

Among the various PBR designs, flat-panel reactors are considered particularly suitable for hydrogen production due to their relatively low cost, high photosynthetic efficiency, effective gas pressure control, and minimal backpressure resulting from reduced hydrogen accumulation (Nyberg et al. 2015). Similarly, rhomboidal photobioreactors help mitigate gas hold-up issues that reduce effective working volume in conventional designs. Multilayer PBR configurations further enhance light distribution but present challenges in maintaining uniform temperature and efficient mixing during hydrogen production (Eloffy et al. 2022). Despite extensive research and technological advancements, the development of commercially viable photobioreactors for large-scale hydrogen production remains a significant challenge. The selection of an appropriate photobioreactor depends on multiple factors, including hydrogen production efficiency, economic feasibility, operational simplicity, and scalability. Each reactor design offers distinct advantages and limitations that must be carefully evaluated to optimize biohydrogen production.

Airlift photobioreactors (PBRs) are widely recognized for their efficient mixing and enhanced gas–liquid mass transfer characteristics. The improved hydrodynamics within airlift systems promote uniform growth conditions throughout the reactor, resulting in higher biomass productivity. Additionally, airlift PBRs are relatively amenable to scale-up. However, their operation is associated with high-energy demands due to continuous air circulation, and their complex design and control requirements pose operational challenges. Bubble column PBRs represent another commonly used configuration due to their simple design, ease of operation, and effective gas–liquid mass transfer. These reactors can be operated in continuous mode and scaled up with relative ease. Nevertheless, at high cell densities, oxygen accumulation becomes a significant limitation, leading to photoinhibition and a consequent decline in hydrogen production. To mitigate this issue, reactor design modifications—such as improved gas stripping or enhanced circulation—are often required to minimize oxygen build-up. Membrane-based photobioreactors offer an efficient means of separating biomass from the culture medium, thereby simplifying harvesting procedures and reducing the risk of contamination. These systems are well suited for continuous operation and high-density microalgal cultures. Despite these advantages, membrane fouling remains a major drawback, increasing operational complexity and costs. Careful selection of membrane materials and appropriate pore sizes is therefore essential to achieve optimal and sustained performance. The selection of an appropriate photobioreactor configuration depends on several factors, including resource availability, spatial constraints, budgetary considerations, and specific operational requirements. Moreover, stringent regulatory and safety considerations impose additional constraints on biohydrogen production systems. These include the need for enclosed reactor designs to ensure efficient hydrogen capture without leakage, ease of sterilization, and high surface-to-volume ratios to maximize light utilization and microalgal growth, thereby enhancing hydrogen recovery. Although significant progress has been made in microalgae-based hydrogen production over recent years, further in-depth research is required to improve photobioreactor designs for commercial-scale applications and to reduce overall production costs. A major bottleneck in microalgal hydrogen production remains

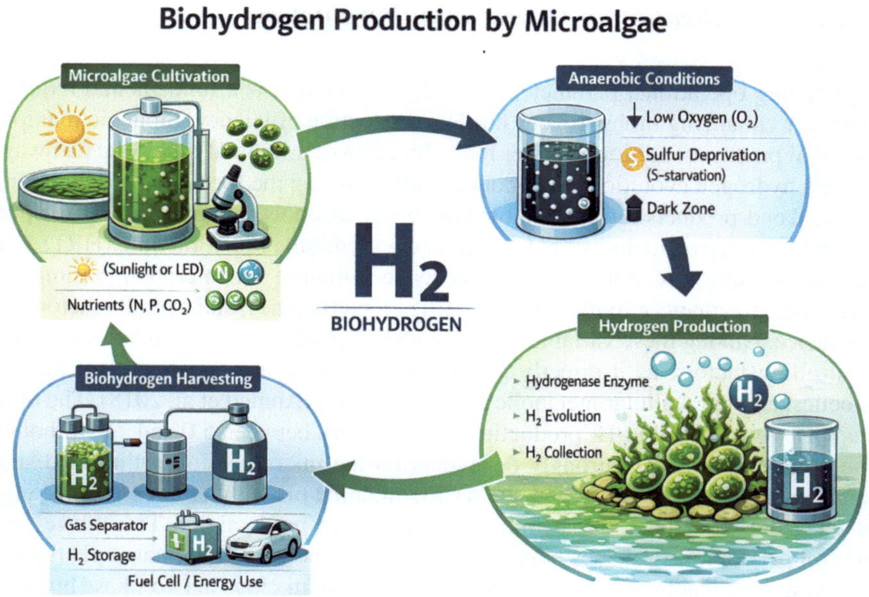

Fig. 2.1 State-of-the-art approaches for biohydrogen production by microalgae

the requirement for large quantities of biomass, which can be cultivated either in photobioreactors or open pond systems. While photobioreactors offer superior control over critical parameters such as pH, light intensity, temperature, dissolved oxygen, and carbon dioxide levels, they are often energy-intensive and costly to operate. In contrast, open pond cultivation systems are comparatively less expensive and require lower energy inputs, although they lack precise environmental control. Figure 2.1 schematically represents state-of-the-art strategies for microalgal biohydrogen production, encompassing photobioreactor cultivation, metabolic switching under oxygen-limited conditions, and downstream hydrogen collection.

2.5 Advanced Strategies to Improve Efficiency of Production

Dark fermentative biohydrogen production has gained considerable attention as a sustainable route for clean energy generation. Despite its potential, the process is constrained by several technical and economic challenges that limit hydrogen yield and overall efficiency, thereby restricting large-scale commercialization. To address these limitations, a range of innovative strategies has been explored to optimize process performance and improve hydrogen productivity. These approaches primarily involve the modification of operational parameters as well as the integration of advanced technologies, including genetic engineering and nanotechnology.

2.5.1 Modification of Operational Parameters

Optimizing operational parameters in dark fermentation is a fundamental requirement for enhancing biohydrogen (bio-H_2) production efficiency. Strategic manipulation of process conditions enables the redirection of microbial metabolic pathways toward hydrogen evolution while concurrently limiting the formation of competing reduced end-products such as volatile fatty acids, alcohols, and methane. Core operational parameters, including pH, temperature, hydraulic retention time (HRT), and organic loading rate (OLR), exert a decisive influence on microbial community structure, enzymatic activity, substrate utilization efficiency, and overall reactor performance. Among these variables, pH is widely regarded as one of the most critical controlling factors, as it directly governs hydrogenase activity, microbial growth kinetics, and intracellular metabolic flux distribution (Ahmed et al. 2018). The optimal pH window for bio-H_2 production generally lies between 5.0 and 7.0, although the precise optimum is strongly dependent on the microbial consortium and substrate characteristics (Tagne et al. 2024). Maintaining pH within this favorable range is therefore essential for sustaining high hydrogen productivity. To achieve this, several pH control strategies have been developed, including automated pH regulation systems capable of accommodating dynamic fluctuations in two-phase bioreactors designed for sequential hydrogen and methane production (Delavar and Wang 2021). In addition, buffering agents—most commonly phosphate-based buffers—are frequently employed to stabilize reactor pH and prevent abrupt inhibitory shifts (Zhao et al. 2010). Substrate preconditioning approaches, which involve adjusting the initial pH prior to inoculation, have also proven effective in promoting early dominance of hydrogen-producing microorganisms. Experimental evidence highlights the strain-specific nature of pH optimization. Yun et al. (2012) manually regulated pH using 3 N KOH and HCl during dark fermentation of *Chlorella vulgaris*, maintaining an optimal pH of 7.4 and achieving a maximum hydrogen yield of 31.2 mL H_2 g^{-1} dry cell weight (DCW). Similarly, Aldaby et al. (2023) reported a hydrogen yield of 3.23 mmol H_2 g^{-1} dry biomass at pH 9 using a phosphate-buffered single-culture system. These findings collectively underscore that optimal pH values are highly microorganism-dependent, necessitating precise, system-specific pH control strategies. Temperature represents another pivotal operational parameter, as it directly affects microbial metabolism, enzymatic reaction rates, and overall biochemical kinetics within the fermentation system (Akhlaghi and Najafpour-Darzi 2020). Dark fermentation can be conducted across a broad temperature spectrum, ranging from mesophilic (25–40 °C) to thermophilic (40–65 °C), extreme thermophilic (65–80 °C), and hyper-thermophilic (>80 °C) regimes (Singh and Das 2019). Temperature manipulation enables selective enrichment of specific microbial populations, thereby influencing hydrogen production pathways. Common optimization strategies include thermophilic operation, temperature phasing, and cyclic temperature control (Okonkwo et al. 2019). Reported optimal temperatures vary significantly depending on feedstock and microbial composition. Usmanbaha et al. (2019) observed a 2.27-fold increase in hydrogen production at 35 °C compared to 55 °C when using microalgal biomass, whereas Sillero et al. (2023) achieved maximum hydrogen yields at 55 °C with a mixed sludge consortium. Ziara et al. (2019) demonstrated optimal hydrogen generation from slaughterhouse wastewater at 45 °C, yielding 0.85 mol H_2 mol^{-1} substrate. Although mesophilic conditions around 37 °C are commonly adopted as a baseline, these studies clearly indicate that temperature optimization must be tailored to the specific microbial system and substrate employed. Hydraulic retention time (HRT) plays a critical role in determining the duration of substrate–microorganism interactions within the reactor. Excessively long HRTs can promote the accumulation of inhibitory metabolites and favor hydrogen-consuming microorganisms, whereas overly short HRTs may result in biomass washout and insufficient metabolic conversion. Various optimization strategies have therefore been proposed, including reducing HRT to suppress hydrogen consumption (Palomo-Briones et al. 2017), decoupling solid and liquid retention through cell immobilization techniques (Boshagh et al. 2024), and applying extended HRTs for the effective degradation of complex or recalcitrant substrates (Mikheeva et al. 2022). Experimental outcomes further highlight the variability of optimal HRTs. Batista et al. (2015) investigated fermentation duration using *Chlorella vulgaris* and *Scenedesmus obliquus* biomass, while Tena et al. (2021) identified an optimal HRT of 0.5 days, achieving a hydrogen yield of 0.90 L H_2 L^{-1}. In contrast, Martins and Surra reported peak hydrogen production at an HRT of 5 days (18.2 L kg^{-1}), with diminished yields observed at both shorter and longer retention times due to biomass limitation and metabolite inhibition, respectively. Conversely, Ri et al. (2019) achieved optimal hydrogen productivity at a markedly shorter HRT of 4 h, yielding 5.1 L H_2 L^{-1}. These divergent results emphasize that optimal HRT is highly dependent on reactor configuration, microbial ecology, and substrate complexity. Organic loading rate (OLR) optimization is equally critical, as it governs substrate availability while preventing process inhibition caused by excessive organic accumulation. Effective OLR management strategies include stepwise loading increases, pulsed or intermittent feeding, and co-digestion approaches to balance nutrient availability and enhance hydrogen productivity (Chaher et al. 2023; Anzola-Rojas et al. 2024). Batista et al. (2014) demonstrated that increasing OLR does not necessarily translate into higher hydrogen yields during fermentation of *Scenedesmus obliquus*, highlighting the limitations of substrate overloading. Species-specific tolerance to organic loading has also been reported; *Enterobacter aerogenes* achieved a maximum hydrogen yield of 58 mL H_2 g^{-1} volatile solids (VS) at OLRs below 5 g L^{-1}, whereas *Clostridium butyricum* reached 113 mL H_2 g^{-1} VS at substantially higher OLRs of 50 g L^{-1}. Overall, the precise and integrated optimization of pH, temperature, HRT, and OLR is essential for establishing favorable conditions that maximize bio-H_2 production in anaerobic dark fermentation systems. Given the strong interdependence among these parameters, optimization strategies must be customized for each specific microbial consortium, substrate type, and reactor configuration. A persistent challenge remains the activity of hydrogen-consuming microorganisms that compete with hydrogen producers; nevertheless, targeted operational and microbial control strategies to suppress these competitors present significant opportunities for further improving biohydrogen production efficiency.

2.5.2 Suppression of Hydrogen Consumers

Methanogenic archaea constitute one of the most significant biological constraints in anaerobic digestion systems designed for biohydrogen (bio-H_2) production. These microorganisms actively compete with hydrogen-producing bacteria by consuming hydrogen, carbon dioxide, and acetate to form methane, thereby substantially diminishing net hydrogen yields (Greening et al. 2019). Suppressing methanogenic activity is therefore a critical prerequisite for maximizing hydrogen recovery in dark fermentation processes. To address this challenge, a wide range of pretreatment and inhibition strategies have been developed to selectively target hydrogen-consuming microorganisms, including physical, chemical, electrochemical, and hybrid approaches. Among these methods, heat pretreatment remains the most widely adopted strategy due to its simplicity, cost-effectiveness, and proven selectivity. Hydrogen-producing bacteria particularly spore-forming genera such as *Clostridium* exhibit relatively high resistance to thermal stress, whereas methanogens are extremely sensitive to elevated temperatures (Mohammadzadeh et al. 2022). As a result, heat pretreatment preferentially enriches hydrogen-producing populations while effectively suppressing methanogenic archaea. Two primary heat-based approaches are commonly employed: heat shock and continuous high-temperature operation. Heat shock treatment typically involves exposing the inoculum to temperatures ranging from 60 to 100 °C for 15–60 min prior to reactor inoculation, which selectively inactivates non-spore-forming microorganisms while preserving hydrogen-producing spores. Alternatively, sustained operation under thermophilic or extreme thermophilic conditions promotes the establishment of temperature-adapted hydrogen-producing communities and creates an unfavorable environment for methanogen survival (Ryue et al. 2020). Chemical inhibition represents another extensively investigated strategy for mitigating hydrogen consumption by methanogens. Methanogen-specific inhibitors, most notably sodium 2-bromoethanesulfonate (2-BES), act by blocking methyl-coenzyme M reductase, a key enzyme responsible for the final step of methane biosynthesis. Additional chemical inhibitors, such as chloroform and lumazine, have also demonstrated efficacy in suppressing methanogenic activity. Kumar et al. (2016) reported a threefold enhancement in hydrogen yield (29.5 mL H_2 g^{-1} volatile solids) following the application of 2-BES during the dark fermentation of microalgal hydrolysate. Furthermore, certain antibiotics, including roxithromycin, have been shown to preferentially inhibit methanogens while exerting minimal effects on hydrolytic and acidogenic bacteria, leading to increased volatile fatty acid (VFA) accumulation and improved hydrogen production. Despite their effectiveness, chemical inhibitors require careful optimization to avoid unintended inhibitory effects on hydrogen-producing microorganisms and to minimize residual toxicity, particularly when microalgae-derived substrates are used. In this context, pH manipulation offers a comparatively simple and effective alternative for methanogen suppression. Acidic operating conditions disrupt methanogenic cell membranes, interfere with proton motive force generation, and impair enzyme functionality, thereby restricting methanogenic metabolism. Similarly, surfactants such as sodium dodecyl sulfate (SDS) have been reported to selectively inhibit methanogens at elevated concentrations by damaging cell membranes and altering surface properties (Takizawa et al. 2020). The use of metal ions, including nickel and cobalt, has also been explored as a methanogenesis inhibition strategy, with reported reductions in methanogenic activity of up to 40%. However, the application of such metals raises significant environmental, economic, and regulatory concerns due to their potential toxicity, accumulation in effluents, and the need for additional downstream treatment. To overcome these limitations, bioelectrochemical systems (BES), which apply controlled electric currents to anaerobic reactors, have emerged as a promising alternative. BES can suppress methanogenic activity while simultaneously enhancing hydrogen recovery by influencing microbial electron transfer pathways and redox balance. Beyond external suppression techniques, intrinsic system factors—including microbial community composition, substrate characteristics, and reactor configuration—play a decisive role in determining the overall effectiveness of methanogen control strategies. While physical, chemical, and electrochemical interventions provide direct means of enhancing hydrogen yields, emerging internal approaches based on microbial and metabolic engineering offer complementary solutions. Genetic modification of hydrogen-producing microorganisms enables targeted redirection of metabolic flux toward hydrogen evolution, reduced formation of competing end-products, and improved tolerance to inhibitory conditions (Singh et al. 2021; Nath et al. 2021). Collectively, the integration of selective inhibition strategies with advanced microbial engineering holds substantial promise for overcoming methanogenic bottlenecks and achieving sustainable, high-efficiency biohydrogen production.

2.5.3 *Genetic Modification*

Genetic modification has emerged as a highly effective strategy for enhancing bio-H_2 production by enabling direct intervention in microbial metabolic networks (Mohanraj et al. 2019). Through precise manipulation of genes and regulatory elements, metabolic fluxes can be redirected toward hydrogen evolution, cellular tolerance to adverse process conditions can be improved, and competing biochemical pathways can be selectively suppressed. However, the success of genetic engineering approaches depends on careful control of gene expression levels, as excessive overexpression or complete gene knockout may impose metabolic burden, disrupt redox balance, or compromise cell viability, thereby reducing overall process stability. One of the most extensively explored genetic strategies involves enhancing hydrogenase activity, which represents the final and rate-limiting step in biological hydrogen production. This can be achieved either through overexpression of native hydrogenase-encoding genes or through heterologous expression and structural modification to improve catalytic efficiency and oxygen tolerance (Zhao et al. 2017). Son et al. (2021) demonstrated this approach by genetically engineering

Clostridium acetobutylicum to express the CA-zwf (pIMP-zwf) and CA-hydrA (pMTL-hydrA) genes, resulting in 1.15- and 1.39-fold increases in hydrogen production, respectively, compared with the wild-type strain. Similarly, the hydrogenase gene cluster *hoxEFUYH* from *Synechocystis* sp. PCC 6803 was successfully introduced into *Enterobacter aerogenes* ATCC 13408, leading to an 88% enhancement in hydrogen yield relative to the native strain. Khanna et al. (2011) further reported that amplification and heterologous expression of the *hydrA* gene from *Enterobacter cloacae* IIT-BT 08 using the pGEX-Kan-*hydrA* vector resulted in a 1.3-fold increase in hydrogenase activity. Beyond hydrogenase-centric approaches, metabolic flux redirection has proven to be a powerful genetic strategy for maximizing hydrogen production. This approach focuses on attenuating or eliminating competing pathways—such as lactate, ethanol, or acetate formation—thereby channeling a greater proportion of reducing equivalents toward hydrogen evolution. In this context, manipulation of intracellular redox balance, particularly the $NADH/NAD^+$ pool, has gained significant attention. Elevated NADH availability enhances ferredoxin reduction, which serves as a critical electron donor for hydrogenase-mediated hydrogen production (Kontur et al. 2012). Seol et al. (2016) demonstrated that overexpression of the *zwf* and *gnd* genes redirected carbon flux from the Embden–Meyerhof–Parnas (EMP) pathway toward the pentose phosphate (PP) pathway, leading to increased NADPH and NADH generation and achieving a hydrogen yield of 1.32 mol H_2 mol^{-1} glucose. Gene knockout strategies provide an additional layer of control by selectively removing nonessential or counterproductive metabolic routes. Jiang et al. (2023) reported that simultaneous knockout of the *nuoE*, *nuoF*, and *nuoG* genes, combined with overexpression of *maeA*, significantly increased intracellular NADH levels and formic acid accumulation, resulting in a 65% improvement in hydrogen production compared with the wild-type strain. Such combinatorial genetic interventions highlight the importance of integrated pathway engineering rather than single-gene modifications. Genetic engineering approaches have also shown considerable promise in microalgal systems. Metabolic modification of *Auxenochlorella protothecoides*, for instance, enhanced intracellular electron transfer efficiency and biomass accumulation, thereby improving its suitability as a feedstock for downstream dark fermentation-based hydrogen production. In filamentous cyanobacteria, Ehira et al. (2014) overexpressed a sucrose-phosphate synthase gene in *Anabaena* sp. PCC 7120, achieving sucrose accumulation levels of up to 10% of dry biomass, which substantially increased the availability of fermentable carbohydrates for subsequent biohydrogen generation. Despite the substantial potential of genetic modification to improve bio-H_2 yields, several technical and societal challenges remain. These include ensuring long-term genetic stability under industrial operating conditions, implementing effective biocontainment strategies to prevent environmental release, navigating complex regulatory approval processes, and achieving public acceptance of genetically modified microorganisms for large-scale applications. Addressing these challenges through robust genetic circuit design, comprehensive risk assessment, and harmonized regulatory frameworks will be essential for translating laboratory-scale genetic engineering advances into commercially viable and sustainable biohydrogen production technologies.

2.5.4 *Microbial Immobilization*

Microbial immobilization, which involves the physical attachment or encapsulation of microorganisms onto or within solid support materials, has emerged as an effective strategy for enhancing bio-H_2 production in dark fermentation systems. By promoting biofilm formation and increasing effective cell density within the reactor, immobilization improves process stability, prevents biomass washout, and enhances resistance to environmental stresses such as high organic loading rates, toxic compounds, and sudden pH fluctuations. In addition, immobilized systems provide localized buffering effects and physical protection, thereby sustaining microbial activity under otherwise inhibitory conditions (Maru et al. 2021). The retention of high concentrations of metabolically active biomass facilitates rapid substrate conversion, improved electron transfer efficiency, and the development of robust, reusable hydrogen-producing microbial consortia. A variety of immobilization techniques have been reported to improve bio-H_2 productivity, including cell entrapment within polymeric or gel matrices, adsorption onto porous carrier materials, and biofilm formation on reactor surfaces, membranes, or electrode materials. Among these, adsorption onto porous supports—such as activated carbon, biochar, ceramic particles, and synthetic polymers—has attracted considerable attention due to its simplicity and scalability. Porous carriers provide extensive surface area for microbial colonization, leading to enhanced biomass retention and improved hydrogen productivity. For instance, a comparative study employing activated carbon, maghemite, and alumina as support materials for a mixed culture of *Enterobacter* sp. H1 and *Citrobacter freundii* H3 demonstrated enhanced hydrogen production across all carriers, with activated carbon yielding the highest hydrogen concentration of 184 mmol L^{-1}. Similarly, Zhang et al. (2017) identified 200 g L^{-1} activated carbon as the optimal immobilization dose for monocultures of *Enterobacter cancerogenus* HG6 2A and *Enterobacter homaechei* 83, achieving a hydrogen yield of 1.2 mol H_2 mol^{-1} glucose—corresponding to a 259% increase compared with conventional suspended-cell systems. Encapsulation-based immobilization represents another effective approach, in which microbial cells are entrapped within gel matrices such as calcium alginate, carrageenan, polyvinyl alcohol, or other synthetic polymers. This technique minimizes cell leakage, reduces contamination risk, and enhances substrate utilization efficiency by maintaining close proximity between cells and substrates. Maru et al. (2021) reported that encapsulation of a mixed culture of *Enterobacter* sp. H1 and *Citrobacter freundii* with silicon dioxide in a synthetic medium resulted in hydrogen production of 174 mmol L^{-1}. Encapsulation also improves microbial tolerance to inhibitory compounds and fluctuating operational conditions, thereby supporting stable long-term reactor performance. Biofilm-based immobilization further enhances bio-H_2 production by enabling strong biomass retention and facilitating synergistic interactions among microbial populations. The formation of stable biofilms supports continuous hydrogen production by maintaining high microbial density and protecting cells from hydraulic disturbances. Kirli and Kapdan (2016) demonstrated that immobilizing acidogenic cultures of anaerobic sludge on a plastic nylon sponge reinforced with metal mesh increased hydrogen yields to 2.1 mol H_2 mol^{-1} glucose, compared with only 0.55 mol H_2 mol^{-1} glucose in a control reactor lacking particle support. The improved performance was attributed to enhanced microbial attachment, reduced biomass loss, and increased biofilm stability. Hybrid immobilization strategies, which combine multiple immobilization mechanisms, have shown superior performance by integrating the advantages of adsorption, encapsulation, and biofilm formation. Kumar et al. (2015) developed a hybrid immobilization matrix composed of 2% (w/v) sodium alginate, 1% (w/v) chitosan, and 2% (w/v) activated carbon, extruded into a 2% (w/v) calcium chloride solution. Using anaerobic granular sludge from a brewery wastewater treatment plant and red microalgae (*Gelidium amansii*) hydrolysate as the substrate, this system achieved complete removal of the inhibitory compound 5-hydroxymethylfurfural (5-HMF) and attained a hydrogen production rate of 1.48 L L^{-1} d^{-1}. The enhanced performance was attributed to improved protection of the microbial inoculum from inhibitory compounds, enhanced nutrient adsorption, and increased effective surface area for microbial attachment. Overall, microbial immobilization significantly improves bio-H_2 production by maintaining high cell density, stabilizing reactor operation, increasing tolerance to inhibitory and fluctuating conditions, and enabling repeated or continuous biomass reuse. These advantages make immobilization a highly promising strategy for the scale-up and long-term operation of dark fermentation-based hydrogen production systems (Sekoai et al. 2018).

2.5.5 Nanotechnological Approaches

The rapid progress of nanotechnology has opened new avenues for intensifying dark fermentative bio-H_2 production by enhancing microbial activity, accelerating biomass conversion, and catalyzing critical biochemical reactions at the nanoscale. As comprehensively reviewed by Biswal et al. (2023), nano-enabled interventions can substantially improve hydrogen yields by increasing substrate bioavailability, promoting interspecies electron transfer, and enhancing the efficiency of redox-driven metabolic pathways. Nanotechnology-assisted bio-H_2 production encompasses a diverse range of strategies, including the application of metal-based nano-cofactors, nanocatalysts, and nanoparticle-assisted biomass pretreatment processes. Among these approaches, metal-based nanoparticles particularly those containing iron and nickel have attracted significant attention due to their direct interaction with hydrogenase enzymes. These nanomaterials act as cofactors that stabilize hydrogenase structure, enhance electron transfer kinetics, and improve catalytic efficiency, thereby facilitating hydrogen evolution. Taherdanak et al. (2015) demonstrated that supplementation of dark fermentation systems with zero-valent iron (Fe^0) and nickel (Ni^0) nanoparticles increased hydrogen yield by nearly 200%, achieving a maximum of 149.8 mL H_2 g^{-1} volatile solids (VS) under optimized conditions. Similarly, the incorporation of Fe_3O_4 nanoparticles during the dark fermentation of *Chlorella* sp. biomass resulted in a 44% increase in hydrogen yield at an optimal nanoparticle concentration of 30 mg L^{-1}, highlighting the dose-dependent nature of nanoparticle effects. Beyond iron-based nanomaterials, noble and transition metal nanoparticles including gold, silver, palladium, copper, and cobalt have demonstrated strong nanocatalytic potential owing to their superior electrical conductivity, redox flexibility, and surface reactivity. These properties enable effective modulation of microbial metabolic pathways involved in hydrogen production. Since hydrogenase-mediated hydrogen evolution requires a continuous supply of reducing equivalents, metal nanoparticles can function as electron shuttles or donors, thereby enhancing enzyme–substrate interactions and accelerating hydrogen generation. For example, El-Sheekh et al. (2023) reported that the addition of magnesium–zinc ferrite ($Mg–ZnFe_2O_4$) nanoparticles to enzymatically pretreated *Oscillatoria* sp. biomass, fermented using a co-culture of *Klebsiella pneumoniae* and *Enterobacter cloacae*, resulted in a hydrogen yield of 3254 mL H_2 L^{-1} hydrolysate. Likewise, the application of Fe_3O_4 nanoparticles during fermentation of *Lyngbya limnetica* red microalgal hydrolysate enhanced cumulative hydrogen production by 37.14% relative to nanoparticle-free controls. Synergistic effects have also been observed when combining multiple nanometals. In one study, cobalt nanoparticles alone produced 5.40 mL H_2 g^{-1} VS, whereas the combined application of cobalt and zero-valent iron nanoparticles increased hydrogen production to 7.0 mL H_2 g^{-1} VS, demonstrating that mixed nanometal systems can further enhance electron transfer efficiency and hydrogenase activity. In addition to directly enhancing enzymatic hydrogen evolution, nanomaterials play a crucial role in biomass pretreatment through nanocatalytic mechanisms that improve the hydrolysis of complex carbohydrates into fermentable sugars. Nanoparticle-assisted pretreatment disrupts cell walls, increases surface area, and facilitates enzymatic access to intracellular polysaccharides. For instance, $Mg–ZnFe_2O_4$ nanoparticles enabled nearly a fourfold increase in sugar release from *Oscillatoria acuminata* biomass compared to thermochemical pretreatment alone, with sugar reductions of 230.48 mg g^{-1} for nanoparticle-assisted biological treatment versus 45.34 mg g^{-1} for conventional thermochemical treatment (El-Sheekh et al. 2023). Similarly, the use of nickel ferrite nanoparticles increased sugar yield by up to 2.5-fold and enhanced hydrogen production rates to 19.92 mL L^{-1} h^{-1} when microalgal biomass was employed as the substrate. Despite these promising laboratory-scale outcomes, the large-scale application of nanotechnology-assisted biohydrogen production remains constrained by several technical and environmental challenges. These include the high cost and energy intensity of nanoparticle synthesis, potential ecotoxicological and human health risks, difficulties associated with nanoparticle recovery and reuse, and the risk of microbial inhibition at elevated nanoparticle concentrations. Addressing these limitations through sustainable nanoparticle design, comprehensive life cycle and risk assessments, and integration of recoverable or magnetic nanomaterials into scalable reactor systems will be essential to unlock the full industrial potential of nano-enabled biohydrogen production (Srivastava et al. 2021).

2.5.6 Detoxification of Inhibitors

Despite substantial progress in bioreactor optimisation, biomass pretreatment, genetic engineering, microbial immobilization, and nanotechnology-assisted process intensification, the presence of inhibitory compounds remains a critical bottleneck limiting bio-H_2 production. These inhibitors are frequently generated during biomass pretreatment, microalgal hydrolysis, and subsequent fermentation stages, and they can severely impair microbial viability, disrupt enzymatic activity, and destabilize metabolic pathways, ultimately leading to reduced hydrogen yields. Consequently, the development and integration of robust detoxification strategies are essential for improving process efficiency, operational stability, and scalability, particularly when microalgae-derived substrates are employed. Detoxification strategies applied in biohydrogen production systems can be broadly categorized into physical, chemical, and biological approaches, each targeting the selective removal or neutralization of inhibitory compounds. Physical detoxification methods primarily rely on adsorption or separation mechanisms to eliminate toxic compounds from hydrolysates or fermentation broths. Commonly used materials include activated carbon, biochar, ion-exchange resins, membrane filtration systems, and simulated moving bed separators (Luo et al. 2022). These materials effectively adsorb a wide range of inhibitors, such as phenolics, furans, organic acids, and heavy metals. Evaporation-based techniques may also be applied to remove volatile inhibitors, including ethanol; however, such approaches can inadvertently remove beneficial volatile organic compounds, potentially compromising downstream fermentation performance. Kim et al. (2022) demonstrated that biochar derived from *Eucheuma spinosum* efficiently adsorbed multiple toxic inhibitors from pretreated microalgal biomass, resulting in an enhanced hydrogen productivity of 1.98 mL H_2 mL^{-1}. Chemical detoxification strategies focus on the neutralization, precipitation, or extraction of inhibitory compounds through targeted chemical reactions. Alkaline reagents such as lime or caustic soda are commonly used to precipitate lignin fragments, denature proteins, and immobilize metal ions, thereby reducing overall inhibitor concentrations. Wang et al. (2018a, b) reported that free-ammonia pretreatment effectively neutralized inhibitory compounds and increased hydrogen yield from 2.5 to 3.8 L H_2 kg^{-1} volatile solids (VS) when microalgae were used as the feedstock. More recently, hydrophobic deep eutectic solvents (DESs) have emerged as environmentally benign alternatives for chemical detoxification. Makos-Chełstowska et al. (2021) achieved removal efficiencies of 86.71% for furfural, 77.92% for 5-hydroxymethylfurfural (HMF), and 70.64% for levulinic acid from microalgal hydrolysate using DES-based extraction. Similarly, Quaid and Reza (2023) identified type-5 hydrophobic DESs, including tetrabutylammonium chloride, capable of selectively absorbing HMF, levulinic acid, and sorbitol, thereby alleviating fermentation inhibition. Sindermann et al. (2016) further demonstrated that a combined solvent–salt system using methyl-isobutyl-ketone (MIBK) and sodium chloride achieved up to 99% extraction efficiency for HMF. Collectively, these chemical detoxification approaches offer adaptable and effective solutions for conditioning microalgal feedstocks prior to fermentative hydrogen production. Biological detoxification strategies exploit the metabolic capabilities of microorganisms to degrade or transform inhibitory compounds into less toxic intermediates or rely on advanced cell retention systems to selectively remove inhibitors while preserving fermentable substrates. Certain microbial strains possess enzymatic pathways capable of metabolizing furfural, HMF, and related toxic compounds, thereby reducing their inhibitory effects. Kumar et al. (2016) reported complete (100%) removal of 5-HMF using immobilized microbial systems, achieving hydrogen production rates of 1.48 L L^{-1} d^{-1} from *Gelidium amansii* hydrolysate. In parallel, membrane-assisted cell retention technologies based on selective pore size, surface charge, and hydrophobicity have enabled efficient inhibitor removal while retaining up to 16.6 g L^{-1} of fermentable sugars, resulting in a threefold increase in hydrogen production. These findings underscore the effectiveness of immobilized and retained microbial systems in maintaining stable and high-rate hydrogen production under inhibitory conditions. Overall, the effective detoxification of inhibitory compounds plays a pivotal role in enhancing bio-H_2 yields by preserving microbial health, sustaining enzymatic activity, and stabilizing fermentative metabolic pathways. Reducing inhibitor toxicity enables prolonged microbial activity and supports higher hydrogen productivity over extended operational periods. Nevertheless, detoxification strategies must be critically evaluated with respect to economic feasibility, energy requirements, environmental sustainability, and potential interference with valuable fermentation products. A balanced integration of detoxification methods into biohydrogen production systems will be essential to ensure their practical viability and successful scale-up for industrial applications.

2.6 Challenges and Future Opportunities

The global energy community increasingly regards hydrogen as a versatile intermediate energy carrier with significant potential for renewable energy storage and long-distance energy transport, as well as a clean fuel for fuel cell technologies capable of delivering near-zero emissions. Despite these advantages, the large-scale commercial deployment of microalgae-based biohydrogen production remains limited, primarily due to the high energy inputs and economic constraints associated with optimizing biomass cultivation, harvesting, and downstream processing technologies. Nevertheless, biohydrogen continues to be recognized as a promising renewable fuel owing to its potential to substantially reduce greenhouse gas emissions in the energy and transportation sectors while decreasing reliance on fossil fuel resources. This chapter critically evaluates recent technological advances in microalgal cultivation strategies and downstream processing techniques that aim to facilitate the sustainable and economically viable production of biohydrogen. While the precise role of biohydrogen within a future hydrogen-based energy economy has yet to be fully defined, its integration as a sustainable energy vector is expected to yield significant environmental and socioeconomic benefits, provided that key scientific, technical, and economic barriers are effectively addressed. Consequently, a comprehensive understanding of microalgae-driven biohydrogen production pathways is essential for the development of efficient, scalable, and robust strategies capable of supporting next-generation clean energy systems.

References

Adessi A, Corneli E, De Philippis R (2017) Photosynthetic purple non sulfur bacteria in hydrogen producing systems: new approaches in the use of well known and innovative substrates. In: Modern topics in the phototrophic prokaryotes: environmental and applied aspects. Springer International Publishing, Cham, pp 321–350

Ahmed ME, Chattopadhyay S, Wang L, Brazzolotto D, Pramanik D, Aldakov D, Fize J, Morozan A, Gennari M, Duboc C, Dey A (2018) Hydrogen evolution from aqueous solutions mediated by a heterogenized [NiFe]-hydrogenase model: low pH enables catalysis through an enzyme-relevant mechanism. Angew Chem Int Ed 57(49):16001–16004

Ahmed SF, Mofijur M, Parisa TA, Islam N, Kusumo F, Inayat A, Le VG, Badruddin IA, Khan TY, Ong HC (2022) Progress and challenges of contaminate removal from wastewater using microalgae biomass. Chemosphere 286:131656

Ahmed SF, Debnath JC, Mehejabin F, Islam N, Tripura R, Mofijur M, Hoang AT, Rasul MG, Vo DVN (2023) Utilization of nanomaterials in accelerating the production process of sustainable biofuels. Sustain Energy Technol Assess 55:102894

Akhlaghi N, Najafpour-Darzi G (2020) A comprehensive review on biological hydrogen production. Int J Hydrog Energy 45(43):22492–22512

Aldaby ES, Mahmoud AH, El-Bery HM, Ali MM, Shoreit AA, Mawad AM (2023) Microalgal upgrading of the fermentative biohydrogen produced from Bacillus coagulans via non-pretreated plant biomass. Microb Cell Factories 22(1):190

Amutha KB, Murugesan AG (2011) Biological hydrogen production by the algal biomass Chlorella vulgaris MSU 01 strain isolated from pond sediment. Bioresour Technol 102(1):194–199

Anwar M, Lou S, Chen L, Li H, Hu Z (2019) Recent advancement and strategy on bio-hydrogen production from photosynthetic microalgae. Bioresour Technol 292:121972

Anzola-Rojas MDP, Fuess LT, Zaiat M (2024) Specific organic loading rate control for improving fermentative hydrogen production. Fermentation 10(4):213

Ardo FM, Lim JW, Ramli A, Lam MK, Kiatkittipong W, Abdelfattah EA, Shahid MK, Usman A, Wongsakulphasatch S, Sahrin NT (2022) A review in redressing challenges to produce sustainable hydrogen from microalgae for aviation industry. Fuel 330:125646

Arimbrathodi SP, Javed MA, Hamouda MA, Aly Hassan A, Ahmed ME (2023) BioH2 production using microalgae: highlights on recent advancements from a bibliometric analysis. Water 15(1):185

Ayub HMU, Nizami M, Qyyum MA, Iqbal N, Al-Muhtaseb AAH, Hasan M (2024) Sustainable hydrogen production via microalgae: technological advancements, economic indicators, environmental aspects, challenges, and policy implications. Environ Res 244:117815

Bahadar A, Kanthasamy R, Sait HH, Zwawi M, Algarni M, Ayodele BV, Cheng CK, Wei LJ (2022) Elucidating the effect of process parameters on the production of hydrogen-rich syngas by biomass and coal Co-gasification techniques: a multi-criteria modeling approach. Chemosphere 287:132052

Barghash H, Al Farsi M, AlRashdi Z, Okedu KE, Al-Alawi A (2024) Achieving decarbonization considering green hydrogen production: case study of Oman. Results Eng 23:102657

Batista AP, Moura P, Marques PA, Ortigueira J, Alves L, Gouveia L (2014) Scenedesmus obliquus as feedstock for biohydrogen production by Enterobacter aerogenes and Clostridium butyricum. Fuel 117:537–543

Batista AP, Ambrosano L, Graça S, Sousa C, Marques PA, Ribeiro B, Botrel EP, Neto PC, Gouveia L (2015) Combining urban wastewater treatment with biohydrogen production–an integrated microalgae-based approach. Bioresour Technol 184:230–235

Benemann JR (2000) Hydrogen production by microalgae. J Appl Phycol 12:291–300. https://doi.org/10.1023/A:1008175112704

Bhatia SK, Jagtap SS, Bedekar AA, Bhatia RK, Rajendran K, Pugazhendhi A, Rao CV, Atabani AE, Kumar G, Yang YH (2021) Renewable biohydrogen production from lignocellulosic biomass using fermentation and integration of systems with other energy generation technologies. Sci Total Environ 765:144429

Biswal T, Shadangi KP, Sarangi PK (2023) Application of nanotechnology in the production of biohydrogen: a review. Chem Eng Technol 46(2):218–233

Boodhun BSF, Mudhoo A, Kumar G, Kim SH, Lin CY (2017) Research perspectives on constraints, prospects and opportunities in biohydrogen production. Int J Hydrog Energy 42(45):27471–27481

Boshagh F, Rostami K, Moazami N (2024) Dark fermentative hydrogen production in packed-bed bioreactor using the Persian Gulf dead coral, ceramic saddle, and ceramic ball as support matrixes. Int J Hydrog Energy 52:447–456

Chaher NEH, Engler N, Nassour A, Nelles M (2023) Effects of co-substrates' mixing ratios and loading rate variations on food and agricultural wastes' anaerobic co-digestion performance. Biomass Convers Biorefinery 13(8):7051–7066

Cheirsilp B, Torpee S (2012) Enhanced growth and lipid production of microalgae under mixotrophic culture condition: effect of light intensity, glucose concentration and fed-batch cultivation. Bioresour Technol 110:510–516

Chen PC, Fan SH, Chiang CL, Lee CM (2008) Effect of growth conditions on the hydrogen production with cyanobacterium Anabaena sp. strain CH3. Int J Hydrog Energy 33(5):1460–1464

Cheng YW, Ng KH, Lam SS, Lim JW, Wongsakulphasatch S, Witoon T, Cheng CK (2019) Syngas from catalytic steam reforming of palm oil mill effluent: an optimization study. Int J Hydrog Energy 44(18):9220–9236

Chong CC, Cheng YW, Ng KH, Vo DVN, Lam MK, Lim JW (2022) Bio-hydrogen production from steam reforming of liquid biomass wastes and biomass-derived oxygenates: a review. Fuel 311:122623

Chowdury KH, Nahar N, Deb UK (2020) The growth factors involved in microalgae cultivation for biofuel production: a review. Comput Water Energy Environ Eng 9(4):185–215

Dalena F, Senatore A, Tursi A, Basile A (2017) Bioenergy production from second-and third-generation feedstocks. In: Bioenergy systems for the future. Woodhead Publishing, pp 559–599

Das SR, Basak N (2021) Molecular biohydrogen production by dark and photo fermentation from wastes containing starch: recent advancement and future perspective. Bioprocess Biosyst Eng 44(1):1–25

Delavar MA, Wang J (2021) Numerical investigation of pH control on dark fermentation and hydrogen production in a microbioreactor. Fuel 292:120355

Deviram G, Mathimani T, Anto S, Ahamed TS, Ananth DA, Pugazhendhi A (2020) Applications of microalgal and cyanobacterial biomass on a way to safe, cleaner and a sustainable environment. J Clean Prod 253:119770

dos Santos MGB, Duarte RL, Maciel AM, Abreu M, Reis A, de Mendonça HV (2021) Microalgae biomass production for biofuels in Brazilian scenario: a critical review. Bioenergy Res 14(1):23–42

Ehira S, Kimura S, Miyazaki S, Ohmori M (2014) Sucrose synthesis in the nitrogen-fixing cyanobacterium Anabaena sp. strain PCC 7120 is controlled by the two-component response regulator OrrA. Appl Environ Microbiol 80(18):5672–5679

El-Dalatony MM, Zheng Y, Ji MK, Li X, Salama ES (2020) Metabolic pathways for microalgal biohydrogen production: current progress and future prospectives. Bioresour Technol 318:124253. https://doi.org/10.1016/j.biortech.2020.124253. Epub 2020 Oct 15. PMID: 33129070

Eloffy MG, Elgarahy AM, Saber AN, Hammad A, El-Sherif DM, Shehata M, Mohsen A, Elwakeel KZ (2022) Biomass-to-sustainable biohydrogen: insights into the production routes, and technical challenges. Chem Eng J Adv 12:100410

El-Sheekh M, Elshobary M, Abdullah E, Abdel-Basset R, Metwally M (2023) Application of a novel biological-nanoparticle pretreatment to Oscillatoria acuminata biomass and coculture dark fermentation for improving hydrogen production. Microb Cell Factories 22(1):34

Fangkum A, Reungsang A (2011) Biohydrogen production from sugarcane bagasse hydrolysate by elephant dung: effects of initial pH and substrate concentration. Int J Hydrog Energy 36(14):8687–8696

Geada P, Vasconcelos V, Vicente A, Fernandes B (2017) Microalgal biomass cultivation. In: Algal green chemistry. Elsevier, pp 257–284

Goswami RK, Mehariya S, Obulisamy PK, Verma P (2021) Advanced microalgae-based renewable biohydrogen production systems: a review. Bioresour Technol 320:124301

Greening C, Geier R, Wang C, Woods LC, Morales SE, McDonald MJ, Rushton-Green R, Morgan XC, Koike S, Leahy SC, Kelly WJ (2019) Diverse hydrogen production and consumption pathways influence methane production in ruminants. ISME J 13(10):2617–2632

Hawkes FR, Dinsdale R, Hawkes DL, Hussy I (2002) Sustainable fermentative hydrogen production: challenges for process optimisation. Int J Hydrog Energy 27(11–12):1339–1347

IEA (2020) World energy balances: overview. IEA, Paris, p 387

Iqbal K, Saxena A, Pande P, Tiwari A, Joshi NC, Varma A, Mishra A (2022) Microalgae-bacterial granular consortium: striding towards sustainable production of biohydrogen coupled with wastewater treatment. Bioresour Technol 354:127203

Javed MA, Zafar AM, Hassan AA, Zaidi AA, Farooq M, El Badawy A, Lundquist T, Mohamed MMA, Al-Zuhair S (2022) The role of oxygen regulation and algal growth parameters in hydrogen production via biophotolysis. J Environ Chem Eng 10(1):107003

Jayaseelan M, Usman M, Somanathan A, Palani S, Muniappan G, Jeyakumar RB (2021) Microalgal production of biofuels integrated with wastewater treatment. Sustainability 13(16):8797

Jiang K, Bai R, Gao T, Lu P, Zhang J, Zhang S, Xu F, Wang S, Zhao H (2023) Optimization of hydrogen production in Enterobacter aerogenes by Complex I peripheral fragments destruction and maeA overexpression. Microb Cell Factories 22(1):137

Jiao H, Tsigkou K, Elsamahy T, Pispas K, Sun J, Manthos G, Schagerl M, Sventzouri E, Al-Tohamy R, Kornaros M, Ali SS (2024) Recent advances in sustainable hydrogen production from microalgae: mechanisms, challenges, and future perspectives. Ecotoxicol Environ Saf 270:115908. https://doi.org/10.1016/j.ecoenv.2023.115908

Jing C, Tao H, Jiang T, Wang Y, Zhai J, Cao L, Su B (2020) Population, urbanization and economic scenarios over the Belt and Road region under the Shared Socioeconomic Pathways. J Geogr Sci 30(1):68–84

Kapdan IK, Kargi F, Oztekin R, Argun H (2009) Bio-hydrogen production from acid hydrolyzed wheat starch by photo-fermentation using different Rhodobacter sp. Int J Hydrog Energy 34(5):2201–2207

Khan MI, Shin JH, Kim JD (2018) The promising future of microalgae: current status, challenges, and optimization of a sustainable and renewable industry for biofuels, feed, and other products. Microb Cell Factories 17(1):36

Khanna N, Dasgupta CN, Mishra P, Das D (2011) Homologous overexpression of [FeFe] hydrogenase in Enterobacter cloacae IIT-BT 08 to enhance hydrogen gas production from cheese whey. Int J Hydrog Energy 36(24):15573–15582

Khetkorn W, Rastogi RP, Incharoensakdi A, Lindblad P, Madamwar D, Pandey A, Larroche C (2017) Microalgal hydrogen production–a review. Bioresour Technol 243:1194–1206

Kim DH, Kim MS (2012) Semi-continuous photo-fermentative H2 production by Rhodobacter sphaeroides: effect of decanting volume ratio. Bioresour Technol 103(1):481–483

Kim SH, Da Yi Y, Kim HJ, Bhatia SK, Gurav R, Jeon JM, Yoon JJ, Kim SH, Park JH, Yang YH (2022) Hyper biohydrogen production from xylose and xylose-based hemicellulose biomass by the novel strain Clostridium sp. YD09. Biochem Eng J 187:108624

Kirli B, Kapdan IK (2016) Selection of microorganism immobilization particle for dark fermentative biohydrogen production by repeated batch operation. Renew Energy 87:697–702

Kontur WS, Noguera DR, Donohue TJ (2012) Maximizing reductant flow into microbial H2 production. Curr Opin Biotechnol 23(3):382–389

Kumar G, Sivagurunathan P, Kobayashi T, Xu KQ, Kim SH (2015) Simultaneous removal of 5-hydroxy methyl furfural (5-HMF) and hydrogen production from acid (H2SO4) pretreated red-algal hydrolysate via hybrid immobilized cells. Algal Res 11:326–333

Kumar G, Zhen G, Sivagurunathan P, Bakonyi P, Nemestóthy N, Bélafi-Bakó K, Kobayashi T, Xu KQ (2016) Biogenic H2 production from mixed microalgae biomass: impact of pH control and methanogenic inhibitor (BESA) addition. Biofuel Res J 3(3):470–474

Kumari S, Nasr M, Kumar S (2017) Technological advances in biohydrogen production from microalgae. In: Gupta S, Malik A, Bux F (eds) Algal biofuels. Springer, Cham. https://doi.org/10.1007/978-3-319-51010-1_17

Luo X, Zeng B, Zhong Y, Chen J (2022) Production and detoxification of inhibitors during the destruction of lignocellulose spatial structure. BioResources 17(1):1939

Makos-Chełstowska P, Słupek E, Kucharska K, Kramarz A, Gębicki J (2021) Efficient extraction of fermentation inhibitors by means of green hydrophobic deep eutectic solvents. Molecules 27(1):157

Maltsev Y, Maltseva K, Kulikovskiy M, Maltseva S (2021) Influence of light conditions on microalgae growth and content of lipids, carotenoids, and fatty acid composition. Biology 10(10):1060

Mandotra SK, Sharma C, Srivastava N, Ahluwalia AS, Ramteke PW (2023) Current prospects and future developments in algal bio-hydrogen production: a review. Biomass Convers Biorefinery 13(10):8575–8592

Maneeruttanarungroj C, Phunpruch S (2017) Effect of pH on biohydrogen production in green alga Tetraspora sp. CU2551. Energy Procedia 138:1085–1092

Marchi B, Zanoni S, Zavanella LE, Jaber MY (2019) Supply chain models with greenhouse gases emissions, energy usage, imperfect process under different coordination decisions. Int J Prod Econ 211:145–153

Maru BT, Lopez F, Medina F, Constantí M (2021) Improvement of biohydrogen and usable chemical products from glycerol by co-culture of Enterobacter spH1 and Citrobacter Freundii H3 using different supports as surface immobilization. Fermentation 7(3):154

Melitos G, Voulkopoulos X, Zabaniotou A (2021) Waste to sustainable biohydrogen production via photo-fermentation and biophotolysis– a systematic review. Renew Energy Environ Sustain 6:45

Mikheeva ER, Katraeva IV, Vorozhtsov DL, Kovalev DA, Kovalev AA, Grigoriev VS, Litti YV (2022) Dark fermentative biohydrogen production from confectionery wastewater in continuous-flow reactors. Int J Hydrog Energy 47(53):22348–22358

Mishra P, Krishnan S, Rana S, Singh L, Sakinah M, Ab Wahid Z (2019) Outlook of fermentative hydrogen production techniques: an overview of dark, photo and integrated dark-photo fermentative approach to biomass. Energ Strat Rev 24:27–37

Mohammadzadeh R, Mahnert A, Duller S, Moissl-Eichinger C (2022) Archaeal key-residents within the human microbiome: characteristics, interactions and involvement in health and disease. Curr Opin Microbiol 67:102146

Mohanraj S, Pandey A, Mohan SV, Anbalagan K, Kodhaiyolii S, Pugalenthi V (2019) Metabolic engineering and molecular biotechnology of biohydrogen production. In: Biohydrogen. Elsevier, pp 413–434

Mona S, Kumar SS, Kumar V, Parveen K, Saini N, Deepak B, Pugazhendhi A (2020) Green technology for sustainable biohydrogen production (waste to energy): a review. Sci Total Environ 728:138481

Morales-Sánchez D, Martinez-Rodriguez OA, Martinez A (2017) Heterotrophic cultivation of microalgae: production of metabolites of commercial interest. J Chem Technol Biotechnol 92(5):925–936

Nagarajan D, Lee DJ, Kondo A, Chang JS (2017) Recent insights into biohydrogen production by microalgae–from biophotolysis to dark fermentation. Bioresour Technol 227:373–387

Nath D, Chakraborty I, Ghangrekar MM (2021) Methanogenesis inhibitors used in bio-electrochemical systems: a review revealing reality to decide future direction and applications. Bioresour Technol 319:124141

Nguyen HY et al (2024) Microalgae-based biohydrogen production. In: Microbial biohydrogen production. Springer. https://doi.org/10.1007/978-3-031-75132-5_1

Nyberg M, Heidorn T, Lindblad P (2015) Hydrogen production by the engineered cyanobacterial strain Nostoc PCC 7120 ΔhupW examined in a flat panel photobioreactor system. J Biotechnol 215:35–43

Okonkwo O, Escudie R, Bernet N, Mangayil R, Lakaniemi AM, Trably E (2019) Impacts of short-term temperature fluctuations on biohydrogen production and resilience of thermophilic microbial communities. Int J Hydrog Energy 44(16):8028–8037

Oncel SVSF, Sukan FV (2011) Effect of light intensity and the light: dark cycles on the long term hydrogen production of Chlamydomonas reinhardtii by batch cultures. Biomass Bioenergy 35(3):1066–1074

Palomo-Briones R, Razo-Flores E, Bernet N, Trably E (2017) Dark-fermentative biohydrogen pathways and microbial networks in continuous stirred tank reactors: novel insights on their control. Appl Energy 198:77–87

Peter AP, Koyande AK, Chew KW, Ho SH, Chen WH, Chang JS, Krishnamoorthy R, Banat F, Show PL (2022) Continuous cultivation of microalgae in photobioreactors as a source of renewable energy: current status and future challenges. Renew Sustain Energy Rev 154:111852

Quaid T, Reza T (2023) COSMO-RS predictive screening of type 5 hydrophobic deep eutectic solvents for selective platform chemicals absorption. J Mol Liq 382:121918

Ramaraj R, Murugesan S, Unpaprom Y (2025) Recent advances in microalgae-based biohydrogen production: a review of innovative strategies, pretreatment, genetic engineering, and bioelectrochemical technologies. Maejo Int J Energy Environ Commun 7(1):42–56

Rastogi RP, Madamwar D, Nakamoto H, Incharoensakdi A (2020) Resilience and self-regulation processes of microalgae under UV radiation stress. J Photochem Photobiol C: Photochem Rev 43:100322

Razu MH, Hossain F, Khan M (2019) Advancement of bio-hydrogen production from microalgae. In: Microalgae biotechnology for development of biofuel and wastewater treatment. Springer Singapore, Singapore, pp 423–462

Ri PC, Kim JS, Kim TR, Pang CH, Mun HG, Pak GC, Ren NQ (2019) Effect of hydraulic retention time on the hydrogen production in a horizontal and vertical continuous stirred-tank reactor. Int J Hydrog Energy 44(33):17742–17749

Ruiz-Marin A, Canedo-López Y, Chávez-Fuentes P (2020) Biohydrogen production by Chlorella vulgaris and Scenedesmus obliquus immobilized cultivated in artificial wastewater under different light quality. AMB Express 10(1):191

Ryue J, Lin L, Kakar FL, Elbeshbishy E, Al-Mamun A, Dhar BR (2020) A critical review of conventional and emerging methods for improving process stability in thermophilic anaerobic digestion. Energy Sustain Dev 54:72–84

Sahrin NT, Khoo KS, Lim JW, Shamsuddin R, Ardo FM, Rawindran H, Hassan M, Kiatkittipong W, Abdelfattah EA, Da Oh W, Cheng CK (2022) Current perspectives, future challenges and key technologies of biohydrogen production for building a carbon–neutral future: a review. Bioresour Technol 364:128088

Sambusiti C, Bellucci M, Zabaniotou A, Beneduce L, Monlau F (2015) Algae as promising feedstocks for fermentative biohydrogen production according to a biorefinery approach: a comprehensive review. Renew Sustain Energy Rev 44:20–36

Sathinathan P, Parab HM, Yusoff R, Ibrahim S, Vello V, Ngoh GC (2023) Photobioreactor design and parameters essential for algal cultivation using industrial wastewater: a review. Renew Sustain Energy Rev 173:113096

Sekoai PT, Awosusi AA, Yoro KO, Singo M, Oloye O, Ayeni AO, Bodunrin M, Daramola MO (2018) Microbial cell immobilization in biohydrogen production: a short overview. Crit Rev Biotechnol 38(2):157–171

Seol E, Sekar BS, Raj SM, Park S (2016) Co-production of hydrogen and ethanol from glucose by modification of glycolytic pathways in Escherichia coli–from Embden-Meyerhof-Parnas pathway to pentose phosphate pathway. Biotechnol J 11(2):249–256

Sharma A, Arya SK (2017) Hydrogen from algal biomass: a review of production process. Biotechnol Rep 15:63

Shuba ES, Kifle D (2018) Microalgae to biofuels: 'promising' alternative and renewable energy, review. Renew Sustain Energy Rev 81:743–755

Sillero L, Solera R, Perez M (2023) Effect of temperature on biohydrogen and biomethane production using a biochemical potential test with different mixtures of sewage sludge, vinasse and poultry manure. J Clean Prod 382:135237

Sindermann EC, Holbach A, de Haan A, Kockmann N (2016) Single stage and countercurrent extraction of 5-hydroxymethylfurfural from aqueous phase systems. Chem Eng J 283:251–259

Singh V, Das D (2019) Potential of hydrogen production from biomass. In: Science and engineering of hydrogen-based energy technologies. Elsevier, pp 123–164

Singh H, Das D (2020) Biohydrogen from microalgae. In: Handbook of microalgae-based processes and products. Academic Press, pp 391–418

Singh NK, Kumari P, Singh R (2021) Intensified hydrogen yield using hydrogenase rich sulfate-reducing bacteria in bio-electrochemical system. Energy 219:119583

Sivaramakrishnan R, Shanmugam S, Sekar M, Mathimani T, Incharoensakdi A, Kim SH, Parthiban A, Geo VE, Brindhadevi K, Pugazhendhi A (2021) Insights on biological hydrogen production routes and potential microorganisms for high hydrogen yield. Fuel 291:120136

Son YS, Jeon JM, Kim DH, Yang YH, Jin YS, Cho BK, Kim SH, Kumar S, Lee BD, Yoon JJ (2021) Improved bio-hydrogen production by overexpression of glucose-6-phosphate dehydrogenase and FeFe hydrogenase in Clostridium acetobutylicum. Int J Hydrog Energy 46(74):36687–36695

Srivastava N, Srivastava M, Mishra PK, Kausar MA, Saeed M, Gupta VK, Singh R, Ramteke PW (2020) Advances in nanomaterials induced biohydrogen production using waste biomass. Bioresour Technol 307:123094

Srivastava N, Hussain A, Kushwaha D, Haque S, Mishra PK, Gupta VK, Srivastava M (2021) Nickel ferrite nanoparticles induced improved fungal cellulase production using residual algal biomass and subsequent hydrogen production following dark fermentation. Fuel 304:121391

Su Y (2021) Revisiting carbon, nitrogen, and phosphorus metabolisms in microalgae for wastewater treatment. Sci Total Environ 762:144590

Sundaram T, Rajendran S, Gnanasekaran L, Rachmadona N, Jiang JJ, Khoo KS, Show PL (2023) Bioengineering strategies of microalgae biomass for biofuel production: recent advancement and insight. Bioengineered 14(1):2252228

Tagne RFT, Costa P, Casella S, Favaro L (2024) Optimization of biohydrogen production by dark fermentation of African food-processing waste streams. Int J Hydrog Energy 49:266–276

Taherdanak M, Zilouei H, Karimi K (2015) Investigating the effects of iron and nickel nanoparticles on dark hydrogen fermentation from starch using central composite design. Int J Hydrog Energy 40(38):12956–12963

Takizawa S, Baba Y, Tada C, Fukuda Y, Nakai Y (2020) Sodium dodecyl sulfate improves the treatment of waste paper with rumen fluid at lower concentration but decreases at higher condition. J Mater Cycles Waste Manage 22(3):656–663

Tang T, Chen Y, Liu M, Du Y, Tan Y (2022) Effect of pH on the performance of hydrogen production by dark fermentation coupled denitrification. Environ Res 208:112663

Tena M, Perez M, Solera R (2021) Effect of hydraulic retention time on hydrogen production from sewage sludge and wine vinasse in a thermophilic acidogenic CSTR: a promising approach for hydrogen production within the biorefinery concept. Int J Hydrog Energy 46(11):7810–7820

Usmanbaha N, Jariyaboon R, Reungsang A, Kongjan P, Chu CY (2019) Optimization of batch dark fermentation of Chlorella sp. using mixed-cultures for simultaneous hydrogen and butyric acid production. Energies 12(13):2529

Vargas SR, dos Santos PV, Zaiat M, do Carmo Calijuri M (2018) Optimization of biomass and hydrogen production by Anabaena sp. (UTEX 1448) in nitrogen-deprived cultures. Biomass Bioenergy 111:70–76

Veeravalli SS, Shanmugam SR, Ray S, Lalman JA, Biswas N (2019) Biohydrogen production from renewable resources. In: Advanced bioprocessing for alternative fuels, biobased chemicals, and bioproducts. Woodhead Publishing, pp 289–312

Vijayaraghavan K, Karthik R, Nalini SK (2009) Hydrogen production by Chlamydomonas reinhardtii under light driven sulfur deprived condition. Int J Hydrog Energy 34(19):7964–7970

Wang Q, Gong Y, Liu S, Wang D, Liu R, Zhou X, Nghiem LD, Zhao Y (2018a) Free ammonia pretreatment to improve bio-hydrogen production from anaerobic dark fermentation of microalgae. ACS Sustain Chem Eng 7(1):1642–1647

Wang H, Xu J, Sheng L, Liu X, Lu Y, Li W (2018b) A review on bio-hydrogen production technology. Int J Energy Res 42(11):3442–3453

Wang AJ, Wang HC, Cheng HY, Liang B, Liu WZ, Han JL, Zhang B, Wang SS (2020) Electrochemistry-stimulated environmental bioremediation: development of applicable modular electrode and system scale-up. Environ Sci Ecotechnol 3:100050–100050

Woon JM, Khoo KS, Akermi M, Alanazi MM, Lim JW, Chan YJ, Goh PS, Chidi BS, Lam MK, Zaini J, Bilad MR (2023) Reviewing biohydrogen production from microalgal cells through fundamental mechanisms, enzymes and factors that engendering new challenges and prospects. Fuel 346:128312

Yun YM, Jung KW, Kim DH, Oh YK, Shin HS (2012) Microalgal biomass as a feedstock for biohydrogen production. Int J Hydrog Energy 37(20):15533–15539

Zhang C, Kang X, Liang N, Abdullah A (2017) Improvement of biohydrogen production from dark fermentation by cocultures and activated carbon immobilization. Energy Fuel 31(11):12217–12222

Zhao T, Xu JL, Chen H, Li YF (2010) Effects of phosphate buffer solution on fermentative biohydrogen production of Biohydrogenbacterium R3 sp. nov. Adv Mater Res 113:2185–2188

Zhao J, Song W, Cheng J, Liu M, Zhang C, Cen K (2017) Improvement of fermentative hydrogen production using genetically modified Enterobacter aerogenes. Int J Hydrog Energy 42(6):3676–3681

Ziara RM, Miller DN, Subbiah J, Dvorak BI (2019) Lactate wastewater dark fermentation: the effect of temperature and initial pH on biohydrogen production and microbial community. Int J Hydrog Energy 44(2):661–673

Chapter 3
Next-Generation Catalysts for Hydrogen Generation: Insights into Synthesis, Characterization, and Evaluation

Abstract This chapter discusses advances in catalysts for hydrogen production, covering traditional noble and transition metal systems alongside emerging technologies. It highlights plasma-assisted and microwave-enhanced catalysis, as well as nanostructured materials including nanoparticles, nanoclusters, COFs, quantum dots, and 2D materials that offer enhanced activity, selectivity, and tunable properties. The chapter emphasizes insights from computational modeling and advanced characterization for rational catalyst design and discusses key challenges such as stability, scalability, and cost. Future directions focus on developing efficient, sustainable, and commercially viable hydrogen generation technologies.

Keywords Hydrogen production · Noble metal catalysts · Transition metal catalysts · Plasma-assisted catalysis · Nanostructured catalysts · Nanoclusters · Quantum dots

3.1 Introduction

The demand for renewable and alternative energy sources has steadily increased since the oil crises of the 1970s. Fossil fuels have remained the dominant global energy source since the Industrial Revolution, whereas earlier societies relied primarily on wood and other biological materials to meet their energy needs. The prolonged and widespread dependence on fossil fuels has led to severe environmental consequences, particularly due to greenhouse gas (GHG) emissions, which have contributed significantly to climate change and environmental degradation. As a result, renewable energy technologies have emerged as promising alternatives and are now increasingly deployed across multiple sectors, currently accounting for more than 20% of global energy consumption. At the beginning of the twenty-first century, fossil fuels were considered the most cost-effective option for electricity generation (Urpelainen and Van de Graaf 2015). However, growing policy commitments to reduce GHG emissions have shifted attention toward renewable energy sources. Continuous research and development efforts have substantially reduced

D. M. Y. Mir et al., *Low-Carbon Hydrogen*, Clean Energy Production Technologies,
https://doi.org/10.1007/978-981-95-8142-9_3

the cost of renewable technologies, in many cases making them more economical than fossil-fuel-based systems. Despite these advancements, the inherent intermittency of renewable energy sources and the challenges associated with long-distance energy transportation remain major obstacles for large-scale deployment. In this context, hydrogen (H_2) has gained recognition as a versatile solution to address both intermittency and energy transport challenges. Hydrogen can function as an efficient energy storage medium and can also facilitate energy transport over long distances. Consequently, hydrogen is increasingly viewed as a key component in completing the renewable energy ecosystem by bridging supply fluctuations and enabling flexible energy delivery. These integrated systems play a crucial role in improving hydrogen production efficiency and reducing overall production costs. Globally, most hydrogen production still relies on fossil-fuel-based technologies such as gasification, steam reforming, and pyrolysis, which collectively account for the majority of current hydrogen output (Tanzil et al. 2021). However, these conventional methods are inherently unsustainable due to their dependence on fossil resources and associated carbon emissions. In light of growing concerns over climate change and international commitments such as the Kyoto Protocol, there is an urgent need to develop hydrogen production routes that do not emit CO_2. To accurately assess whether hydrogen contributes to GHG reduction compared to conventional energy carriers, it is essential to consider the entire life cycle of hydrogen production, distribution, and utilization. In recent years, sustainable hydrogen production methods such as water electrolysis and biomass-based biological processes—including algae, agricultural residues, and industrial or municipal waste—have received increasing attention due to their environmental benefits (Morya et al. 2023). Additionally, hydrogen production through wastewater electrolysis and biological pathways such as dark fermentation, photo-fermentation, microbial electrolysis cells (MECs), and hybrid biological systems has demonstrated significant potential. These approaches not only generate clean energy but also promote waste valorization by utilizing biomass residues and wastewater effluents as feedstocks, thereby reducing environmental burdens (Ndayisenga et al. 2022). Nevertheless, several technical challenges hinder the commercial viability of these sustainable hydrogen production technologies. Biological processes typically suffer from low hydrogen yields, while electrolysis-based methods face issues such as high energy consumption, electrode degradation, and membrane deterioration. Addressing these limitations is critical for improving system efficiency and achieving cost-effective and sustainable hydrogen production. Recent advancements in catalyst development have shown promise in enhancing hydrogen production performance in both electrochemical and biological systems. A wide range of commercially available and newly synthesized catalysts including copper, cobalt, nickel, palladium, platinum, and ternary-based materials have been explored for this purpose (Alpaydın et al. 2020). Furthermore, various catalyst support materials have been investigated, such as carbon-based materials (graphene, graphene oxide, activated carbon, and carbon nanotubes) and metal oxide-based supports (silicon oxide, titanium oxide, copper oxide, molybdenum oxide, among others). Several studies have demonstrated the effectiveness of advanced catalyst-support combinations.

For example, Agrell et al. (2003) reported that the incorporation of ZrO_2 significantly enhanced the activity of Cu/ZnO catalysts in methanol steam reforming, achieving turnover frequencies two to three times higher than ZrO_2-free catalysts while also improving stability and redox resistance. Yang et al. (2013) demonstrated that isolated gold atoms deposited on TiO_2 via UV irradiation exhibited exceptional catalytic activity in the water–gas shift reaction. Similarly, Navarro et al. (2006) showed that Ni-supported catalysts on Ce–La-modified Al_2O_3 outperformed Pt-based catalysts in oxidative reforming of hexadecane, offering higher hydrogen selectivity. Biswas and Kunzru (2007) further reported that a 30% Ni–$Ce_{0.74}Zr_{0.26}O_2$ catalyst displayed high catalytic activity for ethanol reforming at 600 °C. Photocatalytic systems have also demonstrated significant potential. Chen et al. (2014) developed a sandwich-structured CoO_x–MgO–Ta_3N_5 photocatalyst that exhibited remarkably high water oxidation activity under visible light, achieving an oxygen evolution rate 23 times greater than that of pure Ta_3N_5. From an energy perspective, the reference electricity demand for producing one standard cubic meter of hydrogen is approximately 5.5 kWh. Comparative assessments of catalyst performance have been conducted across various experimental studies. For instance, Xiao et al. (2014) evaluated molybdenum carbide nanorods as catalysts and reported enhanced performance in hydrogen production reactions. For various compositions of nickel–molybdenum carbide catalysts, Tafel slopes ranging from 45 to 73 mV dec^{-1} were reported in a 1 M KOH electrolyte, with the highest slope observed for the catalyst containing 10 wt% nickel. In contrast, the maximum exchange current density, reaching 0.27 mA cm^{-2}, was achieved with a 30 wt% nickel–molybdenum carbide composition, indicating enhanced electrocatalytic activity. Similarly, Faber et al. (2014) examined metallic cobalt pyrite catalysts with micro- and nanostructured morphologies for hydrogen evolution reactions (HER) in H_2SO_4 solution. Their results showed Tafel slopes between 51.4 and 58 mV dec^{-1}, along with normalized exchange current densities ranging from 1.97 to 5.27 μA cm^{-2}, highlighting the effectiveness of cobalt pyrite-based catalysts. Zhang et al. (2017a, b) investigated ternary $NiCo_2P_x$ nanowires for hydrogen production and demonstrated that the synergistic interaction between nickel and cobalt significantly enhances HER performance. Additional studies have reported improved hydrogen production efficiencies using nanostructured nickel phosphide, titania–graphene nanosheet composites, molybdenum carbide supported on graphene nanoribbons, and platinum–nickel catalysts, underscoring the importance of nano structuring and catalyst-support interactions in improving catalytic activity (Gao et al. 2016; Xie et al. 2018). More recently, Jin et al. (2022) synthesized selenium vacancy–engineered ultrafine platinum–nickel nanowires to promote hydrogen production under ultralow operating voltages. Using a water-splitting electrolyzer, the system achieved 100% Faradaic efficiency at a current density of 10 mA cm^{-2} and a voltage of only 0.637 V, demonstrating the effectiveness of defect engineering in catalyst design. In another study, earth-abundant and cost-effective IB group metals such as silver and copper were employed as electrocatalysts to enhance hydrogen evolution performance (Yang et al. 2024). The authors reported a volcano-type relationship between the Cu:Ag ratio and HER activity, with a minimum overpotential of 223 mV at a current density of 10 mA cm^{-2} in 0.5 M H_2SO_4 electrolyte. Furthermore, Han et al. (2024) explored an electrochemical redox activation strategy to precisely tune the surface valence states and microstructure of copper foam catalysts. This activation significantly improved electrocatalytic benzyl alcohol oxidation (EBA) while simultaneously enhancing hydrogen production. Compared to untreated copper foam, the activated material exhibited a threefold increase in pure hydrogen production at 1.5 V. Additionally, the anodic current density for EBA oxidation increased substantially from 9.5 to 29.3 mmol cm^{-2}, demonstrating the potential of coupling hydrogen generation with value-added chemical production. To date, numerous review articles have addressed hydrogen energy from various perspectives, including production pathways, process enhancement strategies, and environmental impacts. For example, Dawood et al. (2020) provided a comprehensive overview of hydrogen's role in the global energy transition, covering multiple production routes and hydrogen-based energy storage systems. Similarly, Younas et al. (2022) reviewed the current status, future potential, and limitations of both renewable and non-renewable hydrogen production technologies. However, only a limited number of reviews have comprehensively examined the role of catalysts in hydrogen production, and most existing studies focus on specific catalyst classes or individual processes. For instance, Acar et al. (2014) evaluated heterogeneous photocatalysts for hydrogen production from technical, environmental, and economic perspectives, while Dal Santo et al. (2012) focused on bimetallic heterogeneous catalysts. Sikander et al. (2017) reviewed hydrotalcite-based catalysts, and Ma et al. (2017) discussed molybdenum carbide as an alternative catalyst for hydrogen production. Other studies have concentrated on catalyst effects in specific applications such as proton exchange membrane (PEM) fuel cells, water electrolysis, and photo-reforming of organic compounds (Clarizia et al. 2014; Wang et al. 2021). Overall, existing literature indicates a lack of comprehensive evaluations addressing the influence of diverse catalyst systems across multiple hydrogen production pathways. This gap highlights the urgent need for an integrated and systematic assessment of catalyst roles in enhancing hydrogen production performance. To address this limitation, the present review provides, for the first time, a comprehensive overview of various catalyst types employed in hydrogen production through biological and thermochemical processes. The primary objective of this study is to present an updated and holistic perspective on recent advancements in catalyst-assisted hydrogen production. Despite the growing volume of research and review articles, a unified assessment that considers catalyst performance normalization, by-product reuse, sustainability evaluation, technological barriers, and future opportunities remains largely absent. To the best of our knowledge, this work represents the most comprehensive analysis of catalyst utilization in hydrogen production systems. Furthermore, by normalizing performance metrics across different studies using a defined base-case scenario, this review enables direct and meaningful comparison of catalyst efficiencies in terms of hydrogen production rates. This normalization-based comparative approach distinguishes the present study from existing literature and provides clearer insights into catalyst selection and optimization for future

Table 3.1 Comparative properties of carbon-based catalysts used in hydrogen production

Catalyst material	Key physical/chemical properties	Impact on H_2 production	References
Nitrogen-doped carbon nanotubes (N-CNTs)	1D structural stability and high density of pyridinic nitrogen sites.	Provides "electron highways" and reduces the overpotential for HER.	Liu and Dai (2016)
Graphene and graphene-based materials	High surface area (~2600 m^2/g); tunable bandgap; functionalization increases catalytic sites	HER overpotential: ~200 mV; exchange current density: 0.15 mA/cm^2	Zhang et al. (2017a, b)
Boron, phosphorus, and sulfur doping	Synergistic effects from multiple dopants; improves stability and hydrogen evolution activity	Hydrogen evolution rate: ~10.5 mmol h^{-1} g^{-1}; quantum efficiency: 40%	Carrasco-Jaim et al. (2018)
Polymeric catalysts	Enhanced charge separation; tunable electronic properties	HER overpotential: ~200 mV; stability: ~100 h	Choi et al. (2019)
Carbon nanotubes (CNTs)	High electrical conductivity (~10^4 S/cm); functionalization enhances active sites and stability	HER overpotential: ~230 mV; current density: 10 mA/cm^2	Choi et al. (2019)
Graphitic carbon nitride	Semiconductor properties with a narrow bandgap	Ideal for photocatalytic water splitting under visible light	Liu et al. (2022)
Carbon quantum dots	Ultra-small size, edge-rich surface	Rapid surface redox kinetics and improved light harvesting	Sun et al. (2023a, b, c)
Heteroatom-doped graphene (N, P, S)	High electrical conductivity and optimized Gibbs free energy	Enhances electron transfer; shifts adsorption sites to favor hydrogen release	Gao et al. (2024)
Single-atom catalysts (SACs) on porous carbon	Isolated metal centers (Fe, Co, Ni) anchored to carbon defects	Achieves 100% atom efficiency; mimics performance of noble metals	Samanta and Pradhan (2025)
Biomass-derived porous carbon	High specific surface area and hierarchical porosity	Increases active site density and facilitates rapid gas bubble release	Kumar et al. 2025)

hydrogen technologies. Table 3.1 compiles the structural characteristics and catalytic performance metrics of representative carbon-based catalysts, including graphene, carbon nanotubes, doped carbon materials, and polymeric catalysts, highlighting their relevance in optimizing hydrogen generation through electrochemical, photocatalytic, and reforming processes.

3.2 Traditional Catalysts for Hydrogen Production

Hydrogen production relies on a diverse range of catalysts, including noble metals such as platinum, palladium, and rhodium, alongside non-noble metals like nickel and copper (Choi et al. 2024). To enhance catalytic stability, dispersion, and overall activity, these metals are often supported on substrates such as carbon-based materials, metal oxides, and zeolites (Lee et al. 2024). Recent research has increasingly focused on developing cost-effective and sustainable alternatives to traditional catalysts, with particular attention on high-entropy materials. Composed of five or more metallic elements uniformly distributed within a single phase, high-entropy catalysts exhibit remarkable potential for improved catalytic performance, enhanced durability, and tunable properties, making them highly promising candidates for next-generation hydrogen production technologies.

3.2.1 Noble Metal Catalysts

Noble metal catalysts including Pt, Pd, Ru, Rh, and Ir have long been central to hydrogen production technologies, particularly in electrocatalytic and photocatalytic water-splitting systems. Their favorable electronic structures and exceptional catalytic properties make them benchmark materials for both the hydrogen evolution reaction (HER) and oxygen evolution reaction (OER) in water electrolysis (Niu et al. 2023). However, their high cost and limited natural abundance have motivated extensive research into strategies for improving efficiency, durability, and economic feasibility (Zhang et al. 2024a, b). Recent advances in noble metal catalyst design focus on structural and electronic engineering to enhance performance. Approaches such as crystal phase regulation, alloying, microenvironment modulation, and nanoscale architecture design have been shown to improve charge transfer kinetics, catalytic activity, and stability (Liu et al. 2020). Single-atom catalysts (SACs) have emerged as a particularly effective strategy, maximizing noble metal utilization by creating a high density of accessible active sites while maintaining strong metal support interactions that prevent agglomeration and enhance long-term stability. Computational methods, including density functional theory (DFT) and machine learning (ML), are increasingly applied to guide catalyst development. These tools enable predictive screening of alloy compositions, surface structures, and electronic properties, allowing precise tuning of hydrogen adsorption-free energies and facilitating rational design of next-generation catalysts. Experimental innovations have complemented these computational advances. Wang et al. (2023) developed nitrogen-doped graphene (NPGR) via radio-frequency cold plasma treatment, creating an amphiphilic surface that served as an effective support for Pt nanoparticles in aqueous-phase hydrogenation of cinnamaldehyde. The enhanced anchoring sites and improved metal–support interactions significantly boosted catalytic performance, particularly in Pickering emulsion systems, offering a green and scalable approach for graphene-supported noble metal catalysts. Alloying strategies have also proven effective; Cheng et al. (2024) reported a Pt_3Ni_2 nanowire catalyst (Pt_3Ni_2NWs-S/C) exhibiting superior HER activity compared to pure Pt. This performance was attributed to synergistic Pt–Ni interactions and the preferential exposure of catalytically active Pt(111) facets, highlighting the benefits of electronic structure tuning through alloy engineering.

Carbon-based supports such as graphene and carbon nanotubes have further enhanced noble metal catalysts by providing high surface area, excellent conductivity, and chemical stability. For example, Yuan et al. (2024) developed a graphene chainmail–shelled Cu–Ni alloy ($CuNi_{0.05}$@OC) for selective aqueous-phase hydrogenation, achieving high activity, selectivity, and durability. Strong metal–support interactions, isolated Ni active sites, strain and alloying effects, and the protective graphene shell all contributed to enhanced performance, offering a robust and environmentally friendly catalyst design. Despite these successes, several challenges limit large-scale deployment of noble metal catalysts. High cost and scarcity remain primary barriers, while degradation phenomena such as surface reconstruction, metal leaching, and particle agglomeration compromise long-term stability (Song et al. 2023). Increasingly, non-noble metal catalysts—such as MoS_2, WS_2, and Fe–Ni alloys—have shown comparable catalytic performance at a fraction of the cost, prompting reconsideration of noble metals' long-term role. To address these issues, current research emphasizes hybrid catalyst designs that combine noble metals with transition metals or carbon-based supports to reduce precious metal loading without sacrificing activity (Liu et al. 2021). Stability-enhancing strategies, including surface passivation, defect engineering, and protective coatings, are critical for extending catalyst lifetimes. Scalable synthesis methods, alongside advanced in situ and operando characterization techniques, are also essential for elucidating reaction mechanisms, understanding degradation pathways, and improving commercial feasibility. In summary, noble metal catalysts remain indispensable for hydrogen production due to their unmatched activity and efficiency. However, their sustainable deployment depends on overcoming cost, stability, and scalability challenges through advanced material design, hybridization strategies, and system-level optimization. These innovations will be key to ensuring that noble metals continue to play a leading role in efficient and sustainable hydrogen generation technologies.

3.2.2 Transition Metal Catalysts

Transition metal catalysts have become increasingly important in advancing efficient and sustainable hydrogen production, offering a cost-effective and earth-abundant alternative to conventional noble-metal-based systems (Kazemi et al. 2024). This section reviews recent progress, current challenges, and future prospects in this rapidly evolving field. Among various hydrogen generation strategies, water splitting has been widely recognized as a promising route to address global energy demands. Historically, noble metals particularly platinum have dominated this space due to their exceptional catalytic performance. However, their scarcity and high cost have driven research toward transition metal-based catalysts, which are abundant, economically viable, and highly tunable. Transition metals are notable for their compositional flexibility and adjustable electronic structures, enabling precise tailoring of catalytic properties for the hydrogen evolution reaction (HER)

(Chen et al. 2019). Among these, transition metal nitrides (TMNs) have attracted considerable attention. Nitrogen incorporation into metal lattices modifies the d-band structure by narrowing the d-band width, producing active sites with electronic characteristics resembling those of platinum. Consequently, TMNs exhibit enhanced catalytic activity, improved thermal stability, and superior electrical conductivity, positioning them as highly promising HER catalysts (Wu et al. 2024). Innovative non-noble metal catalyst designs continue to expand the field. Zhang et al. (2024a, b) developed a CoP/Co heterojunction supported on porous g-C_3N_4 nanosheets, synthesized via combined pyrolysis and phosphorus incorporation. This design exposed a high density of active sites and optimized hydrogen and water adsorption free energies, achieving a turnover frequency of 26 min^{-1} at room temperature and maintaining stable performance over five cycles. Similarly, Qiao et al. (2024) fabricated a nanoporous Ni/NiO heterostructure using a sustainable two-step process, where $Ni_{30}Mn_{70}$ alloy sheets were chemically processed to create a 3D porous nickel framework and subsequently oxidized to form a uniform NiO layer. This metal/oxide heterostructure exhibited exceptional HER activity and durability in alkaline media, highlighting the benefits of synergistic interactions at the metal–oxide interface. Further progress has been achieved with advanced sulfide-based systems. Xie et al. (2024) engineered metallic tetra-coordinated W_2S_3 nanosheets via precise stoichiometric growth and tested them in a flow-type anion exchange membrane electrolyzer with commercial IrO_2 as the anode. The W_2S_3 catalyst demonstrated platinum-like HER activity across both acidic and alkaline media, exhibiting low overpotentials, favorable Tafel slopes, and high bifunctionality and stability. This work underscores the potential for non-noble metal catalysts to rival traditional platinum-based systems in practical hydrogen production. Understanding structure–activity relationships is critical for rational catalyst design, and advanced characterization tools such as X-ray absorption spectroscopy (XAS) and in situ/operando techniques have become essential. These methods allow real-time observation of local electronic structure, oxidation states, and coordination environments, revealing that active sites undergo dynamic transformations during HER—challenging traditional static models. Single-atom catalysts (SACs) represent another promising avenue, dispersing isolated transition metal atoms on suitable supports. SACs maximize metal utilization, offer precise control of electronic structure at the atomic level, and often achieve high catalytic activity with minimal material usage. Despite these advances, key challenges remain. Catalyst stability under harsh operating conditions, particularly in alkaline environments, continues to be a concern, with long-term degradation reported in some systems. Additionally, the exact nature of active sites remains under debate, with ongoing research required to clarify the relative roles of metal centers, coordination environments, and support materials. Addressing these challenges is essential to enable the rational design, scale-up, and commercial deployment of transition metal catalysts for sustainable hydrogen production.

3.3 Emerging Technologies

3.3.1 Plasma-Assisted Catalysis

Plasma-assisted catalysis (PAC) represents an emerging hybrid approach in which non-thermal plasma (NTP) is synergistically coupled with catalytic materials to intensify reaction kinetics while improving overall energy efficiency. Unlike conventional thermocatalytic processes that rely on high operating temperatures and energy inputs often derived from fossil fuels, PAC leverages energetic electrons, ions, radicals, and excited species generated within the plasma to activate chemical bonds under comparatively mild thermal conditions. This unique activation mechanism enables reactions to proceed at lower bulk temperatures, thereby reducing total energy demand and associated greenhouse gas emissions. Owing to these advantages, PAC has gained recognition as a promising pathway for sustainable hydrogen production. Initially, plasma–catalyst systems were investigated primarily for environmental remediation, volatile organic compound abatement, and ammonia synthesis. However, recent advances have shifted attention toward hydrogen production, where PAC has demonstrated the ability to enhance reaction selectivity, conversion efficiency, and process flexibility (Whitehead 2019). The growing interest in this field has been further stimulated by the integration of advanced catalyst design and data-driven optimization strategies. For instance, Cai et al. (2024) developed a hybrid machine-learning framework combining regression trees, support vector regression, and artificial neural networks to optimize methane dry reforming over Ni/Al_2O_3 catalysts in a dielectric barrier discharge reactor. By coupling this framework with a genetic algorithm for hyperparameter optimization, the study identified optimal operational conditions—namely, a discharge power of 20 W, a CO_2/CH_4 ratio of 1.5, and a nickel loading of 7.8 wt%—resulting in enhanced energy output. Importantly, the analysis highlighted gas flow rate as a critical parameter governing plasma–catalyst synergy and overall reactor performance. Comparable progress has been reported in warm plasma reactor configurations. Lian et al. (2024) demonstrated that $Ni/CeO_2/Al_2O_3$ catalysts integrated into thermally insulated warm plasma reactors could achieve energy efficiencies approaching 85%, with specific energy consumption as low as 1.16 kWh Nm^{-3}. This corresponds to approximately 13.6 kWh kg^{-1} H_2, underscoring the significant impact of thermal management and catalyst selection on PAC performance. Collectively, these results indicate that PAC systems can attain moderate to high energy efficiencies, effectively bridging the gap between highly efficient but carbon-intensive thermal processes and environmentally benign yet energy-intensive alternatives. More unconventional PAC-related concepts have also been explored. Gravity-driven piezocatalysis, for example, has been reported to simulate energy utilization equivalent to 19,150 kWh for the daily production of 290.9 kmol of hydrogen, corresponding to roughly 66 kWh kg^{-1} H_2 (Lin and Wu 2024). While such findings highlight the conceptual versatility and innovation within the PAC domain, they also reveal that current energy requirements remain relatively high compared with mature hydrogen production technologies. By comparison, conventional water electrolysis systems—including proton exchange membrane (PEM) and alkaline electrolyzers—typically consume 50–65 kWh kg^{-1} H_2, with their sustainability strongly dependent on the carbon intensity of the electricity source (Tao et al. 2022). Thermocatalytic methane cracking, often referred to as turquoise hydrogen production, has emerged as a lower-emission alternative, requiring approximately 20–40 kWh kg^{-1} H_2 while simultaneously producing valuable solid carbon as a by-product (Sanyal et al. 2024). The energy performance of PAC systems is governed by several tightly coupled factors. The choice of plasma source is critical: microwave plasmas tend to deliver more uniform energy distributions, whereas gliding arc discharges can achieve higher reaction rates but demand careful control of thermal effects. Reactor design also plays a pivotal role, with effective thermal insulation shown to reduce heat losses, improve catalyst utilization, and lower net energy consumption. On the materials side, nickel- and cerium-based catalysts have consistently demonstrated superior hydrogen yields by strengthening plasma–catalyst interactions and facilitating surface-mediated reaction pathways. Recent PAC research has increasingly focused on ammonia (NH_3) as a hydrogen carrier, motivated by its high hydrogen density, established infrastructure, and carbon-free combustion. Non-thermal plasma-assisted ammonia decomposition has enabled efficient on-site hydrogen generation with substantially reduced carbon footprints compared to fossil-based routes (Awad et al. 2024). These systems have shown marked improvements in energy efficiency and hydrogen selectivity, representing an important step toward scalable and decentralized hydrogen supply chains. Advances in understanding plasma catalyst interactions have further accelerated progress, with tailored catalyst architectures such as metal-supported and nanostructured materials proving particularly effective in enhancing conversion efficiencies and suppressing undesired by-products. Innovation in experimental design and diagnostics has been equally influential. Optimized reactor configurations, including microreactors, allow precise control over key parameters such as electron density, gas residence time, and local temperature, enabling fine tuning of reaction efficiency and selectivity (Yadav et al. 2023). Meanwhile, advanced characterization techniques, including in situ spectroscopy and high-resolution electron microscopy, have provided molecular-level insights into active sites, charge transfer processes, and reaction mechanisms, thereby supporting rational catalyst and reactor design. In parallel, contact electrocatalysis (CEC) has emerged as a complementary concept to PAC, exploiting electron transfer generated through contact electrification to enhance catalytic activity. Early investigations suggest that CEC can offer valuable insights into charge-driven activation mechanisms that may be harnessed to further optimize plasma-assisted catalytic systems (Chen et al. 2024). Despite its substantial promise, PAC still faces notable challenges. The intrinsic complexity of plasma-driven systems where gas-phase chemistry and surface reactions occur simultaneously complicates the identification of dominant reaction pathways and can promote undesired side reactions that reduce hydrogen selectivity. Long-term catalyst stability remains a critical issue, as prolonged plasma exposure can induce particle sintering, surface reconstruction, and structural degradation. Although recent advances in robust catalyst formulations and protective coatings have yielded encouraging results, universally applicable solutions have yet to be realized (Tian et al. 2024). In summary, plasma-assisted catalysis offers a compelling route toward sustainable hydrogen production by combining enhanced reaction efficiency with reduced environmental impact. However, the successful transition of PAC from laboratory scale demonstrations to industrial deployment will depend on overcoming challenges related to catalyst durability, process complexity, and system level optimization. Continued progress in catalyst engineering, reactor design, and fundamental understanding of plasma catalyst interactions will be essential for establishing PAC as a cornerstone technology in the future hydrogen economy.

3.3.2 Microwave-Enhanced Catalysis

Microwave-enhanced catalysis (MWC) has emerged as an innovative and increasingly attractive route for hydrogen production, capitalizing on the distinctive interactions between microwave radiation and catalytic materials. In contrast to conventional thermal processes, which rely on external heating of the reactor walls and subsequent heat transfer to the reacting medium, MWC enables selective and volumetric heating. Microwave energy couples directly with catalysts and reaction interfaces, resulting in rapid, localized heating without raising the temperature of the entire reactor volume. This unique heating mode accelerates reaction kinetics, improves energy utilization, and supports scalable hydrogen generation, making MWC well aligned with the objectives of clean and sustainable energy systems. Although microwave-assisted methods were initially explored to shorten reaction times and reduce energy consumption in chemical processing, their relevance to hydrogen production has grown substantially in recent years in response to the global energy transition. Considerable progress has been made in applying MWC to hydrogen-relevant reactions. For example, microwave-assisted reforming of methane and low-carbon alcohols has demonstrated significantly enhanced reaction rates and increased hydrogen yields, primarily due to the rapid and selective heating effects induced by microwave irradiation (Zhao et al. 2023). In parallel, microwave plasma-assisted approaches have enabled efficient conversion of hydrogen-rich feedstocks such as plastics and biomass, where the high-energy plasma environment promotes bond activation while suppressing the formation of undesired byproducts. Catalyst innovation has played a central role in advancing MWC-based hydrogen production. Increasing attention has been directed toward nanostructured and microwave-absorbing materials, including ferrites, carbon nanotubes, and conductive ceramics. These materials exhibit strong dielectric or magnetic losses under microwave irradiation, leading to enhanced microwave absorption and improved catalytic performance. As a result, hydrogen production efficiency is significantly increased compared to conventional catalysts that interact weakly with microwave fields (Vuppaladadiyam et al. 2022). In addition, sustainable strategies such as microwave-enhanced catalytic biomass pyrolysis have demonstrated strong potential for green hydrogen generation by utilizing renewable feedstocks while minimizing emissions and process inefficiencies. Advances in reactor engineering have further strengthened the viability of MWC systems. Modern reactor designs increasingly emphasize safety, energy efficiency, and scalability, with microreactor-based configurations enabling precise control over temperature profiles, reaction residence times, and electromagnetic field distribution. Such designs improve heat management and make MWC more suitable for industrial-scale deployment. Nevertheless, several challenges continue to hinder widespread adoption. Extended exposure to microwave radiation can accelerate catalyst deactivation through sintering, phase transformation, or structural degradation, leading to losses in activity and long-term stability. Achieving uniform microwave field distribution also remains a major engineering challenge, as localized hotspots can negatively affect reaction selectivity and catalyst integrity. While microwave heating itself reduces thermal losses during chemical reactions, the overall environmental sustainability of MWC systems depends strongly on the electricity source used to generate microwave radiation. If powered by fossil-based electricity, the environmental advantages may be significantly diminished. Moreover, the high cost of specialized catalysts and microwave-compatible reactor materials poses an economic barrier to large-scale implementation. To address these concerns, the integration of microwave-enhanced catalytic systems with renewable energy sources particularly solar energy has emerged as a promising pathway. Microwave-driven catalytic processes offer rapid, selective, and volumetric heating, often outperforming conventional thermal approaches in terms of reaction kinetics and energy efficiency (Sun et al. 2023a, b, c). When powered by solar energy, either directly through solar-thermal systems or indirectly via photovoltaic-generated electricity, MWC platforms can provide viable solutions for clean energy conversion and decentralized hydrogen production (Domínguez-Saldaña et al. 2024). The feasibility of this integration is supported by advances in power electronics and the modular architecture of modern microwave reactors, which allow flexible operation under variable renewable energy inputs (Yao et al. 2025). Material-level innovations further reinforce the synergy between solar energy and MWC. For instance, doped ceria-based catalysts have demonstrated enhanced microwave-driven redox cycling, enabling efficient water splitting at reduced operating temperatures. Tailoring their electronic structure improves microwave coupling, increases hydrogen yields, and simplifies system design (Domínguez-Saldaña et al. 2024). Similarly, emerging concepts such as photothermal hydrogels illustrate the potential for integrating solar energy harvesting with catalytic processes, enabling simultaneous water evaporation and hydrogen generation under solar irradiation (Horikoshi et al. 2021). From an energy-efficiency standpoint, microwave-assisted catalytic systems are particularly well suited for solar-powered operation. Unlike conventional heating methods, which suffer from high thermal inertia and non-uniform temperature gradients, microwave heating allows rapid, targeted energy delivery with minimal losses. Comparative studies indicate that microwave-based systems can achieve substantially higher energy efficiencies than traditional thermal processes, especially in biomass-to-hydrogen conversion pathways (Lyu et al. 2022). When coupled with clean electricity from solar photovoltaics, MWC systems support modular, off-grid hydrogen production, making them especially attractive for deployment in remote or infrastructure-limited regions. Despite these advantages, several barriers continue to limit the widespread deployment of solar-powered MWC technologies. Technical challenges include the intermittency of solar energy, the need for effective energy storage or real-time power modulation, and the development of catalysts with strong and durable microwave absorption characteristics. Scaling up microwave reactors introduces additional complexities related to electromagnetic field uniformity and long-term catalyst degradation, necessitating sophisticated reactor engineering solutions. Economic factors are equally important, as the high capital costs associated with microwave generators, power electronics, and solar integration components must be justified by gains in efficiency, durability, and operational flexibility (Zhao et al. 2023). The future role of MWC in hydrogen production remains an active area of discussion within the scientific community. While some researchers advocate for rapid industrial adoption based on its efficiency, modularity, and compatibility with renewable energy, others emphasize unresolved challenges related to catalyst longevity, energy sourcing, and system integration. Comparisons with other emerging hydrogen technologies, such as plasma-assisted catalysis and solid oxide electrolysis cells, further underscore the need for rigorous benchmarking under realistic operating conditions. To overcome existing limitations, future research should focus on developing durable and cost-effective microwave-responsive catalysts, hybrid systems that integrate MWC with complementary energy optimization strategies, and renewable-powered microwave generation. Advances in computational modeling, combined with pilot-scale demonstrations, will be critical for optimizing reactor designs and assessing scalability. Overall, microwave-enhanced catalysis holds substantial potential to transform hydrogen production by uniting high efficiency with environmental sustainability. With continued innovation and targeted research, MWC could emerge as a key enabling technology in the global transition toward cleaner, more resilient energy systems.

3.3.3 Novel Nanostructured Catalysts

The advent of nanostructured catalysts has revolutionized the field of catalysis, significantly influencing energy conversion technologies, environmental sustainability, and chemical manufacturing processes. Traditional bulk catalysts, while widely used, are inherently limited by low surface-area-to-volume ratios and restricted tunability, which constrain their catalytic efficiency and limit their adaptability to diverse reaction conditions. In contrast, nanostructured materials including nanoparticles, nanoclusters, quantum dots, metal–organic frameworks (MOFs), and covalent organic frameworks (COFs) have redefined catalyst design by offering markedly enhanced activity, selectivity, and reactivity. These nanoscale architectures provide exceptionally high surface areas and afford precise control over chemical composition, particle morphology, and electronic properties at atomic and molecular scales.

Such control enables tailored surface chemistry, optimized active sites, and improved interaction with reactants, thereby unlocking new possibilities in reaction engineering and catalysis optimization. Moreover, the tunability of nanostructured catalysts allows them to address specific challenges in energy conversion, such as hydrogen production, carbon dioxide reduction, and biomass valorization, while also facilitating environmentally sustainable chemical processes. This section highlights recent advances in the design and synthesis of nanostructured catalysts, examines the ongoing challenges including stability, scalability, cost, and outlines prospective research directions across various classes of nanomaterials. Collectively, these developments underscore the transformative potential of nanostructured catalysts and their pivotal role in shaping the future of high-performance, sustainable catalytic technologies.

3.3.4 Nanoparticles and Nanoclusters

Recent research has increasingly focused on the precise synthesis of nanoparticles and nanoclusters, emphasizing control over size, morphology, and composition to optimize catalytic performance. For example, Shi et al. (2024) developed a nanostructured Pt–NiFe oxide catalyst that exhibited outstanding activity for the hydrogen evolution reaction (HER) in alkaline membrane water electrolysis. In this system, HER proceeds primarily through hydrogen adsorption at Pt edge sites, facilitating efficient H_2 formation. Density functional theory (DFT) calculations revealed that substituting Ni and Fe and their interaction with $Ni(OH)_2$ significantly altered the electronic structure of the catalyst, enhancing its overall performance. Remarkably, the Pt–NiFe oxide catalyst achieved nearly four times the specific activity and more than double the mass activity of commercial Pt/C under alkaline conditions. This performance was attributed to a synergistic interaction between Pt and the Ni–Fe oxide phase, which promoted water dissociation and provided moderate hydrogen binding energies, thereby accelerating hydrogen adsorption–desorption kinetics and improving overall HER rates. These findings underscore the value of rational catalyst design strategies for high-performance alkaline HER applications. Catalytic activity in nanoparticle systems is strongly dictated by particle size and morphology. Smaller nanoparticles generally exhibit a higher density of low-coordination surface atoms, which act as highly active sites. The integration of active learning frameworks has further accelerated catalyst discovery by combining DFT simulations with machine learning (ML) models to predict HER activity in bimetallic and trimetallic nanoparticles. Liu et al. (2023) demonstrated that ML-guided selection of candidate compositions allowed rapid experimental validation, producing catalysts with mass activities exceeding those of Pt/C in alkaline electrolyzers. This computational experimental synergy has significantly shortened development cycles and highlights the expanding role of artificial intelligence in catalyst innovation. Advances in synthesis techniques such as wet chemical reduction, thermal decomposition, and vapor deposition now allow precise control over nanoparticle shapes, including cubes, rods, and octahedra, each exhibiting distinct catalytic properties (Pan et al. 2023). However, nanoparticle aggregation during operation remains a major limitation, often causing loss of surface area and diminished catalytic efficiency. Nanoclusters, typically smaller than 2 nm, represent a promising class of catalysts due to their discrete electronic states and pronounced quantum size effects, which confer unique reactivity and selectivity. For instance, ruthenium nanoclusters have shown superior activity for ammonia synthesis under milder conditions compared with conventional catalysts. Despite their potential, challenges persist in achieving uniform nanocluster synthesis and stable immobilization on supports. A notable advancement in scalable catalyst design was reported by Li et al. (2024), who developed a freestanding, hierarchically structured carbon electrode for industrial hydrogen production. Natural wood was carbonized at 900 °C to generate a porous framework with aligned channels, which then served as a scaffold for N-doped carbon nanotubes encapsulating Ni nanoparticles. This architecture enhanced electrolyte diffusion and gas transport, improving HER performance. The resulting Ni@NCW electrode achieved an industrially relevant current density of 1000 mA cm^{-2} at a low overpotential of 401 mV and sustained stable operation for over 100 h. When integrated into an anion exchange membrane water electrolyzer (AEMWE), the electrode enabled ampere-level hydrogen production at a reduced cell voltage of 2.43 V without noticeable degradation, demonstrating its strong potential for scalable, durable, and efficient hydrogen evolution applications. This body of work highlights the critical importance of nanoscale control, advanced synthesis strategies, and hierarchical architectures in the development of high-performance catalysts, illustrating the transformative role of nanoparticles and nanoclusters in modern hydrogen production technologies.

3.3.5 Covalent Organic Frameworks (COFs)

Covalent organic frameworks (COFs) have recently emerged as a highly promising class of catalysts, offering a complementary and in some cases advantageous alternative to metal–organic frameworks (MOFs). While both COFs and MOFs share modular, tunable architectures, COFs are composed entirely of light elements connected via robust covalent bonds. This all-organic construction imparts exceptional chemical stability, superior structural tunability, and precise atomic-level design capabilities, enabling high surface areas, intrinsic porosity, and controlled incorporation of catalytic functionalities. Gottwald et al. (2024) demonstrated the potential of imine-linked COFs for water-splitting applications. Their study focused on DCNA-X and DNCA-X materials, where conduction and valence band edges were analyzed relative to electrochemical potentials, with and without overpotential corrections. The alignment confirmed their suitability for both the hydrogen evolution reaction (HER) and oxygen evolution reaction (OER) across a broad pH range. Density functional theory (DFT) calculations were used to explore the impact of protonation, imine linkage orientation, and solvation effects on electronic properties. Notably, protonation was found to significantly narrow the band gap, while

COFs with donor groups positioned closer to the nitrogen of the imine linkage exhibited smaller band gaps compared with donors nearer the carbon atom. Reaction energy calculations for HER and OER further highlighted that variations in hydrogen evolution rates are closely tied to charge distribution within the framework. These insights underscore the essential role of DFT in predicting band alignment and clarifying structure–property relationships in COF-based catalysts. Advances in multiscale DFT simulations have enabled the rational design of imine-linked COFs by concurrently optimizing electronic structure and proton-transfer pathways. Experimental studies in electrochemical cells have confirmed that hydrogen evolution rates closely match theoretical predictions, validating DFT-guided design strategies (Liu et al. 2023). Beyond water splitting, COFs have exhibited remarkable catalytic activity across a wide range of reactions, including oxidation, reduction, and carbon–carbon coupling. Functionalization with metal centers or catalytically active organic groups further enhances their performance. Particularly notable is the integration of porphyrin and phthalocyanine units, which mimic enzymatic active sites and impart biomimetic catalytic behavior. Despite these advancements, several challenges remain. Achieving precise control over pore size and geometry, as well as ensuring uniform distribution of active sites throughout the framework, continues to be difficult. Addressing these limitations will require ongoing innovation in synthetic strategies, including refined bottom-up assembly methods and post-synthetic modifications (Xia 2024). Overall, COFs represent a versatile and tunable platform for next-generation catalysis, combining high stability, design flexibility, and the potential for molecular-level control over catalytic activity.

3.3.6 Quantum Dots and Two-Dimensional (2D) Materials

Quantum dots and two-dimensional (2D) materials, such as molybdenum disulfide (MoS_2) and phosphorene, have emerged as highly promising classes of nanostructured catalysts due to their unique electronic structures and high surface areas. Quantum dots, with pronounced quantum confinement effects, exhibit size-dependent and tunable electronic properties, making them particularly effective for photocatalytic and electrocatalytic applications. For instance, cadmium sulfide (CdS) quantum dots have demonstrated efficient photocatalytic hydrogen production, leveraging their adjustable band gaps and large surface-to-volume ratios to enhance light absorption and facilitate rapid charge separation (Sun et al. 2023a, b, c). A notable example of 2D material catalysis is the study by Xiong et al. (2023), which investigated hydrogen evolution reaction (HER) activity on reconstructed and double-stacked black phosphorene (BP) edge structures using density functional theory (DFT) and molecular dynamics simulations. These simulations showed that total energy and temperature converged within 1–2 ps, indicating thermal stability of the BP edges at room temperature with minimal distortion. Five bilayer BP edge configurations displayed higher HER activity than Pt, with two specific edges exhibiting particularly low energetic barriers for the Tafel reaction. The superior performance of bilayer BP compared to monolayers was attributed to unique edge geometries and increased electronic density following hydrogen adsorption. While BP is susceptible to oxidation, mitigation strategies such as de-aerated water immersion or the formation of $MoSe_2$ heterojunctions can preserve stability. These findings offer valuable design principles for noble-metal-free HER catalysts based on BP. Beyond phosphorene, other 2D materials like MoS_2 and graphene are widely recognized for their catalytic potential due to atomically thin structures and high surface areas (Sahoo 2021). MoS_2 has garnered particular attention for HER, where catalytic activity is typically concentrated at nanosheet edges. Xu et al. (2022) introduced an innovative strategy to activate the basal planes by creating Frenkel defects, which involve Mo atoms vacating lattice positions to form vacancy-interstitial pairs. Using microelectrochemical devices to isolate basal-plane activity, they demonstrated that Frenkel-defected monolayer MoS_2 exhibited significantly enhanced HER performance, surpassing both pristine monolayers and Pt single-atom-doped MoS_2. This work highlights the critical role of point defects in modulating electronic structure and establishes defect engineering as a powerful tool for optimizing catalytic activity. Advances in exfoliation, heteroatom doping, and defect engineering have further expanded the versatility of MoS_2 for diverse electrochemical and photocatalytic reactions. Across quantum dots and 2D catalysts, DFT-guided studies have been instrumental in rationalizing edge structures, defect configurations, and electronic properties, directly informing experimental synthesis. Device-level testing consistently validates these predictions, confirming that computational modeling can accelerate the discovery of high-performance HER catalysts (Cerone et al. 2024). Despite their promise, quantum dots and 2D materials face challenges in scalability and long-term stability. Controlled, cost-effective, and reproducible synthesis remains difficult, and maintaining structural and chemical integrity under harsh electrochemical or high-temperature conditions is a persistent hurdle. Continued research into stabilization strategies and degradation-resistant architectures will be essential for practical applications. In summary, nanostructured catalysts including covalent organic frameworks (COFs), quantum dots, and 2D materials represent powerful platforms for advancing hydrogen production technologies. By improving active site availability, charge transfer, and reaction kinetics, these materials enable enhanced performance across electrochemical, photocatalytic, and reforming pathways. Figure 3.1 provides an overview of advanced catalytic systems for hydrogen generation, encompassing noble-metal, carbon-based, single-atom, framework-based, and bio-inspired catalysts, illustrating their complementary roles in sustainable hydrogen evolution.

3.4 Challenges and Future Directions

Recent advances in hydrogen production catalysts have delivered substantial improvements in performance across a wide range of material classes, each offering unique advantages and encountering distinct limitations. Key progress has been

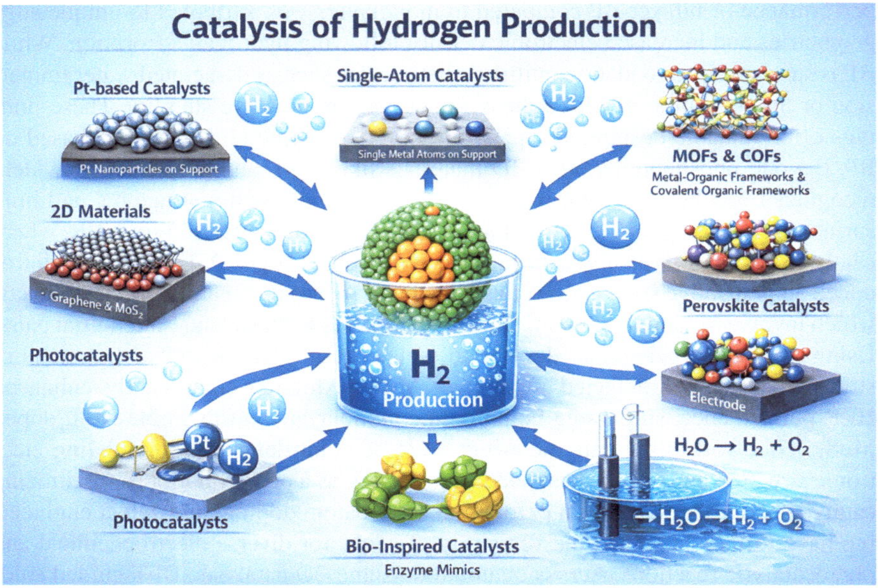

Fig. 3.1 Schematic representation of various advanced catalysts used in hydrogen production

driven by the development of efficient and scalable synthesis methods, including sol–gel processing, electrodeposition, and advanced nanofabrication techniques. Despite these advances, several persistent challenges remain, such as enhancing catalyst stability under harsh operational conditions, reducing reliance on scarce and costly elements, and translating laboratory-scale results into commercially viable technologies. Achieving truly sustainable hydrogen production therefore requires a holistic approach that integrates scientific innovation, interdisciplinary collaboration, and supportive policy frameworks. While innovations in catalyst design have significantly boosted activity and selectivity, the broader potential of hydrogen as a clean energy carrier depends on addressing both technical and non-technical barriers. Machine learning (ML) and data-driven methodologies are emerging as particularly powerful tools for accelerating catalyst discovery and optimization. By combining ML algorithms with high-throughput experimental data and density functional theory (DFT) calculations, researchers can rapidly predict material properties, identify key performance descriptors, and efficiently explore vast chemical design spaces. This approach reduces the reliance on trial-and-error experimentation, enabling targeted identification of promising compositions and guiding the rational design of next-generation hydrogen production catalysts. Environmental considerations are also increasingly being integrated into early-stage catalyst development. Life-cycle assessment (LCA) provides a systematic framework for evaluating environmental impacts, from raw material extraction to end-of-life disposal. This is especially critical for nanostructured catalysts, which may offer superior activity but can also pose ecological and health risks. Incorporating LCA allows informed material selection, balances performance with sustainability, and helps minimize overall environmental footprints. The development of standardized LCA protocols and open-access impact databases will be instrumental in promoting cross-disciplinary evaluation and responsible innovation. Computational chemistry and hybrid catalytic systems have further reshaped the landscape of sustainable hydrogen production. The combination of DFT, ML, and multiscale modeling has yielded deep insights into reaction mechanisms, catalyst structures, and performance optimization. Computational screening has been particularly effective in elucidating structure–activity relationships and guiding experimental synthesis. For example, Mai et al. (2022) demonstrated the use of ML to explore material chemistry and understand electrocatalytic and photocatalytic reaction pathways, while Liu et al. (2023) employed DFT to investigate van der Waals layered materials for HER, showcasing the predictive power of adsorption free-energy descriptors. Multiscale modeling, bridging atomistic simulations with continuum-level behavior, has enhanced predictive accuracy for complex catalytic systems, including functionalized 2D materials (Zaeem et al. 2024). Hybrid catalytic architectures are also gaining traction as platforms for sustainable hydrogen production. Photoelectrochemical systems integrate light harvesting with electrochemical catalysis to improve solar-to-hydrogen conversion efficiencies, while bio-electrocatalytic systems employ biological catalysts under mild conditions to enhance selectivity and sustainability. These hybrid approaches exemplify a paradigm shift toward multifunctional, synergistic catalyst designs that combine complementary functionalities for superior performance. Despite these advances, gaps remain in optimizing and deploying hydrogen production technologies. Noble-metal, transition-metal, and bio-inspired catalysts continue to be pivotal for achieving high efficiency and long-term durability. Overcoming challenges related to stability, scalability, and environmental impact will be critical for translating emerging catalytic innovations into practical, large-scale hydrogen production systems. In summary, continued progress in computational modeling, materials design, and integrated system development will be central to advancing hydrogen technologies toward industrial-scale implementation and supporting global clean energy transitions.

References

Acar C, Dincer I, Zamfirescu C (2014) A review on selected heterogeneous photocatalysts for hydrogen production. Int J Energy Res 38:1903–1920

Agrell J, Birgersson H, Boutonnet M, Melián-Cabrera I, Navarro RM, Fierro JLG (2003) Production of hydrogen from methanol over Cu/ZnO catalysts promoted by ZrO2 and Al2O3. J Catal 219:389–403

Alpaydın CY, Gülbay SK, Colpan CO (2020) A review on the catalysts used for hydrogen production from ammonia borane. Int J Hydrog Energy 45:3414–3434

Awad OI, Zhou B, Kadirgama K, Chen Z, Mohammed MN (2024) Nonthermal plasma-assisted catalysis NH3 decomposition for COx-free H2 production: a review. Int J Hydrog Energy 56:452–470

Biswas P, Kunzru D (2007) Steam reforming of ethanol for production of hydrogen over Ni/CeO2–ZrO2 catalyst: effect of support and metal loading. Int J Hydrog Energy 32:969–980

Cai Y, Mei D, Chen Y, Bogaerts A, Tu X (2024) Machine learning-driven optimization of plasma-catalytic dry reforming of methane. J Energy Chem 96:153–163

Carrasco-Jaim OA, Torres-Martínez LM, Moctezuma E (2018) Enhanced photocatalytic hydrogen production of AgMO3 (M= Ta, Nb, V) perovskite materials using CdS and NiO as co-catalysts. J Photochem Photobiol A Chem 358:167–176

Cerone N, Zimbardi F, Contuzzi L, Tosti S, Fabbiano L, Zito GD, Carnevale MO, Valerio V (2024) Pre-pilot scale study of hydrogen production from biomass syngas via water-gas shift in Pd–Ag catalytic membrane reactor and dedicated hydrogen permeation unit. Int J Hydrog Energy 95:1204–1214

Chen S, Qi Y, Liu G, Yang J, Zhang F, Li C (2014) A wide visible-light-responsive tunneled MgTa 2 O 6− x N x photocatalyst for water oxidation and reduction. Chem Commun 50:14415–14417

Chen Z, Duan X, Wei W, Wang S, Ni B-J (2019) Recent advances in transition metal-based electrocatalysts for alkaline hydrogen evolution. J Mater Chem A 7:14971–15005

Chen Z, Lu Y, Hong R, Liang Z, Wen L, Liu X, Liu Q (2024) Recent progress of solid–liquid interface-mediated contact-electro-catalysis. Langmuir 40(11):5557–5570

Cheng T, Wang Z, Fang S, Jin H, Zhu C, Zhao S, Zhuang G, Chen Q, Zhu Y (2024) Fcc/hcp PtNi heterostructured alloy nanocrystals with ultrathin Pt shell for enhanced catalytic performance towards hydrogen evolution reaction. Nano Res 17:9822–9829

Choi MJ, Yang E, Yu HW, Kim IS, Oh SE, Chae KJ (2019) Transition metal/carbon nanoparticle composite catalysts as platinum substitutes for bioelectrochemical hydrogen production using microbial electrolysis cells. Int J Hydrog Energy 44(4):2258–2265

Choi J, Nguyen QT, Park S, Ghule BG, Park JH, Park JR, Nakate UT, Jang J-H, Kim D-W, Park S (2024) Interfacially engineered palladium nanoparticle-decorated nickel oxide nanostructured electrocatalysts for high-performance hydrogen evolution reaction. Chem Eng J 497:154407

Clarizia L, Spasiano D, Di Somma I, Marotta R, Andreozzi R, Dionysiou DD (2014) Copper modified-TiO2 catalysts for hydrogen generation through photoreforming of organics. A short review. Int J Hydrog Energy 39:16812–16831

Dal Santo V, Gallo A, Naldoni A, Guidotti M, Psaro R (2012) Bimetallic heterogeneous catalysts for hydrogen production. Catal Today 197:190–205

Dawood F, Anda M, Shafiullah GM (2020) Hydrogen production for energy: an overview. Int J Hydrog Energy 45:3847–3869

Domínguez-Saldaña A, Navarrete L, Balaguer M, Carrillo AJ, Santos J, García-Baños B, Plaza-González P, Catalán-Martínez D, Catalá-Civera JM, Serra JM (2024) Enhanced hydrogen production in microwave-driven water-splitting redox cycles by engineering ceria properties. Adv Energy Mater 14(38):2401443

Faber MS, Dziedzic R, Lukowski MA, Kaiser NS, Ding Q, Jin S (2014) High-performance electrocatalysis using metallic cobalt pyrite (CoS2) micro-and nanostructures. J Am Chem Soc 136:10053–10061

Gao W, Shi Y, Zhang Y, Zuo L, Lu H, Huang Y, Fan W, Liu T (2016) Molybdenum carbide anchored on graphene nanoribbons as highly efficient all-pH hydrogen evolution reaction electrocatalyst. ACS Sustain Chem Eng 4:6313–6321

Gao C, Yao H, Wang P, Zhu M, Shi XR, Xu S (2024) Carbon-based composites for oxygen evolution reaction electrocatalysts: design, fabrication, and application. Materials 17(10):2265. https://doi.org/10.3390/ma17102265

Gottwald F, Penschke C, Saalfrank P (2024) Water splitting at imine-linked covalent organic frameworks. Phys Chem Chem Phys 26(32):21821–21831

Han Y, Yu C, Huang H, Wei Q, Dong J, Chen L, Qiu J (2024) Controllable surface reconstruction of copper foam for electrooxidation of benzyl alcohol integrated with pure hydrogen production. SmartMat 5:e1206

Horikoshi S, Takahashi L, Sueishi K, Tanizawa H, Serpone N (2021) Microwave-driven hydrogen production (MDHP) from water and activated carbons (ACs). Application to wastewaters and seawater. RSC Adv 11(50):31590–31600

Jin Y, Zhang Z, Yang H, Wang P, Shen C, Cheng T, Huang X, Shao Q (2022) Boosting hydrogen production with ultralow working voltage by selenium vacancy-enhanced ultrafine platinum–nickel nanowires. SmartMat 3:130–141

Kazemi A, Manteghi F, Tehrani Z (2024) Metal electrocatalysts for hydrogen production in water splitting. ACS Omega 9:7310–7335

Kumar N, Aepuru R, Lee SY, Park SJ (2025) Advances in catalysts for hydrogen production: a comprehensive review of materials and mechanisms. Nanomaterials 15(4):256. https://doi.org/10.3390/nano15040256

Lee J-H, Kim J-W, Lee S-Y, Park S-J (2024) Facile production of highly porous graphitic nanosheets for enhanced hydrogen storage. Chem Eng J 486:149988

Li D, Cheng H, Hao X, Yu G, Qiu C, Xiao Y, Huang H, Lu Y, Zhang B (2024) Wood-derived freestanding carbon-based electrode with hierarchical structure for industrial-level hydrogen production. Adv Mater 36(4):2304917

Lian HY, Sun ZW, Ru Y, Liu JL, Li XS, Zhu AM (2024) Warm plasma catalytic coreforming of dilute bioethanol and methane for hydrogen production. Plasma Process Polym 21(1):2300062

Lin HY, Wu JM (2024) Revolutionary energy harvesting: gravity-driven Piezocatalysts for sustainable hydrogen production in MoS2@ Mo2CTx systems. Adv Energy Mater 14(39):2402164

Liu X, Dai L (2016) Carbon-based metal-free catalysts. Nat Rev Mater 1(11):1–12

Liu M, Zhou L, Luo X, Wan C, Xu L (2020) Recent advances in noble metal catalysts for hydrogen production from ammonia borane. Catalysts 10:788

Liu Y, Zhou D, Deng T, He G, Chen A, Sun X, Yang Y, Miao P (2021) Research progress of oxygen evolution reaction catalysts for electrochemical water splitting. ChemSusChem 14:5359–5383

Liu X, Deng Y, Zheng L, Kesama MR, Tang C, Zhu Y (2022) Engineering low-coordination single-atom cobalt on graphitic carbon nitride catalyst for hydrogen evolution. ACS Catal 12(9):5517–5526

Liu J, Wang Z, Kou L, Gu Y (2023) Mechanism exploration and catalyst design for hydrogen evolution reaction accelerated by density functional theory simulations. ACS Sustain Chem Eng 11(2):467–481

Lyu X, Li H, Xiang H, Mu Y, Ji N, Lu X, Fan X, Gao X (2022) Energy efficient production of 5-hydroxymethylfurfural (5-HMF) over surface functionalized carbon superstructures under microwave irradiation. Chem Eng J 428:131143

Ma Y, Guan G, Hao X, Cao J, Abudula A (2017) Molybdenum carbide as alternative catalyst for hydrogen production—a review. Renew Sust Energ Rev 75:1101–1129

Mai H, Le TC, Chen D, Winkler DA, Caruso RA (2022) Machine learning for electrocatalyst and photocatalyst design and discovery. Chem Rev 122(16):13478–13515

Morya R, Andrianantenaina FH, Singh S, Pandey AK, Kim G-B, Verma JP, Kumar G, Raj T, Kim S-H (2023) Exploring rice straw as substrate for hydrogen production: critical challenges and opportunities. Environ Technol Innov 31:103153

Navarro RM, Álvarez-Galván MC, Rosa F, Fierro JLG (2006) Hydrogen production by oxidative reforming of hexadecane over Ni and Pt catalysts supported on Ce/Ladoped Al2O3. Appl Catal A Gen 297:60–72

Ndayisenga F, Yu Z, Wang B, Wu G, Zhang H, Phulpoto IA, Zhao J, Yang J (2022) Thermophilic-operating environment promotes hydrogen-producing microbial growth in a lignocellulose-fed DF-MEC system for enhanced biohydrogen evolution. Process Saf Environ Prot 167:213–224

Niu H, Wang Q, Huang C, Zhang M, Yan Y, Liu T, Zhou W (2023) Noble metal-based heterogeneous catalysts for electrochemical hydrogen evolution reaction. Appl Sci 13:2177

Pan L, Parnière A, Dunseath O, Fongalland D, Nicolau G, Weber CC, Lu J, Klingenhof M, Arinchtein A, Oh HS, Blanchard PY (2023) Enhancing the performance of shape-controlled octahedral rhodium-doped PtNi nanoalloys inside hydrogen–air fuel cell cathodes using a rational design of catalysts, supports, and layering. ACS Catal 14(1):10–20

Qiao L, Xi C, Li C, Zhang K, Li Q, Han J, Ding Y (2024) Self-limited formation of nanoporous nickel heterostructure catalyst for electrochemical hydrogen production. Adv Funct Mater 34:2402286

Sahoo AP (2021) An empirical study on movement of stock market of BRIC economies-are they co-integrated? ComFin Res 9(4):11–16

Samanta S, Pradhan A (2025) Metal-free carbon-based porous materials, a promising electrocatalyst for hydrogen fuel production. Chem Commun. https://doi.org/10.1039/D5CC01359C

Sanyal A, Malalasekera W, Bandulasena H, Wijayantha KGU (2024) Review of the production of turquoise hydrogen from methane catalytic decomposition: optimising reactors for sustainable hydrogen production. Int J Hydrog Energy 72:694–715

Shi G, Tano T, Tryk DA, Iiyama A, Uchida M, Terao K, Osada H, Yamaguchi M, Tamoto K, Kakinuma K (2024) Nanostructured Pt-NiFe oxide catalyst for hydrogen evolution reaction in alkaline electrolyte membrane water electrolyzers. ACS Catal 14(12):9460–9468

Sikander U, Sufian S, Salam MA (2017) A review of hydrotalcite based catalysts for hydrogen production systems. Int J Hydrog Energy 42:19851–19868

Song R, Wang X, Ge J (2023) Recent progress of noble metal-based single-atom electrocatalysts for acidic oxygen evolution reaction. Curr Opin Electrochem 42:101379

Sun H, Lee SY, Park SJ (2023a) Bimetallic CuPd alloy nanoparticles decorated ZnO nanosheets with enhanced photocatalytic degradation of methyl orange dye. J Colloid Interface Sci 629:87–96

Sun J, Shang H, Miao C, Yang J, Liao Y (2023b) Microwave enhanced hydrogen production from liquid organic hydrogen carriers: a review. Chem Eng Process Process Intensif 190:109432

Sun P, Xing Z, Li Z, Zhou W (2023c) Recent advances in quantum dots photocatalysts. Chem Eng J 458:141399

Tanzil AH, Brandt K, Wolcott M, Zhang X, Garcia-Perez M (2021) Strategic assessment of sustainable aviation fuel production technologies: yield improvement and cost reduction opportunities. Biomass Bioenergy 145:105942

Tao M, Azzolini JA, Stechel EB, Ayers KE, Valdez TI (2022) Engineering challenges in green hydrogen production systems. J Electrochem Soc 169(5):054503

Tian F, Zhou N, Chen W, Zhan J, Tang L, Wu M (2024) Progress in green ammonia synthesis technology: catalytic behavior of ammonia synthesis catalysts. Adv Sustainable Syst 8(8):2300618

Urpelainen J, Van de Graaf T (2015) The international renewable energy agency: a success story in institutional innovation? Int Environ Agreem Polit Law Econ 15:159–177

Vuppaladadiyam AK, Vuppaladadiyam SSV, Awasthi A, Sahoo A, Rehman S, Pant KK, Murugavelh S, Huang Q, Anthony E, Fennel P, Bhattacharya S (2022) Biomass pyrolysis: a review on recent advancements and green hydrogen production. Bioresour Technol 364:128087

Wang S, Lu A, Zhong C-J (2021) Hydrogen production from water electrolysis: role of catalysts. Nano Convergence 8:4

Wang P, Yi X, Lu M, Huang P, Wang Y, Gao D, Wei Y, Xie X, Zhu J (2023) An amphiphilic nitrogen-doped graphene prepared by ammonia cold plasma and its supported platinum nanocatalyst for aqueous-phase hydrogenation of cinnamaldehyde. Appl Catal A Gen 663:119286

Whitehead JC (2019) Plasma catalysis: challenges and future perspectives. In: Plasma catalysis: fundamentals and applications. Springer International Publishing, Cham, pp 343–348

Wu L-W, Yao Y-F, Xu S-Y, Cao X-Y, Ren Y-W, Si L-P, Liu H-Y (2024) Electrocatalytic hydrogen evolution of transition metal (Fe, co and cu)–Corrole complexes bearing an imidazole group. Catalysts 14:5

Xia N (2024) Persian Sasanian silver coins recently unearthed in Turfan, Xinjiang. In: Studies in silk road archaeology. Singapore, Springer Nature Singapore, pp 29–37

Xiao P, Yan Y, Ge X, Liu Z, Wang J-Y, Wang X (2014) Investigation of molybdenum carbide nanorod as an efficient and durable electrocatalyst for hydrogen evolution in acidic and alkaline media. Appl Catal B Environ 154:232–237

Xie L, Liu Q, Shi X, Asiri AM, Luo Y, Sun X (2018) Superior alkaline hydrogen evolution electrocatalysis enabled by an ultrafine PtNi nanoparticle-decorated Ni nanoarray with ultralow Pt loading. Inorg Chem Front 5:1365–1369

Xie L, Wang L, Liu X, Zhao W, Liu S, Huang X, Zhao Q (2024) Tetra-coordinated W2S3 for efficient dual-pH hydrogen production. Angew Chem Int Ed 63:e202316306

Xiong J, Gong Q, Feng T, Wang M, Zhang X, Liu G, Qiao G, Xu Z (2023) Enhance hydrogen evolution reaction performance via double-stacked edges of black phosphorene. Inorg Chem 62(51):21115–21127

Xu J, Shao G, Tang X, Lv F, Xiang H, Jing C, Liu S, Dai S, Li Y, Luo J, Zhou Z (2022) Frenkel-defected monolayer MoS2 catalysts for efficient hydrogen evolution. Nat Commun 13(1):2193

Yadav D, Lu X, Vishwakarma CB, Jing D (2023) Advancements in microreactor technology for hydrogen production via steam reforming: a comprehensive review of experimental studies. J Power Sources 585:233621

Yang M, Allard LF, Flytzani-Stephanopoulos M (2013) Atomically dispersed Au–(OH) x species bound on titania catalyze the low-temperature water-gas shift reaction. J Am Chem Soc 135:3768–3771

Yang T, Cheng C, Xiao L, Wang M, Zhang F, Wang J, Yin P, Shen G, Yang J, Dong C (2024) A descriptor of IB alloy catalysts for hydrogen evolution reaction. SmartMat 5:1204

Yao L, Li H, Song Z, Wang W, Mao Y (2025) Mechanism of microwave-enhanced catalytic of plastic cracking for hydrogen production over iron catalysts: a reactive molecular dynamics study. Fuel 389:134646

Younas M, Shafique S, Hafeez A, Javed F, Rehman F (2022) An overview of hydrogen production: current status, potential, and challenges. Fuel 316:123317

Yuan H, Hong M, Huang X, Qiu W, Dong F, Zhou Y, Chen Y, Gao J, Yang S (2024) Graphene chainmail shelled dilute Ni–Cu alloy for selective and robust aqueous phase catalytic hydrogenation. Adv Sci 11:2304349

Zaeem MA, Thomas S, Kavousi S, Zhang N, Mukhopadhyay T, Mahata A (2024) Multiscale computational modeling techniques in study and design of 2D materials: recent advances, challenges, and opportunities. 2D Mater 11(4):042004

Zhang J, Li X, Chen H, Qi M, Zhang G, Hu H, Ma X (2017a) Hydrogen production by catalytic methane decomposition: carbon materials as catalysts or catalyst supports. Int J Hydrog Energy 42(31):19755–19775

Zhang R, Wang X, Yu S, Wen T, Zhu X, Yang F, Sun X, Wang X, Hu W (2017b) Ternary NiCo2Px nanowires as pH-universal electrocatalysts for highly efficient hydrogen evolution reaction. Adv Mater 29:1605502

Zhang H, Han G, Liu Y, Zhao L, Zhang W, Tahir Khalil M, Wei H, Wang C, Liu T, Guo X et al (2024a) CoP/co heterojunction on porous g-C3N4 nanosheets as a highly efficient catalyst for hydrogen generation. J Colloid Interface Sci 658:22–31

Zhang Y, Yan R, Xu X, He C, Zhou M, Zeng Z, Ma T, Wang M, Li S (2024b) Next generation Noble metal-engineered catalysts: from structure evolution to structure-reactivity correlation in water splitting. Adv Funct Mater 34:2308813

Zhao J, Wang D, Zhang L, He M, Ma W, Zhao S (2023) Microwave-enhanced hydrogen production: a review. RSC Adv 13(22):15261–15273

Chapter 4
Toward Clean Hydrogen: Emerging Carbon Capture and Conversion Technologies in the Energy Transition

Abstract The extensive use of fossil fuels for electricity generation, heating, and transportation remains the dominant source of greenhouse gas emissions, significantly driving global climate change. Transitioning toward cleaner fossil fuel utilization, particularly through low-carbon hydrogen production, represents a critical pathway in the energy transition. This chapter presents a comprehensive review of emerging carbon capture and conversion technologies that can enable clean hydrogen systems. It begins by examining the Intergovernmental Panel on Climate Change (IPCC) roadmap toward net-zero emissions and contextualizes potential decarbonization pathways for energy sector. Subsequently, advanced CO_2 capture and removal technologies including ionic-liquid-based physical absorption, chemical looping, and cryogenic separation are critically assessed in terms of technological maturity, capture efficiency, CO_2 purity, scalability, and implementation challenges. The chapter further categorizes CO_2 conversion routes into chemical approaches and biological pathways evaluating their potential for producing value-added products. The chapter concludes by highlighting integrated CCC hydrogen frameworks as a viable strategy for decarbonizing energy sector, while identifying key technical barriers, research gaps, and future directions required to accelerate large-scale deployment.

Keywords CO_2 capture and conversion · Global greenhouse gas emissions · Decarbonization · Hydrogen production

4.1 Introduction

Global warming has emerged as one of the most pressing challenges of the twenty-first century, prompting an unprecedented collective response from the international community. To address the accelerating impacts of climate change, 195 countries have committed to reducing greenhouse gas (GHG) emissions and pursuing carbon neutrality by mid-century (Horowitz 2016; Jafari et al. 2022). Among all GHGs, carbon dioxide (CO_2) remains the principal contributor to anthropogenic climate

D. M. Y. Mir et al., *Low-Carbon Hydrogen*, Clean Energy Production Technologies,
https://doi.org/10.1007/978-981-95-8142-9_4

change due to its high emission volumes and long atmospheric lifetime. Despite the implementation of mitigation policies and cleaner energy initiatives, global atmospheric CO_2 concentrations have continued to rise, increasing by nearly 1.7% worldwide (Regufe et al. 2021). A temporary deviation from this trend occurred in 2020, when the International Energy Agency (IEA) reported a 5.8% reduction in global CO_2 emissions, largely driven by reduced oil and coal consumption during the COVID-19 pandemic (International Energy Agency 2021). However, this decline proved short-lived, as total emissions from global energy use still reached approximately 31.5 Gt CO_2 almost 50% higher than pre-industrial levels—highlighting the structural dependence of the global energy system on fossil fuels. Coal continues to play a dominant role, supplying about 27% of global primary energy while accounting for nearly 44% of CO_2 emissions (Bui et al. 2018). The expected resurgence in coal consumption in 2021 alone was projected to add roughly 640 Mt. of CO_2 to global emissions. Concurrently, natural gas demand is expanding at an annual rate of approximately 1.9% and is projected to overtake coal by 2030, owing to its greater availability and comparatively lower carbon intensity. Overall, CO_2 emissions are primarily concentrated in three sectors—power generation, industry, and surface transportation with the power sector alone responsible for nearly 46% of total emissions worldwide (International Energy Agency 2022). In response to these trends, the United Nations Framework Convention on Climate Change (UNFCCC) established a global policy framework in 2015 aimed at limiting the increase in average global temperature to well below 2 °C, while striving to cap warming at 1.5 °C (Fragkos 2020). Building on the IPCC Fifth Assessment Report (AR5) (Stocker et al. 2013), early estimates suggested that remaining within the 1.5 °C threshold would require a cumulative carbon budget of approximately 200–415 $GtCO_2$ between 2011 and 2100, with energy-related emissions capped at around 400 $GtCO_2$. These limits were later revised in the IPCC Special Report on 1.5 °C, which expanded the allowable budget to 570–770 $GtCO_2$, corresponding to 67% and 50% probabilities of meeting the temperature target, respectively (IPCC 2018). Achieving such ambitious goals necessitates the deployment of integrated mitigation strategies, including large-scale renewable energy deployment, hydrogen-based energy systems, and carbon dioxide removal, capture, and utilization technologies capable of delivering net-negative emissions beyond 2050. Within this global decarbonization landscape, Qatar occupies a strategically significant position as one of the world's leading energy suppliers. The country holds the second-largest proven natural gas reserves and produces approximately 77 million tonnes of liquefied natural gas annually. In 2020, Qatar's total CO_2 emissions were estimated at 106.65 million tonnes per year (Nations United Programme 2020). While its absolute emissions are relatively low compared to other Gulf Cooperation Council (GCC) countries, Qatar records the highest per capita emissions in the region, at approximately 37.02 tonnes. Emissions are dominated by the oil and gas sector (64.7%), followed by power generation (22%) and other industrial and societal activities (32.2%). Recognizing the environmental and economic implications of this profile, Qatar has intensified its decarbonization efforts, particularly within the industrial sector, through increased investments in renewable energy. These efforts, however, continue to face challenges related to intermittency, grid integration, and system reliability. To complement renewable deployment, Qatar is actively pursuing alternative low-carbon energy carriers, notably hydrogen, as a cornerstone of its long-term energy transition strategy (Rajabloo et al. 2022). Hydrogen can be produced from diverse feedstocks, including fossil fuels, water electrolysis, and biomass. Nevertheless, conventional natural-gas-based hydrogen production (grey hydrogen) remains highly carbon-intensive, with emissions of approximately 89.1 g CO_2/MJ (Birol 2019). The integration of carbon capture technologies into this process gives rise to blue hydrogen, which can reduce carbon intensity to around 22.4 g CO_2/MJ (Regulator 2020) and is considered broadly consistent with pathways aligned to the 1.5 °C target (IEA 2017). In support of this transition, Qatar has made substantial progress in carbon capture and sequestration (CCS), with plans to expand CO_2 storage capacity beyond 5 million tonnes per year by 2025 (Zeynep Beyza Kilic 2019). Several flagship initiatives underscore this commitment. In 2012, Qatar Petroleum, in collaboration with Shell, Imperial College London, and Qatar Science & Technology Park, launched a $70 million research program to establish the Qatar Carbonates and Carbon Storage Research Centre (QCCSRC). Additionally, the Qatar Fertilizer Company (QAFCO) is spearheading the world's largest blue ammonia project, incorporating an integrated CCS facility designed to capture approximately 11 million tonnes of CO_2 annually (Ingram 2022). The North Field East (NFE) expansion project, with an investment of $28.75 billion, has also embedded CCS as a core environmental component (Up et al. 2016). Despite these advances, Qatar's limited domestic geological storage capacity presents long-term constraints for large-scale CO_2 sequestration. In this context, carbon capture and utilization (CCU) emerges as a practical and complementary solution to CCS. CCU pathways enable the conversion of captured CO_2 into value-added products, thereby addressing both storage limitations and economic feasibility challenges. Recent studies, including Al-Yaeeshi et al. (2018), have demonstrated the technical viability, economic competitiveness, and environmental benefits of deploying CCU technologies within Qatar's industrial clusters. Their findings indicate that retrofitting existing industrial facilities with CCU systems can substantially reduce emissions while supporting sustainable industrial development. Consequently, CCU represents a promising pathway for Qatar to reconcile continued industrial growth with global climate objectives and to contribute meaningfully to achieving the 1.5 °C temperature stabilization target.

4.2 Advances in CO_2 Capture and Conversion Technologies

The rapid and excessive accumulation of carbon dioxide (CO_2) in the atmosphere is a major contributor to global climate change, presenting serious risks to ecosystems and the stability of the biosphere. As industrial activities and modern societies seek to curb their carbon emissions, the development and deployment of innovative, efficient, and sustainable mitigation strategies have become increasingly critical. CO_2

Table 4.1 Power generation from carbon capture and storage (CCUS)-based power plants under the IEA Sustainable Development Scenario 2025–2040 (IEA 2020)

Technology	2025	2030	2035	2040
Coal with CCUS	14 TWh	246 TWh	674 TWh	994 TWh
Gas with CCUS	4 TWh	220 TWh	560 TWh	915 TWh

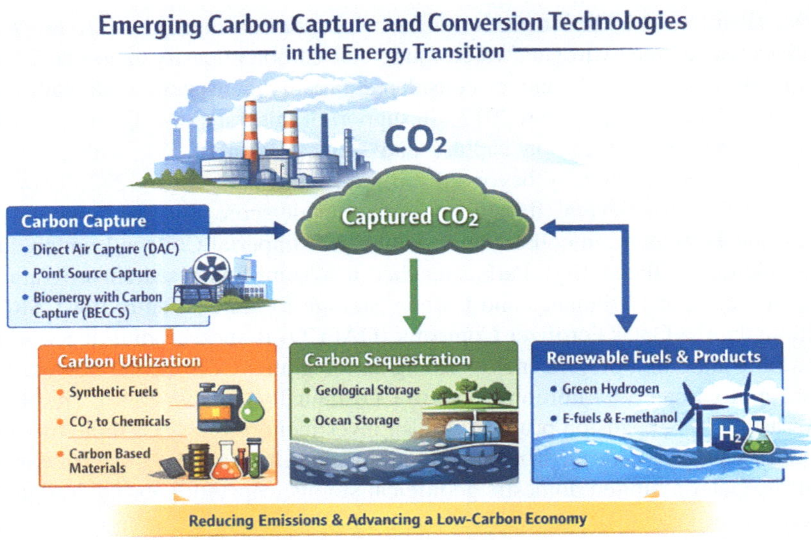

Fig. 4.1 Integrated carbon capture and conversion technologies supporting a low-carbon energy transition

can be separated and captured from the atmosphere or industrial emission streams through two principal approaches, namely, direct and indirect capture methods (Kar et al. 2019). These strategies constitute an integral component of global climate mitigation efforts aimed at lowering the concentration of this dominant greenhouse gas. The IEA Sustainable Development Scenario projects a rapid scale-up of CCUS-integrated power generation between 2025 and 2040, reflecting the critical role of carbon capture in decarbonizing fossil-based electricity systems (Table 4.1).

Carbon capture technologies are generally classified into three main categories based on the stage of fuel conversion at which CO_2 is separated: pre-combustion capture, post-combustion capture, and oxy-fuel combustion capture (Zhao et al. 2024). Each of these approaches offers distinct technical advantages and challenges, depending on the application and operating conditions. Figure 4.1 illustrates the role of emerging carbon capture and conversion (CCC) technologies within the energy transition framework. Carbon dioxide emissions from industrial and energy-generation sources are captured using advanced approaches such as point-source capture, direct air capture, and bioenergy with carbon capture and storage. The captured CO_2 is then directed toward multiple pathways, including carbon utilization for the production of synthetic fuels, chemicals, and carbon-based materials, as well as long-term carbon sequestration through geological and ocean storage. In parallel, captured carbon supports the generation of renewable fuels and products, such as green hydrogen and e-fuels, enabling sector coupling across energy, industry, and transport. Collectively, these integrated carbon capture, utilization, and storage pathways contribute to significant emissions reduction and support the development of a resilient, low-carbon energy system.

4.2.1 Direct Capture of CO_2

Direct CO_2 capture refers to the extraction of carbon dioxide directly from the ambient atmosphere or from concentrated emission streams at point sources, such as power plants and industrial facilities, without reliance on intermediate conversion processes. This approach encompasses two principal categories: direct air capture, which targets dilute atmospheric CO_2, and point-source CO_2 capture, which focuses on emissions released from specific industrial operations.

4.2.2 Direct Air Capture

Direct air capture (DAC) employs large-scale engineered systems to extract CO_2 directly from ambient air through physicochemical processes, most commonly using amine-based materials or other selective sorbents to bind and separate CO_2 from the atmosphere (Vargas et al. 2013). DAC technologies are generally classified into two dominant categories: solid sorbent–based systems and liquid solvent–based systems. In solid sorbent configurations, CO_2 is adsorbed onto a solid material through chemical interactions and subsequently released via thermal regeneration or pressure swing processes, allowing for downstream storage or utilization. In contrast, liquid solvent–based systems rely on the absorption of CO_2 into a liquid medium, followed by separation through heating or alternative regeneration techniques. Despite its high capital and energy requirements, DAC is widely recognized as a pivotal negative-emissions technology. Its ability to remove CO_2 independently of emission point sources makes it particularly valuable for achieving net-zero and net-negative emissions targets, especially in regions lacking concentrated industrial CO_2 sources (Pekdemir 2015).

4.2.3 Point-Source CO_2 Capture

Point-source CO_2 capture refers to the separation and recovery of carbon dioxide directly from concentrated emission streams at specific industrial facilities, including power generation units, cement plants, steel manufacturing sites, chemical processing facilities, and natural gas treatment plants (Shen et al. 2012). The principal technologies employed for point-source CO_2 capture are pre-combustion capture, oxy-fuel combustion capture, and post-combustion capture. Although point-source capture systems are associated with high energy demand and significant capital and operational costs, they are generally more technically and economically viable when deployed at large-scale industrial installations. In such settings, the high concentration of CO_2 in flue gases and the economies of scale can partially mitigate the overall costs, making point-source capture a practical near- to medium-term solution for reducing industrial carbon emissions (Shen et al. 2012).

4.2.4 Pre-combustion Capture

Pre-combustion capture involves the removal of CO_2 from a fuel stream prior to combustion, most commonly through gasification processes that convert solid or gaseous fuels into a synthesis gas composed primarily of hydrogen and carbon monoxide (Zhao et al. 2024). This intermediate gas mixture enables the separation of CO_2, after which the remaining hydrogen-rich fuel can be combusted to generate energy with significantly reduced carbon emissions (Zhao et al. 2024). Ongoing research and development efforts are directed toward enhancing the efficiency and economic feasibility of CO_2 capture technologies, particularly those associated with flue gas treatment and upstream capture processes (Wilberforce et al. 2019). Nevertheless, despite notable technological progress, the large-scale deployment of pre-combustion capture remains constrained by substantial energy requirements and high capital and operating costs. While several commercial-scale CO_2 capture facilities are currently in operation worldwide, broader adoption is still limited by these economic and technical challenges.

4.2.5 Oxy-Fuel Combustion

Oxy-fuel combustion (OFC) is a carbon capture enabled combustion approach in which atmospheric air is substituted with a controlled mixture of recycled flue gas and high-purity oxygen, typically containing about 95 mol% O_2. This configuration fundamentally alters the combustion environment and enables inherent CO_2 capture by simplifying downstream separation processes (Koohestanian and Shahraki 2021). Under OFC operation, nitrogen is largely excluded from the oxidant stream, resulting in a flue gas composed primarily of CO_2 and water vapor. Following moisture condensation, a highly concentrated CO_2 stream is obtained, which can be efficiently purified and compressed using downstream units. Compared with other CCS routes, OFC-based CCS is widely regarded as a cost-competitive option, particularly due to its potential for retrofitting existing air-fired combustion systems with relatively limited structural modifications (Yadav and Mondal 2022). This retrofit capability makes OFC especially attractive for decarbonizing legacy power plants and industrial facilities. However, the overall performance and economic feasibility of OFC systems are strongly influenced by two key auxiliary units: the ASU and the CPU. The ASU is responsible for supplying high-purity oxygen by removing nitrogen and other inert gases from ambient air. At present, cryogenic air separation remains the most mature and commercially established technology for large-scale oxygen production, but it is also highly energy-intensive and contributes substantially to the efficiency penalty of OFC systems. As a result, alternative oxygen production technologies, such as CLP-based air separation and membrane separation, have gained increasing research attention due to their potential to reduce energy consumption and improve overall system efficiency (Seddighi et al. 2018). The CPU plays a critical role in recovering and purifying CO_2 from the OFC flue gas stream. Although OFC inherently produces a CO_2-rich exhaust, impurities such as residual O_2, N_2 leakage, argon, and acid gases must be removed to meet transport, utilization, or storage specifications. The compression and refrigeration requirements associated with the CPU impose a significant energy burden, making it one of the dominant contributors to efficiency losses in OFC-based CCS systems. Consequently, reducing the auxiliary energy demand of both the ASU and CPU is essential for improving net plant efficiency and enabling large-scale commercial deployment of OFC technology (Koohestanian and Shahraki 2021). From an operational perspective, OFC demonstrates a greater potential for reducing CO_2 capture costs than both pre-combustion and post-combustion capture approaches, particularly in energy-intensive industrial applications (Koohestanian and Shahraki 2021). The direct production of a concentrated CO_2 stream minimizes separation complexity and avoids solvent-based capture systems, thereby offering a more energy- and cost-efficient pathway for CCS integration. CLP has emerged as a promising enhancement within OFC systems, especially in the context of oxygen production. In CLP-based air separation, oxygen is transferred indirectly from air to fuel via a solid oxygen carrier, eliminating the need for energy-intensive cryogenic separation. The oxygen carrier is a critical component of this process, as it governs oxygen transfer capacity, reaction kinetics, cyclic stability, and long-term operational reliability (Wu et al. 2018). Recent advances in oxygen carrier materials and reactor design have further improved the technical viability and efficiency of CLP-assisted OFC systems (Koohestanian and Shahraki 2021). To fully quantify the benefits of integrating OFC with CLP-based air separation, comprehensive environmental and techno-economic life cycle assessments are required. Such analyses provide valuable insights into system-level efficiency, emissions reduction potential, and long-term sustainability (Yadav and Mondal 2022). Beyond power generation, OFC has also attracted significant interest due to its operational flexibility and compatibility with a broad range of industrial processes, including oil refining and petrochemical applications (Wu et al. 2018). Collectively, these attributes position OFC particularly when integrated with CLP as a highly promising pathway for industrial decarbonization and large-scale CCS deployment.

4.2.6 Post-combustion Capture

A wide range of alternative methods and materials for carbon capture and sequestration in post-combustion systems are being actively investigated as potential replacements for conventional chemical absorption technologies (Valluri and Kawatra 2021). These emerging approaches include the application of metal oxides in chemical-looping combustion, membrane-based separation, physical absorption, gas hydrate formation, cryogenic distillation, and solid adsorption processes employing pressure swing adsorption (PSA) or temperature swing adsorption (TSA) techniques. Despite these advancements, amine-based absorption remains one of the most widely implemented post-combustion CO_2 capture technologies. Post-combustion CO_2 capture has been applied at an industrial scale for several decades and offers multiple practical benefits. These include (a) straightforward integration into existing power and industrial plants without substantial alterations to plant infrastructure or combustion systems, (b) applicability to gas-fired power plants and other low-pressure exhaust streams, and (c) high operational flexibility, allowing capture systems to be adjusted or maintained without requiring complete plant shutdowns (Valluri and Kawatra 2021). Post-combustion CO_2 capture techniques are extensively utilized within the chemical processing industry; however, continued research and development are necessary to improve system efficiency, reduce energy penalties, and optimize fixed capture configurations for specific industrial applications. Recent studies addressing global warming mitigation have emphasized advances in CO_2 absorption from flue gas streams (Deng et al. 2015). In addition to absorption-based approaches, CO_2 separation technologies such as membrane separation, wet and dry absorption, centrifugal separation, and cryogenic distillation have received considerable attention (Zhu et al. 2014). The ability to efficiently and selectively separate CO_2 from flue gas is widely recognized as a critical technological advancement in the global effort to mitigate climate change and reduce anthropogenic greenhouse gas emissions (Deng et al. 2015).

4.2.7 Absorption Processes

Absorption-based CO_2 capture is fundamentally distinct from adsorption processes, which rely on surface interactions between gas molecules and solid sorbents, as absorption involves the physical or chemical dissolution of CO_2 into the bulk phase of a liquid solvent (Trivedi et al. 2016). This transfer typically occurs at relatively low operating temperatures, generally between 40 and 60 °C, where CO_2 migrates from the gas phase into the liquid phase through either physical solubility or chemical reaction pathways. The CO_2-loaded solvent is then routed to a regeneration unit, commonly a thermal stripper, where elevated temperatures in the range of 120–160 °C at near-atmospheric pressure are applied to release the absorbed CO_2. After regeneration, the lean solvent is recycled back to the absorber, enabling continuous cyclic operation. Within the stripper, CO_2 is separated from the solvent and can be recovered at purities reaching up to 99% (Qin et al. 2011). When suitable solvents are employed—particularly amine-based systems—absorption technology can effectively treat flue gas streams containing CO_2 concentrations of up to approximately 20%. Over time, a wide range of absorption-based capture processes have been developed using diverse classes of absorbents, including alkanolamines, aqueous ammonia, dual-alkali systems, sodium carbonate (Na_2CO_3) slurry absorption, and chilled ammonia processes (Lee et al. 2007). Among these options, amine-based absorption has attracted the greatest industrial and academic attention due to its strong chemical affinity for CO_2, high absorption capacity, and well-established process design. This approach has been extensively implemented for decades in natural gas processing and in the production of food- and beverage-grade CO_2, demonstrating a high level of technological maturity and reliability (Harris et al. 2009). In chemical absorption systems, alkaline solvents react with CO_2 through acid–base neutralization reactions to form carbonate or bicarbonate species (Halliday et al. 2019). Both strong and weak alkaline absorbents are commonly applied, including potassium hydroxide and sodium hydroxide–based solutions. However, the high corrosivity associated with strong alkaline solvents introduces significant material compatibility and operational challenges. Similarly, widely used alkanolamines such as MEA, DEA, TEA, and MDEA suffer from limitations related to high regeneration energy requirements, which adversely affect the overall energy efficiency and operating cost of the capture process (Yu et al. 2024). In contrast, physical solvents capture CO_2 primarily through differences in gas solubility rather than through chemical reactions (Shu et al. 2020). Because of their relatively weak interactions with acid gases, physical solvents are most effective under conditions of high CO_2 or H_2S partial pressures. However, their tendency to co-absorb heavier hydrocarbons restricts their use in certain applications, such as natural gas sweetening, and instead makes them more suitable for syngas purification and related processes (Lee et al. 2007). Commonly used physical solvents include DEPG (Selexol), MeOH (Rectisol), NMP (Purisol), and PC (Fluor Solvent) (Burr and Lyddon 2008). In recent years, ionic liquids have emerged as a promising class of advanced absorbents for acid gas capture. These materials are typically liquid below 100 °C and consist of tunable cation–anion combinations, providing exceptional flexibility for tailoring solvent properties to specific separation requirements (Xu et al. 1992). Their adjustable chemical structure enables reversible capture and release of CO_2, SO_x, and NO_x through a combination of physical dissolution and chemical interactions (Ramdin et al. 2012). Closely related to ionic liquids are deep eutectic solvents (DES), which have gained increasing attention as alternative absorbents. Although

DES are not composed solely of ionic species and are often biodegradable, they exhibit solvent behaviors similar to ionic liquids and offer the potential for lower lifecycle costs, enhancing their attractiveness for large-scale applications (Trivedi et al. 2016). Amino acid–based sorbents represent another emerging category of absorption materials. These ionic, low-volatility compounds show reduced susceptibility to oxidative degradation compared to conventional amines and retain favorable CO_2 selectivity characteristics (Burr and Lyddon 2008). Nevertheless, their broader industrial adoption has been constrained by higher costs, potential solid precipitation during operation, slower reaction kinetics, and susceptibility to thermal degradation (McDonald et al. 2012). Despite ongoing advances in alternative solvents, amines remain the most mature and widely deployed materials for CO_2 absorption and are frequently used as benchmark sorbents for evaluating emerging technologies. Their extensive use in natural gas sweetening continues to demonstrate their industrial robustness and reliability. In parallel, alkaline earth metal oxides have attracted growing interest due to their high CO_2 uptake capacities at elevated temperatures. While their solid-state nature avoids issues related to solvent volatility and corrosion, challenges associated with sorbent handling, regeneration energy demand, and long-term cyclic stability remain significant (Binti Wan Ramli 2017). As a result, sustained research efforts are focused on alkaline earth metals, alkali metals, and a wide range of synthetic and naturally derived metal oxides to develop next-generation sorbents capable of delivering efficient, durable, and scalable CO_2 capture solutions (Binti Wan Ramli 2017).

4.2.7.1 Adsorption: Activated Carbons and Metal-Organic Frameworks

Activated carbons are among the most extensively studied and widely applied solid sorbents for CO_2 capture, owing to their effective adsorption performance under ambient temperature and pressure conditions. These materials are typically derived from carbon-rich precursors that possess high energy content and low ash fractions. To achieve efficient CO_2 adsorption, precursor materials must meet several key criteria, including favorable activation behavior, thermal stability, and resistance to excessive structural degradation during processing (Tong et al. 2017). The resulting physicochemical characteristics of activated carbons—such as specific surface area, pore size distribution, and adsorption capacity—are strongly governed by the intrinsic hydrogen (H_2) and oxygen (O_2) content of the precursor material, which influence pore development and surface chemistry (Deng et al. 2015). The choice of precursor plays a decisive role in determining the adsorption performance of activated carbons. Biomass-derived precursors, particularly agricultural residues such as almond shells and olive stones, have demonstrated remarkable CO_2 uptake capacities of up to 15.6 mol kg^{-1}. This high performance is largely attributed to their favorable elemental composition and the formation of well-developed microporous structures during activation (González et al. 2013). Such findings highlight the potential of low-cost, renewable waste materials as sustainable feedstocks for high-performance CO_2 sorbents. Activated carbon production generally follows either physical or chemical activation routes. Physical activation involves a two-step process comprising initial carbonization (or pyrolysis) followed by activation. During carbonization, volatile components are removed under an inert nitrogen (N_2) atmosphere at temperatures approaching 800 °C, which is considered optimal for devolatilization and the formation of stable carbon frameworks (Üner and Bayrak 2018). Subsequent activation further develops porosity through controlled gasification, typically using steam or CO_2. Chemical activation, in contrast, involves impregnating the precursor with a dehydrating or activating agent either prior to or during carbonization (Üner and Bayrak 2018). Common activating agents include acidic, alkaline, and salt-based chemicals (Burr and Lyddon 2008). This approach has gained increasing attention due to its lower energy demand, shorter processing time, and higher carbon yield compared to physical activation. Chemically activated carbons are usually produced at temperatures between 600 and 800 °C, where short residence times promote enhanced pore formation and improved adsorption capacity (Heidari et al. 2014). In many cases, chemical activation is carried out as a single-step process, with the activating agent present throughout carbonization and activation, facilitating dehydration and oxidation reactions that generate highly porous carbon structures (Heidari et al. 2014). Beyond activated carbons, metal–organic frameworks (MOFs) have emerged as a highly promising class of sorbent materials for gas capture applications. MOFs are crystalline, highly porous solids constructed from transition metal ions or clusters coordinated with organic ligands. The extensive diversity of metal nodes and organic linkers allows precise tuning of pore size, surface functionality, and adsorption behavior, making MOFs particularly attractive for the capture of CO_2, SO_x, NO_x, and H_2S (Sathre and Masanet 2013). Their exceptionally high surface areas and structural versatility offer significant advantages over conventional sorbents, although challenges related to stability and cost remain. Zeolites represent another important category of solid sorbents, characterized by their microporous aluminosilicate frameworks with uniform pore networks. These materials enable selective diffusion and weak interactions between acid gases and metal oxide sites within the zeolite structure. To improve CO_2 affinity, zeolites have been functionalized with amine groups; however, their relatively small pore sizes can restrict amine loading, thereby limiting overall adsorption capacity (Sathre and Masanet 2013). Among transition metal oxides, cadmium oxide (CdO) has been identified as an effective CO_2 sorbent, exhibiting adsorption behavior comparable to that of magnesium oxide (MgO) (Sathre and Masanet 2013). The CO_2 capture mechanism of CdO closely resembles that of MgO, with molten salts acting as reaction promoters to facilitate carbonate formation. However, the high toxicity of cadmium severely constrains the large-scale deployment of CdO-based sorbents, despite their favorable adsorption performance. In addition to metal oxides, a wide range of solid supports including silica, clays, and polymeric materials have been functionalized with amines, amino acids, and ionic liquids to enhance chemical interactions with acid gases. Among these, silica remains the most widely used support due to its low cost, high surface area, mechanical robustness, and tunable porosity. Nevertheless, alternative supports continue to be actively investigated to overcome limitations related to stability, capacity, and long-term cyclic performance (Üner and Bayrak 2018).

4.2.8 Cryogenic Carbon Capture

Cryogenic carbon capture (CCC) is a physical separation approach that exploits differences in boiling points and desublimation characteristics of individual gas components to selectively separate CO_2 from mixed gas streams (Font-Palma et al. 2021). The effectiveness of CCC relies heavily on accurate phase equilibrium data, which define the temperature and pressure windows under which CO_2 transitions between gaseous, liquid, and solid states. Based on these phase transitions, CCC processes are generally categorized into two operational regimes: solid–vapor separation, which employs non-conventional separation concepts, and liquid–vapor separation, which is based on well-established cryogenic distillation principles (Font-Palma et al. 2021). Recent studies have focused on CCC configurations that utilize partial condensation followed by phase separation of liquid and vapor streams, typically integrated with multi-stage compression at operating pressures of around 15 and 30 bar (Karayil et al. 2024). To enhance CO_2 recovery and achieve higher purity levels, low-boiling components such as O_2, N_2, and Ar are progressively removed through additional flash separation steps, coupled with careful optimization of operating pressures and temperatures. For applications that demand ultra-high-purity CO_2, including EOR, an additional stripping column is commonly incorporated to reduce residual oxygen concentrations to below 100 ppm (Font-Palma et al. 2021). Efficient refrigeration is a central requirement for CCC systems, as extremely low temperatures are necessary to induce CO_2 condensation or solidification. Stirling coolers are frequently recommended for this purpose due to their high efficiency and reliability at cryogenic temperatures. A typical CCC configuration employs three Stirling cooling units: the first unit pre-cools the inlet gas stream to remove moisture, the second functions as a freezing tower where CO_2 is desublimated into the solid phase, and the third unit is dedicated to the cryogenic storage of captured CO_2 (Font-Palma et al. 2021). As illustrated in Fig. 4.2, CCC can be synergistically integrated with downstream energy conversion processes. In this configuration, industrial exhaust gases undergo CCC, enabling the separation and storage of CO_2 at ultra-low temperatures (approximately −120 °C). The remaining clean gas stream, composed primarily of N_2 and water vapor, is subsequently routed to hydrogen production via electrolysis powered by renewable energy sources such as wind and solar. The produced hydrogen is then purified, liquefied, and stored for use in fuel cells and transportation applications. This integrated CCC hydrogen framework supports a circular and low-carbon energy system by simultaneously minimizing greenhouse gas emissions and accelerating the transition toward green hydrogen as a carbon-neutral energy carrier.

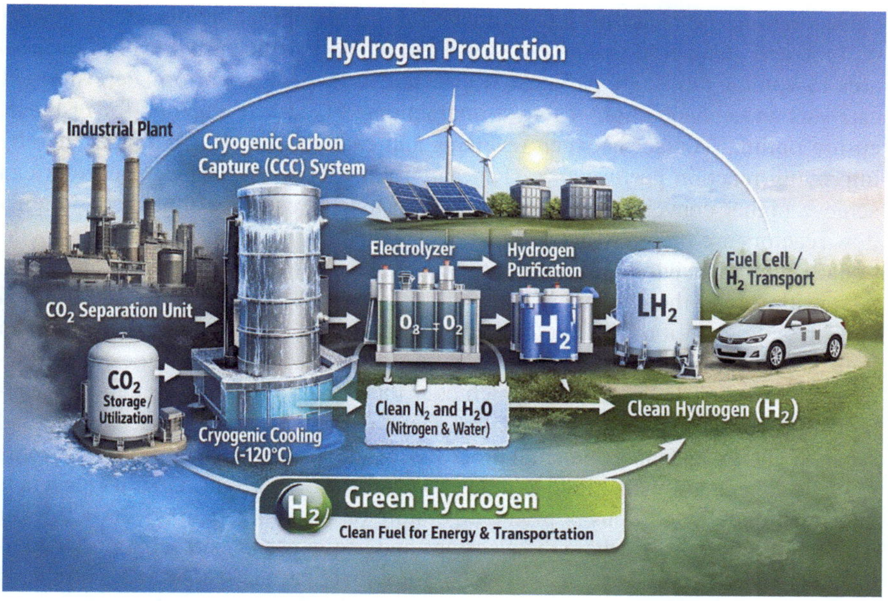

Fig. 4.2 Cryogenic carbon capture and hydrogen production pathway for clean energy transition

4.3 Chemical Conversion Technologies

4.3.1 Hydrogenation

Photo-thermal hydrogenation has emerged as a cutting-edge approach for utilizing solar energy to drive hydrogenation reactions. In this technique, incoming solar radiation is converted into thermal energy, which raises the local temperature of the system and accelerates reaction kinetics. This temperature enhancement not only boosts the rate of redox reactions but also improves the overall catalytic efficiency. Photo-thermal processes achieve this by combining two synergistic mechanisms: electronic excitation caused by ultraviolet (UV) light absorption in semiconductor materials and thermal energy released from exothermic reactions occurring on the catalyst surface. Essentially, the catalyst operates as a photothermal converter, absorbing photons and transforming their energy into heat, creating conditions conducive for hydrogenation reactions (Fan and Tahir 2022). Despite these advantages, the efficiency of CO_2 conversion in photo-thermal hydrogenation is still limited compared to traditional thermal hydrogenation methods. Current reports indicate CO_2 yields in the range of micro-moles per gram of catalyst per hour (μmol g^{-1} h^{-1}), highlighting the need for further improvements. Several challenges restrict the broader adoption of this technology, including suboptimal light absorption by catalysts and the high cost associated with sourcing renewable hydrogen (Tahir and Tahir 2020). Designing an effective photo-thermal reactor involves careful consideration of three critical aspects: the reactor configuration, the source of heat or light, and the operational mode. Structured reactors, which provide high surface-area-to-volume ratios, are particularly effective as they allow uniform deposition of catalysts and enhance light–catalyst interactions. Fixed-bed reactors, for instance, ensure optimal contact between reactants and catalysts under irradiation, thereby improving reaction performance (Fan and Tahir 2022). While batch reactors can achieve higher conversion efficiencies due to longer residence times, continuous-flow reactors are preferred for industrial-scale applications because of their operational stability, scalability, and ability to maintain consistent reaction conditions. Alongside photo-thermal approaches, plasma-assisted hydrogenation has gained considerable attention as a versatile alternative for CO_2 conversion. Plasma is an ionized gas containing a mixture of electrons, ions, radicals, molecules, and electronically excited species, all of which contribute to its exceptional chemical reactivity. This property has enabled plasma technologies to be applied across a broad range of energy and environmental processes (Snoeckx and Bogaerts 2017). Plasma systems are generally classified as either thermal or non-thermal. Thermal plasma is generated under conditions of high temperature (approximately 4000–20,000 K) or elevated gas pressures, where the electrons and heavy particles reach thermal equilibrium (Snoeckx and Bogaerts 2017). Non-thermal plasma, in contrast, is produced by applying strong electric fields across electrodes, causing gas breakdown and accelerating electrons while keeping the bulk gas close to ambient temperature. In non-thermal plasma, energetic electron collisions with molecules induce ionization, excitation, and dissociation events, leading to the formation of reactive species, emission of light, and the synthesis of novel chemical products (Snoeckx and Bogaerts 2017). Collectively, photo-thermal and plasma-assisted hydrogenation represent complementary strategies for CO_2 conversion, with each offering unique mechanisms, advantages, and challenges that can be optimized for sustainable chemical production.

4.3.2 Electrochemical Reduction

Electrochemical CO_2 reduction has emerged as a highly promising strategy for converting carbon dioxide into valuable chemicals and fuels. This process typically takes place in an electrochemical cell composed of a cathode, an anode, and an electrolyte, with an external electrical power source driving the reactions (Kamkeng et al. 2021). At the anode, water undergoes oxidation, producing oxygen, protons, and electrons, while at the cathode, CO_2 is reduced to generate a range of products such as carbon monoxide, formic acid, hydrocarbons, and alcohols. Despite its potential for sustainable chemical production, the widespread deployment of electrochemical CO_2 reduction faces significant technical and economic hurdles. Chief among these are low achievable current densities, decreased energy efficiency due to pH gradients within the cell, and the high electricity costs required to sustain the reaction (Price 2015). To overcome these challenges, research has increasingly focused on the development of advanced materials and optimized device architectures. Bipolar membranes (BPMs) have emerged as an effective solution to the pH gradient problem. By promoting water auto-dissociation at the membrane interface, BPMs maintain near-constant pH levels at both the anode and cathode, thereby enhancing reaction selectivity, reducing energy losses, and improving overall cell efficiency (Cells et al. 2016). In addition, continuous-flow reactor designs have been explored to address mass-transfer limitations and the low solubility of CO_2 in aqueous electrolytes. These systems ensure efficient transport of reactants to the electrode surface and facilitate removal of products, particularly in gas-phase or gas-diffusion electrode configurations, leading to significantly improved reaction performance (Liang et al. 2020). Several advanced electrochemical reactor designs are currently under investigation for their scalability and industrial applicability. These include microfluidic electrochemical cells, zero-gap electrolyzers, and multilayer electrolyzer stacks, all of which aim to combine high current densities with enhanced energy efficiency and operational stability. Collectively, these innovations represent a pathway toward commercially viable electrochemical CO_2 conversion technologies. Detailed discussions and comparative analyses of these device architectures are available in Price (Price 2015).

4.3.3 Photochemical Reduction

Photocatalytic CO_2 reduction has emerged as a highly promising strategy for addressing both energy and environmental challenges by emulating natural photosynthesis through an artificial photosynthetic pathway. In this approach, semiconductor-based photocatalysts absorb light energy, generating charge carriers that drive the conversion of CO_2 into high-energy, value-added chemicals (Kamkeng et al. 2021). The photocatalytic reduction process can be broken down into five key steps: (1) absorption of photons by the photocatalyst, leading to the formation of electron–hole (e^-/h^+) pairs; (2) separation and migration of these charge carriers to active sites on the catalyst surface; (3) adsorption of CO_2 molecules onto the photocatalyst; (4) redox reactions, where photogenerated electrons reduce CO_2 into desired products while the corresponding holes oxidize water to release oxygen; and (5) desorption of the reaction products, completing the catalytic cycle (Wu et al. 2017). Significant progress in heterogeneous photocatalysis has been achieved through the rational design of catalysts, with performance largely governed by several critical factors: efficient adsorption and activation of CO_2 on the catalyst surface, effective separation and transport of charge carriers, broad-spectrum light absorption, and the availability of suitable active sites. Optimizing these parameters is essential not only for increasing conversion efficiency but also for enhancing product selectivity. Achieving high selectivity toward specific hydrocarbons or oxygenates is particularly important for economic feasibility, as it reduces downstream separation and purification costs. Despite its considerable potential, photocatalytic CO_2 reduction faces challenges that limit its practical deployment for sustainable chemical production. Chief among these are low conversion efficiencies, limited catalyst stability under prolonged operation, and difficulties in scaling up laboratory-scale systems to industrially relevant sizes. Overcoming these obstacles will require continued research focused on developing robust, highly efficient photocatalysts and reactor designs capable of sustained operation (Wu et al. 2017).

4.4 Biological Conversion Technologies

4.4.1 Photosynthetic CO_2 Fixation

Photosynthetic CO_2 fixation is the fundamental process through which plants and microalgae capture carbon dioxide and convert it into energy-rich compounds such as glucose and biomass using sunlight, water, and inorganic nutrients (Kamkeng et al. 2021). Microalgae, in particular, demonstrate high CO_2 assimilation efficiency while transforming light energy into chemical energy, making them attractive biological platforms for carbon capture and utilization (CCU). Algal cultivation is typically conducted in two main types of systems: open raceway ponds (RPs) and closed photobioreactors (PBRs). Open RPs are shallow ponds or lagoons that provide growth media for algae, offering the advantage of relatively low capital costs. However, they are challenged by several practical limitations, including large land requirements, vulnerability to environmental fluctuations, water loss due to evaporation, and difficulties in situating ponds near concentrated CO_2 sources (Baena-Moreno et al. 2019; Bhatia et al. 2019). In contrast, closed photobioreactors provide a controlled environment that enables precise regulation of key parameters such as temperature, light intensity, nutrient supply, and CO_2 concentration. These systems offer multiple benefits over open ponds, including minimized water loss, reduced contamination and predator growth, improved light utilization due to larger illumination surfaces, and enhanced capacity for high-density algal cultivation or immobilization (Razzak et al. 2013). Among closed systems, tubular photobioreactors are widely adopted for outdoor applications because they allow efficient sunlight exposure while providing dedicated channels for gas exchange and photosynthetic activity. Nonetheless, tubular PBRs also have notable limitations, such as the formation of pH gradients along the reactor length, high construction costs associated with durable materials, and significant land requirements for large-scale operations. Overall, both open and closed systems offer trade-offs between cost, operational control, and productivity, and optimizing these systems remains critical for the sustainable implementation of microalgal CO_2 fixation technologies.

4.4.2 *Non-photosynthetic CO_2 Fixation*

This technique employs a source of high-energy electrons in conjunction with microorganisms to facilitate the conversion of CO_2 into value-added bio-products. The process can operate under either aerobic or anaerobic conditions. In aerobic systems, microorganisms are directly exposed to oxygen from the surrounding environment, whereas in anaerobic configurations, oxygen is deliberately excluded from the system to enable oxygen-sensitive metabolic pathways (Hawkins et al. 2013). A key advantage of non-photosynthetic CO_2 fixation methods is their ability to generate a wide range of bio-products, including microbial biomass and methane, while operating under relatively mild temperature and pressure conditions.

4.5 Challenges for Commercialization

Effective mitigation of climate change requires the large-scale conversion of CO_2 into value-added products. However, most carbon conversion technologies currently face significant economic barriers, primarily due to high electricity costs, substantial capital expenditure (CAPEX), and uncertain market demand resulting from insufficient regulatory support. While many technical challenges associated with CO_2 conversion can be addressed through continued research and development—particularly for technologies at lower technology readiness levels (TRLs) economic feasibility remains a major concern (Lamberts-Van Assche et al. 2022). Alternative system and reactor designs for emerging technologies may help alleviate some scale-up challenges; nevertheless, the existing literature on such approaches remains limited. From a commercialization perspective, mineralization emerges as the most mature CO_2 conversion pathway, accounting for approximately 43% of current commercial projects. In contrast, no commercial-scale projects currently exist for electrochemical, non-photosynthetic, photochemical, or plasma-based CO_2 conversion technologies, largely due to their low TRLs. Among these, plasma catalysis—typically at TRLs between 1 and 3—has gained attention for its potential economic advantages, including relatively low capital and operating costs (Snoeckx and Bogaerts 2017). However, scalability remains a critical limitation for plasma-based systems. As reactor size increases, CO_2 conversion efficiency often declines due to confinement of the plasma region near the reactor center, resulting in reduced interaction between reactant molecules and the active plasma zone (Bogaerts and Centi 2020). A techno-economic assessment by Lamberts-Van Assche et al. (Lamberts-Van Assche et al. 2022) examined plasma-catalytic CO_2 conversion across various reactor configurations and found that extended space-time and the inclusion of packing materials significantly increase both operational electricity costs (approximately €35 million OPEX) and power-supply-related capital costs (approximately €80 million CAPEX). Similarly, electrochemical reduction (ER) technologies remain at the pilot scale, with TRLs ranging from 3 to 5, and require further technological advancements to achieve commercial viability and long-term sustainability. Several studies have evaluated the techno-economic feasibility of large-scale electrochemical CO_2 reduction. Rumayor et al. (2019) assessed the conversion of CO_2 to formic acid and reported strong environmental performance; however, the process was deemed economically uncompetitive under current market conditions. Agarwal et al. (2011) compared large-scale ER of CO_2 to formic acid and formate salts with conventional carbon capture and storage (CCS). Their analysis showed that ER requires approximately 40% lower CAPEX (USD 2.9 million) compared to CCS (USD 4.7 million) but incurs significantly higher operating costs (USD 1150 per tonne of CO_2 mitigated) relative to CCS (USD 390 per tonne of CO_2 mitigated). Both approaches yielded negative net present values over a 25-year plant lifetime, indicating limited profitability under prevailing economic conditions. Despite these challenges and uncertainties, CCC technologies are expected to play an increasingly important role in future clean energy systems, particularly in enabling low-carbon fuel and hydrogen production pathways. Table 4.2 summarizes

Table 4.2 Comparative analysis of emerging carbon capture and conversion approaches and their sectoral applications

S. No	Technology focus	Key findings	Applications	Key challenges	References
1	Integrated CO_2 capture and conversion	Modular DAC and electrochemical conversion systems transforming CO_2 into synthetic fuels	Renewable fuels, green hydrogen, chemical manufacturing	High energy input, limited catalyst stability, and CO_2 purity issues	Masoumilari et al. (2026)
2	Techno-economic frameworks for CCU	Proposes economic models for scalable CO_2 mineralization and solid carbon products	Cement and construction, carbon-negative materials	High capital costs; uncertain carbon market incentives	Ansari, (2026)
3	AI-enhanced CCUS systems	Integrates AI-driven discovery and predictive modeling to optimize CCUS efficiency	Smart manufacturing, digital twins for emissions control	Data scarcity and integration complexity between AI and real-world processes	Mazhar et al. (2026)
4	Biohybrid carbon sequestration	Uses engineered cyanobacteria for CO_2-to-biomass and biofuel pathways	Bioenergy, sustainable agriculture	Low productivity under variable conditions scalability issues.	Nawaz et al. (2026)
5	Chemical looping for CO_2 utilization	Uses redox cycles for efficient synthesis of NH_3 and methanol from CO_2	Fertilizer, petrochemical sectors	Complex reactor design and oxygen carrier degradation	Ghorbani et al. (2026)
6	Industrial CCUS pathways	Integrates capture with mineralization for deep industrial decarbonization	Steel, cement, heavy industries	High retrofitting costs; lack of supportive infrastructure	Singh (2026)

recent advancements in emerging CCC technologies relevant to the global energy transition. It highlights the major research domains, industrial applications, and key challenges identified in recent studies. The technologies span from AI-enhanced CCUS systems and biohybrid carbon sequestration to chemical looping illustrating the multidisciplinary progress toward carbon utilization and net-zero goals.

4.6 Production of Blue Hydrogen

Blue hydrogen is based on coupling conventional fossil-fuel-based hydrogen production pathways with carbon capture, utilization, and storage (CCUS) technologies to reduce associated greenhouse gas (GHG) emissions (AlHumaidan et al. 2023; Massarweh et al. 2023). This approach is often perceived as a cost-effective transitional alternative to green hydrogen; however, the deployment of CCUS introduces substantial technological complexity, high capital expenditure (CAPEX), and challenges related to public acceptance and social license to operate (Sovacool et al. 2024). The technology readiness levels (TRLs) of blue hydrogen pathways are relatively advanced, ranging from TRL 7 for coal gasification with CCUS to TRL 8 for steam methane reforming (SMR) coupled with CCUS (Noussan et al. 2020). There is currently no universally accepted definition of the CO_2 capture rate required to classify grey hydrogen as blue hydrogen. Depending on the production route and capture point, reported capture efficiencies typically range between 70% and 95%. For example, (Arcos and Santos 2023) demonstrated CO_2 capture from three major emission streams, achieving approximately 90% removal efficiency from pressure swing adsorption (PSA) tail gas and steam reforming flue gas, and over 99% capture from high-pressure raw hydrogen streams. In addition to CO_2 emissions, methane (CH_4) leakage during upstream natural gas extraction and transport must be carefully considered, as it can significantly influence the overall climate performance of blue hydrogen systems (Noussan et al. 2020). To harmonize definitions across Europe, the CertifHy Steering Group proposed a benchmark for low-carbon (blue) hydrogen in 2019, requiring at least a 60% reduction in GHG emissions relative to a conventional SMR-based reference process. This corresponds to an emission intensity threshold of 36.4 g CO_2 e/MJ (131 gCO_2e/kWh), compared with a baseline of 91 gCO_2e/MJ (328 gCO_2e/kWh) for standard SMR hydrogen production (Noussan et al. 2020). Beyond emissions, water consumption represents an often-overlooked environmental impact of blue hydrogen pathways. While electrolysis is commonly associated with high water demand, studies indicate that fossil-based hydrogen production with CCUS can exhibit even higher water consumption rates. (Simoes et al. 2021) reported that water use is frequently underestimated in blue hydrogen assessments. Life cycle inventory analyses show that water consumption can reach up to 24 L per kg of H_2 for SMR and approximately 38 L per kg of H_2 for coal gasification routes (Rafique et al. 2025).

4.6.1 Production of Blue Hydrogen by Steam Reforming of Methane and Light Hydrocarbons

The steam methane reforming (SMR) process involves three principal reactions: (i) steam reforming of hydrocarbons, (ii) the water–gas shift (WGS) reaction, and (iii) methane formation. These reactions are represented by Eqs. (4.1)–(4.3), respectively.

$$CH + nH_2O \rightarrow nCO + (n + m/2)H_2 \tag{4.1}$$

$$CO + H_2O \rightarrow CO_2 + H_2 \tag{4.2}$$

$$CO + 3H_2 \rightarrow CH_4 + H_2O \tag{4.3}$$

An important parameter defining the steam methane reforming (SMR) process is the hydrogen-to-carbon (H:C) ratio of the feedstock. Higher H:C ratios result in lower carbon dioxide formation. Methane, with an H:C ratio of 4:1, produces the lowest CO_2 emissions among hydrocarbon feedstocks, approximately 7 kg CO_2 per kg H_2 produced (Wismann 2019). Product distribution in SMR is governed by thermodynamic constraints: methane reforming is favored at relatively low temperatures (~623 K), whereas higher hydrogen yields are achieved at elevated temperatures of up to ~1273 K (Wang et al. 2024). Nickel-based catalysts are predominantly used in industrial SMR due to their high activity and cost-effectiveness (Yusuf et al. 2024). Catalytic performance depends on the available metal surface area and the ability to withstand severe operating conditions, including temperatures between 600 and 1300 K and steam partial pressures of up to 35 bar (Massarweh et al. 2023). Under industrial conditions, typical Ni catalysts exhibit turnover frequencies of approximately 0.5 s^{-1} at 723 K, corresponding to methane conversions of about 10% (Navarro et al. 2015). However, equilibrium limitations restrict high methane conversion to temperatures exceeding 1170 K. In conventional reformers, reaction kinetics are rarely rate-limiting; instead, performance is constrained by mass and heat transfer limitations within pelletized catalysts, such as Ni/Al_2O_3, where effectiveness factors are often below 10% (Pashchenko 2023). Industrial SMR systems achieve thermal efficiencies in the range of 80–85% (Ma et al. 2022). SMR plant comprises feed gas pre-treatment, pre-reforming, primary reforming, high-temperature water–gas shift (HTWGS), low-temperature water–gas shift (LTWGS), and hydrogen purification units. In the pre-reformer, higher hydrocarbons are converted into C_1 species (CH_4, CO, and CO_2) at relatively low temperatures (673–823 K) (Didenko et al. 2023). Heating the pre-reformed gas to temperatures up to 1073 K prior to entering the main reformer mitigates the risk of carbon formation via thermal cracking. However, the low operating temperature renders pre-reforming catalysts susceptible to carbon deactivation (Gupta et al. 2024). These catalysts typically consist of high nickel loadings (20–30 wt%) supported on alkaline materials such as MgO (60–70 wt%) with high surface area (Massarweh et al. 2023). The water–gas shift (WGS) process comprises two catalytic stages—high-temperature shift (HTS) and low-temperature shift (LTS) to convert residual CO into additional hydrogen, as shown in Eq. (4.2). The HTS reactor operates near equilibrium at 623–693 K, reducing CO concentrations from 8–10 vol% to approximately 4 vol%, while the LTS reactor operates at 453–613 K to further lower CO levels to 0.4–0.8 vol% (Fowles and Carlsson 2021). Industrial HTS and LTS catalysts are typically based on Fe_2O_3–Cr_2O_3 and Cu–ZnO–Al_2O_3 formulations, respectively (Jeong et al. 2013). Despite WGS processing, reformate gas still contains impurities such as unconverted CH_4 and trace CO (Chakrabarti et al. 2015). Pressure swing adsorption (PSA) is widely employed to purify hydrogen, producing H_2 with purities up to 99.99 vol% (Kim and Yang 2025; Krótki et al. 2025). PSA utilizes adsorbents such as zeolites, activated carbon, and silica gel to selectively remove CO_2, CO, CH_4, and N_2 from syngas streams, while hydrogen passes through as the product gas (Pangestu et al. 2024). Regeneration is achieved through cyclic pressure variation, enabling continuous production of high-purity hydrogen (Oh and Lee 2024). Methane leakage represents a critical challenge for blue hydrogen production via SMR (Bauer et al. 2022). Methane is a potent greenhouse gas, and approximately 22% of global annual CH_4 emissions (~80 Tg $yr.^{-1}$) originate from oil and gas supply chains (Davids et al. 2025). Leakage rates as low as 1–3% can negate the emission reductions achieved through CCUS deployment. Methane emissions occur during natural gas extraction, transportation, and reforming operations. Without effective mitigation, blue hydrogen pathways may fail to deliver climate benefits and could exceed the lifecycle emissions of conventional fossil fuels (Shirizadeh et al. 2023). Reducing methane leakage requires integrated technological, operational, and regulatory strategies. Advanced leak detection and repair (LDAR) programs employing infrared imaging, continuous monitoring sensors, and satellite-based surveillance can significantly reduce fugitive emissions. Equipment upgrades, improved pipeline integrity, and best operational practices further mitigate leakage risks. Additionally, on-site hydrogen production at extraction facilities can minimize transport-related emissions (Ruokonen et al. 2021). Regulatory frameworks mandating rigorous monitoring, reporting, and verification (MRV), coupled with economic incentives or penalties, are essential to ensure compliance. Collectively, these measures enhance the climate credibility of blue hydrogen and align its deployment with global net-zero objectives (Pettersen et al. 2022).

4.7 Role of CCC for Deployment of Large-Scale Hydrogen Production

The urgent need to reduce carbon intensity from industrial sectors, including large-scale hydrogen production, is a key driver for achieving global clean energy transition targets by 2050. Within this context, Qatar's commitment to lowering carbon emissions and advancing toward a low-carbon energy system underscores the strategic importance of investing in carbon capture and conversion (CCC) technologies as part of its clean energy portfolio (Shirizadeh et al. 2023). Integrating CCC technologies into the national hydrogen production strategy represents a critical step aligned with the Qatar National Climate Change Action Plan 2030, supporting broader goals of sustainable economic growth and diversification. Coupling CCC technologies with large-scale blue hydrogen production offers dual benefits: reducing dependence on long-term CO_2 sequestration and offsetting hydrogen production costs through the generation of value-added chemicals and fuels. To realize these benefits, systematic studies are necessary to identify viable CCC pathways that leverage Qatar's existing industrial infrastructure, alongside a comprehensive techno-economic and environmental evaluation of integrated blue hydrogen and CCC systems (Oh and Lee 2024). Comparative analyses indicate that the steam methane reforming (SMR) process is particularly well-suited for hydrogen production in Qatar due to its technological maturity, scalability, and cost-effectiveness. Integrating appropriate CCC technologies with SMR-based hydrogen production can therefore serve as an efficient mid-term strategy to accelerate the nation's transition toward a low-carbon economy. From a long-term perspective, with national carbon neutrality targets set within the next 15–30 years, future research should adopt a holistic approach. This includes evaluating commercially available CO_2 capture technologies, emerging carbon conversion routes, and their associated environmental impacts through comprehensive life cycle assessment (LCA) frameworks. Deploying such integrated solutions is essential for ensuring a sustainable and resilient energy future. By systematically examining both established and novel technologies within an LCA-based evaluation, policymakers and industry stakeholders can make informed decisions that facilitate the large-scale implementation of CO_2 mitigation strategies (Davids et al. 2025).

4.8 Conclusion and Future Prospective

Amid growing concerns over global warming, decarbonizing the energy sector is crucial for mitigating industrial CO_2 emissions and achieving carbon neutrality. Qatar offers several promising pathways for decarbonization, including low-carbon hydrogen production and the deployment of solar energy systems. This paper has systematically reviewed these pathways, with particular emphasis on CO_2 capture, CO_2 conversion, and the potential integration of CCC technologies with low-carbon hydrogen production. Through this analysis, several technological gaps and critical research needs that must be addressed to support the energy transition have been identified. The key conclusions of this study are summarized as follows:

4.8.1 CO_2 Capture Technologies

Physical absorption using ionic liquids (ILs) has emerged as a highly promising approach for CO_2 capture due to their negligible volatility, high thermal stability, and tunable chemical properties. These features make ILs attractive alternatives to conventional amine-based solvents. However, the vast structural diversity and tunability of ILs also present a significant challenge, as identifying the optimal IL for efficient CO_2 capture requires screening thousands of potential combinations. Chemical looping combustion (CLC) is another advanced CO_2 capture strategy, widely regarded for its efficiency and environmentally friendly operation in fuel-based power plants. By employing a solid oxygen carrier to transfer oxygen from air to fuel, CLC inherently produces a CO_2-rich flue stream suitable for capture. Nevertheless, practical implementation faces challenges related to the mechanical stability of oxygen carriers and their consistent performance during repeated oxidation–reduction cycles. In parallel, emerging vapor–solid cryogenic CO_2 capture technologies have shown promise for reducing energy consumption compared to traditional vapor–liquid cryogenic processes. By leveraging selective solid sorbents and low-temperature separation, these approaches offer the potential for more energy-efficient CO_2 recovery in industrial applications.

4.8.2 CO_2 Conversion Technologies

CO_2 conversion technologies represent a rapidly expanding research area due to their potential environmental and economic benefits. Nonetheless, emerging approaches such as photo-thermal and plasma hydrogenation require further development to address challenges associated with limited performance data and low CO_2 conversion efficiencies. Electrochemical CO_2 reduction faces scalability constraints, primarily due to low energy efficiency arising from pH gradients across the cell; innovative solutions such as bipolar membranes (BPMs) have been proposed to mitigate these limitations. Photocatalytic CO_2 reduction remains a promising route for addressing energy and environmental challenges, but further efforts are required to improve product selectivity and conversion yields. In biological conversion pathways, closed cultivation systems have demonstrated advantages over open systems, including reduced water evaporation and improved conditions for algae immobilization, although climatic variability remains a major limitation for large-scale deployment.

4.8.3 Hydrogen Production Pathways

Hydrogen is widely regarded as a key energy carrier for decarbonizing the energy sector and meeting future energy demand. Despite its potential, extensive research and technological advancement are required for hydrogen to play a central role in future energy systems. Steam methane reforming (SMR) remains the most widely deployed hydrogen production technology. However, cleaner and more sustainable hydrogen production routes including coal gasification, methane pyrolysis, water electrolysis, and photocatalysis have also been critically reviewed. Gasification and pyrolysis, in particular, demonstrate economic competitiveness and the potential to serve as alternatives to conventional hydrogen production technologies, although several technical and scalability challenges currently hinder their large-scale implementation.

References

Agarwal AS, Zhai Y, Hill D, Sridhar N (2011) The electrochemical reduction of carbon dioxide to formate/formic acid: engineering and economic feasibility. ChemSusChem 4:1301–1310. https://doi.org/10.1002/cssc.201100220

AlHumaidan FS, Absi Halabi M, Rana MS, Vinoba M (2023) Blue hydrogen: current status and future technologies. Energy Convers Manag 283:116840

Al-Yaeeshi AA, Al-Ansari T, Govindan R (2018) The potential for carbon dioxide capture and utilization within the State of Qatar. In: Computer aided chemical engineering. Elsevier Masson SAS. https://doi.org/10.1016/B978-0-444-64235-6.50261-8

Ansari MB (2026) A techno-economic climate-tech manufacturing framework for permanent carbon capture and conversion into stable carbon-based products. Available at SSRN 6006835

Arcos JMM, Santos DMF (2023) The hydrogen color spectrum: techno-economic analysis of the available technologies for hydrogen production. Gases 3:25–46

Baena-Moreno FM, Rodríguez-Galán M, Vega F, Alonso-Farinas ˜B, Vilches Arenas LF, Navarrete B (2019) Carbon capture and utilization technologies: a literature review and recent advances. Energy Sources Part A Recover Util Environ Eff 41:1403–1433. https://doi.org/10.1080/15567036.2018.1548518

Bauer C, Treyer K, Antonini C, Bergerson J, Gazzani M, Gencer E, Gibbins J, Mazzotti M, McCoy ST, McKenna R, Pietzcker R (2022) On the climate impacts of blue hydrogen production. Sustain Energy Fuels 6(1):66–75

Bhatia SK, Bhatia RK, Jeon JM, Kumar G, Yang YH (2019) Carbon dioxide capture and bioenergy production using biological system – a review. Renew Sust Energ Rev 110:143–158. https://doi.org/10.1016/j.rser.2019.04.070

Binti Wan Ramli WK (2017) Exsolved base metal catalyst systems with anchored nanoparticles for carbon monoxide (CO) and nitric oxides *(NO_x) oxidation* (Doctoral dissertation). Newcastle University

Birol F (2019) The future of hydrogen – analysis – IEA [WWW document]. Technol. Rep. URL. https://www.iea.org/reports/the-future-of-hydrogen, 3.13.21

Bogaerts A, Centi G (2020) Plasma technology for CO2 conversion: a personal perspective on prospects and gaps. Front Energy Res 8:1–23. https://doi.org/10.3389/fenrg.2020.00111

Bui M, Adjiman CS, Bardow A, Anthony EJ, Boston A, Brown S, Fennell PS, Fuss S, Galindo A, Hackett LA, Hallett JP, Herzog HJ, Jackson G, Kemper J, Krevor S, Maitland GC, Matuszewski

M, Metcalfe IS, Petit C, Puxty G, Reimer J, Reiner DM, Rubin ES, Scott SA, Shah N, Smit B, Trusler JPM, Webley P, Wilcox J, Mac Dowell N (2018) Carbon capture and storage (CCS): the way forward. Energy Environ Sci 11:1062–1176

Burr B, Lyddon L (2008) A comparison of physical solvents for acid gas removal. In: Gas processors' association convention, Grapevine

Cells ME, Li YC, Zhou D, Yan Z, Gonc RH, Salvatore DA, Berlinguette CP, Mallouk TE (2016) Electrolysis of CO 2 to Syngas in Bipolar. https://doi.org/10.1021/acsenergylett.6b004

Chakrabarti D, De Klerk A, Prasad V, Gnanamani MK, Shafer WD, Jacobs G, Sparks DE, Davis BH (2015) Conversion of CO2 over a CO-based Fischer-Tropsch catalyst. Ind Eng Chem Res 54:1189–1196

Davids D, Grant N, Mittal S, Hawkes A, Oluleye G (2025) Impact of methane leakage rate and carbon capture rate on blue hydrogen sustainability using combined warming index. Appl Energy 394:125888

Deng S, Hu B, Chen T, Wang B, Huang J, Wang Y, Yu G (2015) Activated carbons prepared from peanut shell and sunflower seed shell for high CO2 adsorption. Adsorption 21(1):125–133

Didenko LP, Babak VN, Sementsova LA, Dorofeeva TV, Chizhov PE, Gorbunov SV (2023) Steam conversion of ethane and methane–ethane mixtures in a membrane reactor with a foil made of a Pd–Ru Alloy. Membr Membr Technol 5:69–79

Fan WK, Tahir M (2022) Recent developments in photothermal reactors with understanding on the role of light/heat for CO2 hydrogenation to fuels: a review. Chem Eng J 427:131617. https://doi.org/10.1016/j.cej.2021.131617

Font-Palma C, Cann D, Udemu C (2021) Review of cryogenic carbon capture innovations and their potential applications. C 7(3):58

Fowles M, Carlsson M (2021) Steam reforming of hydrocarbons for synthesis gas production. Top Catal 64:856–875

Fragkos P (2020) Global energy system transformations to 1.5 °C: the impact of revised intergovernmental panel on climate change carbon budgets. Energ Technol 8:1–13. https://doi.org/10.1002/ente.202000395

Ghorbani B, Zendehboudi S, Mahinpey N (2026) Chemical looping ammonia production. In: Mahinpey N (ed) Chemical looping processes. Elsevier, pp 307–375. https://doi.org/10.1016/B978-0-443-26659-1.00011-X

González AS, Plaza MG, Rubiera F, Pevida C (2013) Sustainable biomass-based carbon adsorbents for post-combustion CO2 capture. Chem Eng J 230:456–465

Gupta A, Kumar R, Sharma JP, Ahmadi MH, Najser J, Blazek V, Prokop L (2024) The role of catalyst in hydrogen production: a critical review. J Therm Anal Calorim 149:14517–14534

Halliday C, Harada T, Hatton TA (2019) Toward a mechanistic understanding and optimization of molten alkali metal borates (A x B1–x O1. 5–x) for high-temperature CO2 capture. Chem Mater 32(1):348–359

Harris F, Kurnia KA, Mutalib MIA, Thanapalan M (2009) Solubilities of carbon dioxide and densities of aqueous sodium glycinate solutions before and after CO2 absorption. J Chem Eng Data 54(1):144–147

Hawkins AS, McTernan PM, Lian H, Kelly RM, Adams MWW (2013) Biological conversion of carbon dioxide and hydrogen into liquid fuels and industrial chemicals. Curr Opin Biotechnol 24:376–384. https://doi.org/10.1016/j.copbio.2013.02.017

Heidari A, Younesi H, Rashidi A, Ghoreyshi AA (2014) Evaluation of CO2 adsorption with eucalyptus wood based activated carbon modified by ammonia solution through heat treatment. Chem Eng J 254:503–513

Horowitz CA (2016) Paris agreement. Int Leg Mater:740–755. https://doi.org/10.1017/s0020782900004253

IEA (2017) Key Wold energy statistic 2017. International Energy Agency

IEA (2020) Power generation from carbon capture technologies in the IEA sustainable development scenario, IEA, Paris. https://www.iea.org/data-and-statistics/charts/power-generation-from-carbon-capture-technologies-in-the-iea-sustainable-development-scenario

Ingram J (2022) Qatar launches 'world's largest' blue ammonia project [WWW document]. https://www.mees.com/2022/9/2/refining-petrochemicals/qatar-launches-worlds-largest-blue-ammonia-project/dcc199b0-2ab0-11ed-a8a5-913ffd3e63d0. Accessed 4 Dec 2023

International Energy Agency (2021) Global energy review 2021. Global Energy Review 2020

International Energy Agency (2022) Global energy review: CO2 emissions in 2021 – analysis – IEA [WWW Document]. Glob. Energy Rev. CO2 Emiss. 2021. URL. https://www.iea.org/reports/global-energy-review-co2-emissions-in-2021-2, 8.30.23

IPCC (2018) Global warming of 1.5°C. Ipcc – Sr15

Jafari M, Botterud A, Sakti A (2022) Decarbonizing power systems: a critical review of the role of energy storage. Renew Sust Energ Rev 158:112077. https://doi.org/10.1016/j.rser.2022.112077

Jeong DW, Subramanian V, Shim JO, Jang WJ, Seo YC, Roh HS, Gu JH, Lim YT (2013) High-temperature water gas shift reaction over Fe/Al/Cu oxide based catalysts using simulated waste-derived synthesis gas. Catal Lett 143:438–444

Kamkeng ADN, Wang M, Hu J, Du W, Qian F (2021) Transformation technologies for CO2 utilisation: current status, challenges and future prospects. Chem Eng J. https://doi.org/10.1016/j.cej.2020.128138

Kar S, Goeppert A, Prakash GS (2019) Integrated CO2 capture and conversion to formate and methanol: connecting two threads. Acc Chem Res 52(10):2892–2903

Karayil A, Elseragy A, Aliyu AM (2024) An assessment of CO2 capture technologies towards global carbon net neutrality. Energies 17(6):1460

Kim Y, Yang H (2025) Hydrogen purity: influence of production methods, purification techniques, and analytical approaches. Energies 18:741

Koohestanian E, Shahraki F (2021) Review on principles, recent progress, and future challenges for oxy-fuel combustion CO2 capture using compression and purification unit. J Environ Chem Eng 9(4):105777

Krótki A, Bigda J, Spietz T, Ignasiak K, Matusiak P, Kowol D (2025) Performance evaluation of pressure swing adsorption for hydrogen separation from syngas and water–gas shift syngas. Energies 18:1887

Lamberts-Van Assche H, Thomassen G, Compernolle T (2022) The early-stage design of plasma for the conversion of CO2 to chemicals: a prospective techno-economic assessment. J CO2 Util 64:102156. https://doi.org/10.1016/j.jcou.2022.102156

Lee S, Song HJ, Maken S, Park JW (2007) Kinetics of CO2 absorption in aqueous sodium glycinate solutions. Ind Eng Chem Res 46(5):1578–1583

Liang S, Altaf N, Huang L, Gao Y, Wang Q (2020) Electrolytic cell design for electrochemical CO 2 reduction. J CO_2 Util 35:90–105

Ma J, Jiang B, Gao Y, Yu K, Lv Z, Si-ma W, Yuan D, Tang D (2022) A compact and high-efficiency electrified reactor for hydrogen production by methane steam reforming. Int J Hydrog Energy 47:41421–41431

Masoumilari S, Masoumi Z, Shamsabadi AM, Kyung D, Tayebi M (2026) From emissions to assets: sustainable technologies for CO2 capture, conversion, and integrated strategies. Int J Mol Sci 27(2):847. https://doi.org/10.3390/ijms27020847

Massarweh O, Al-khuzaei M, Al-Shafi M, Bicer Y, Abushaikha AS (2023) Blue hydrogen production from natural gas reservoirs: a review of application and feasibility. J CO2 Util 70:102438

Mazhar S, Mumtaz MW, El Oirdi M, Mukhtar H, Raza MA, Farhan M, Aatif M, Muteeb G (2026) Synergizing advanced materials and artificial intelligence for next-generation carbon capture, utilization, and storage (CCUS): a review. RSC Adv 16(3):2621–2651

McDonald TM, Lee WR, Mason JA, Wiers BM, Hong CS, Long JR (2012) Capture of carbon dioxide from air and flue gas in the alkylamine-appended metal–organic framework mmen-Mg2 (dobpdc). J Am Chem Soc 134(16):7056–7065

Nations United Programme (2020) UNITED NATIONS DEVELOPMENT PROGRAMME - ANNUAL REPORT 2020

Navarro RM, Guil R, Fierro JLG (2015) Introduction to hydrogen production. In: Compendium of hydrogen energy. Woodhead Publishing, London, UK, pp 21–61

Nawaz T, Faiz S, Fahad S, Joshi N, Saud S, Yousaf J, Aqil M, Adnan M, Amjad M, Khan I (2026) Photosynthetic adaptations and light harvesting in bloom-forming cyanobacteria. In: Cyanobacterial blooms: ecology, evolution and biogeochemical impacts: microbial dynamics and global implications. Springer Nature Switzerland, Cham, pp 337–370

Noussan M, Raimondi PP, Scita R, Hafner M (2020) The role of green and blue hydrogen in the energy transition—a technological and geopolitical perspective. Sustainability 13:298

Oh T, Lee S (2024) Optimisation of steam-methane reforming process using PSA off gas. Int J Hydrog Energy 95:902–915

Pangestu MRG, Malaibari Z, Muhammad A, Al-Rowaili FN, Zahid U (2024) Comprehensive review on methane pyrolysis for sustainable hydrogen production. Energy Fuel 38:13514–13538

Pashchenko D (2023) Intra-particle diffusion limitation for steam methane reforming over a Ni-based catalyst. Fuel 353:129205

Pekdemir T (2015) Integrated capture and conversion. In: Carbon dioxide utilisation. Elsevier, pp 253–272

Pettersen J, Steeneveldt R, Grainger D, Scott T, Holst LM, Hamborg ES (2022) Blue hydrogen must be done properly. Energy Sci Eng 10:3220–3236

Price DJ (2015) Carbon dioxide. In: Hamilton and Hardy's industrial toxicology, 6th edn, pp 305–308. https://doi.org/10.1002/9781118834015.ch42

Qin F, Wang S, Kim I, Svendsen HF, Chen C (2011) Heat of absorption of CO2 in aqueous ammonia and ammonium carbonate/carbamate solutions. Int J Greenh Gas Con 5(3):405–412

Rafique N, Khalil FK, Ullah IUI, Abbas MT, Safdar M, Aziz U (2025) Hydrogen fuel cell technology: a sustainable future for mechanical power systems. Spectr Eng Sci 3:214–224

Rajabloo T, De Ceuninck W, Van Wortswinkel L, Rezakazemi M, Aminabhavi T (2022) Environmental management of industrial decarbonization with focus on chemical sectors: a review. J Environ Manage 302:114055. https://doi.org/10.1016/j.jenvman.2021.114055

Ramdin M, de Loos TW, Vlugt TJ (2012) State-of-the-art of CO2 capture with ionic liquids. Ind Eng Chem Res 51(24):8149–8177

Razzak SA, Hossain MM, Lucky RA, Bassi AS, De Lasa H (2013) Integrated CO2 capture, wastewater treatment and biofuel production by microalgae culturing – a review. Renew Sust Energ Rev 27:622–653. https://doi.org/10.1016/j.rser.2013.05.063

Regufe MJ, Pereira A, Ferreira AFP, Ribeiro AM, Rodrigues AE (2021) Current developments of carbon capture storage and/or utilization–looking for net-zero emissions defined in the Paris agreement. Energies 14. https://doi.org/10.3390/en14092406

Regulator CE (2020) Market snapshot: how hydrogen has the potential to reduce the CO2 emissions of natural gas [WWW document]. Canada energy Regul. URL. https://www.cer-rec.gc.ca/en/data-analysis/energy-markets/market-snapshots/2020/market-snapshot-hydrogen-potential.html, 11.19.22

Rumayor M, Dominguez-Ramos A, Perez P, Irabien A (2019) A techno-economic evaluation approach to the electrochemical reduction of CO2 for formic acid manufacture. J CO2 Util 34:490–499. https://doi.org/10.1016/j.jcou.2019.07.024

Ruokonen J, Nieminen H, Dahiru AR, Laari A, Koiranen T, Laaksonen P, Vuokila A, Huuhtanen M (2021) Modelling and cost estimation for conversion of green methanol to renewable liquid transport fuels via olefin oligomerisation. Processes 9:1046

Sathre R, Masanet E (2013) Prospective life-cycle modeling of a carbon capture and storage system using metal–organic frameworks for CO 2 capture. RSC Adv 3(15):4964–4975

Seddighi S, Clough PT, Anthony EJ, Hughes RW, Lu P (2018) Scale-up challenges and opportunities for carbon capture by oxy-fuel circulating fluidized beds. Appl Energy 232:527–542

Shen W, He Y, Zhang S, Li J, Fan W (2012) Yeast-based microporous carbon materials for carbon dioxide capture. ChemSusChem 5(7):1274–1279

Shirizadeh B, Villavicencio M, Douguet S, Trüby J, Bou Issa C, Seck GS, D'herbemont V, Hache E, Malbec LM, Sabathier J, Venugopal M (2023) The impact of methane leakage on the role of natural gas in the European energy transition. Nat Commun 14(1):5756

Shu Q, Legrand L, Kuntke P, Tedesco M, Hamelers HV (2020) Electrochemical regeneration of spent alkaline absorbent from direct air capture. Environ Sci Technol 54(14):8990–8998
Simoes SG, Catarino J, Picado A, Lopes TF, di Berardino S, Amorim F, Gírio F, Rangel CM, Ponce de Leão T (2021) Water availability and water usage solutions for electrolysis in hydrogen production. J Clean Prod 315:128124
Singh H (2026) CCUS: green technological solutions and research paths. In: Industrial Decarbonization and the energy transition. Elsevier, pp 221–264
Snoeckx SR, Bogaerts A (2017) Plasma technology – a novel solution for CO2 conversion. Chem Soc Rev 46:5805–5863. https://doi.org/10.1039/c6cs00066e
Sovacool BK, Del Rio DF, Herman K, Iskandarova M, Uratani JM, Griffiths S (2024) Reconfiguring European industry for net-zero: a qualitative review of hydrogen and carbon capture utilisation and storage benefits and implementation challenges. Energy Environ Sci 17:3523–3569
Stocker TF, Qin D, Plattner GK, Tignor MMB, Allen SK, Boschung J, Nauels A, Xia Y, Bex V, Midgley PM (2013) Climate change 2013 the physical science basis: working group I contribution to the fifth assessment report of the intergovernmental panel on climate change, climate change 2013 the physical science basis: working group I contribution to the fifth assessment report of the intergovernmental panel on climate change. https://doi.org/10.1017/CBO978110
Tahir M, Tahir B (2020) Constructing a stable 2D/2D heterojunction of oxygen-cluster modified Ti3AlC2MAX cocatalyst with proton-rich C3N4for highly efficient photocatalytic CO2Methanation. Ind Eng Chem Res 59:9841–9857. https://doi.org/10.1021/acs.iecr.0c00193
Tong L, Yue T, Zuo P, Zhang X, Wang C, Gao J, Wang K (2017) Effect of characteristics of KI-impregnated activated carbon and flue gas components on Hg0 removal. Fuel 197:1–7
Trivedi TJ, Lee JH, Lee HJ, Jeong YK, Choi JW (2016) Deep eutectic solvents as attractive media for CO 2 capture. Green Chem 18(9):2834-2842
Üner O, Bayrak Y (2018) The effect of carbonization temperature, carbonization time and impregnation ratio on the properties of activated carbon produced from Arundo donax. Microporous Mesoporous Mater 268:225–234
Up S, Newsletter FOR., Stand BA, Will HOW, Sharing THE, Shape E, Commerce F, Summit R (2016) In: News details 2015–2016
Valluri S, Kawatra SK (2021) Use of frothers to improve the absorption efficiency of dilute sodium carbonate slurry for post combustion CO2 capture. Fuel Process Technol 212:106620
Vargas DP, Giraldo L, Erto A, Moreno-Piraján JC (2013) Chemical modification of activated carbon monoliths for CO2 adsorption. J Therm Anal Calorim 114(3):1039–1047
Wang N, Otor HO, Rivera-Castro G, Hicks JC (2024) Plasma catalysis for hydrogen production: a bright future for Decarbonisation. ACS Catal 14:6749–6798
Wilberforce T, Baroutaji A, Soudan B, Al-Alami AH, Olabi AG (2019) Outlook of carbon capture technology and challenges. Sci Total Environ 657:56–72
Wismann ST (2019) Electrically heated steam methane reforming. Technical University of Denmark, Lyngby, Denmark
Wu J, Huang Y, Ye W, Li Y (2017) CO2 reduction: from the electrochemical to photochemical approach 1700194, vols. 1–29. https://doi.org/10.1002/advs.201700194
Wu F, Argyle MD, Dellenback PA, Fan M (2018) Progress in O2 separation for oxy-fuel combustion–A promising way for cost-effective CO2 capture: A review. Prog Energy Combust Sci 67:188–205
Xu Y, Schutte RP, Hepler LG (1992) Solubilities of carbon dioxide, hydrogen sulfide and sulfur dioxide in physical solvents. Can J Chem Eng 70(3):569–573
Yadav S, Mondal SS (2022) A review on the progress and prospects of oxy-fuel carbon capture and sequestration (CCS) technology. Fuel 308:122057
Yu H, Shan C, Li J, Hou X, Yang L (2024) Alkaline absorbents for SO2 and SO3 removal: a comprehensive review. J Environ Manage 366:121532

Yusuf BO, Umar M, Kotob E, Abdulhakam A, Taialla OA, Awad MM, Hussain I, Alhooshani KR, Ganiyu SA (2024) Recent advances in bimetallic catalysts for methane steam reforming in hydrogen production: current trends, challenges, and future prospects. Chem Asian J 19:e202300641

Zeynep Beyza Kilic (2019) Qatar to Store More than 5M Tons of CO2 a Year by 2025. https://www.aa.com.tr/en/energy/projects/qatar-to-store-more-than-5m-tons-of-co2-a-year-by-2025/26924

Zhao L, Hu HY, Wu AG, Terent'ev AO, He LN, Li HR (2024) CO2 capture and in-situ conversion to organic molecules. J CO2 Util 82:102753

Zhu XL, Wang PY, Peng C, Yang J, Yan XB (2014) Activated carbon produced from paulownia sawdust for high-performance CO2 sorbents. Chin Chem Lett 25(6):929–932

Chapter 5
Advanced Metal–Organic Frameworks for Selective Hydrogen Separation and Purification

Abstract Hydrogen is a key energy carrier for a sustainable and low-carbon future, but its widespread deployment requires efficient purification and isotope separation technologies. Achieving high-purity hydrogen and separating isotopes such as deuterium and tritium remain challenging due to their similar physicochemical properties and low natural abundance. Metal–organic frameworks (MOFs), with their highly tunable pore structures and chemical functionalities, have emerged as advanced materials for selective hydrogen separation and purification. This chapter presents a comprehensive overview of MOF-based strategies for hydrogen and hydrogen isotope separation, emphasizing molecular sieving, kinetic quantum sieving, and chemical affinity quantum sieving mechanisms. Design principles, structure–property relationships, and representative experimental and theoretical studies are discussed to highlight the potential of advanced MOFs in addressing key challenges in hydrogen purification and isotope separation.

Keywords Metal–organic frameworks · Hydrogen separation · Hydrogen isotope separation · Molecular sieving

5.1 Introduction

The intensifying global energy crisis, coupled with a rapid rise in energy demand, has driven an urgent quest for sustainable and environmentally friendly energy alternatives. Among these, hydrogen (H_2) stands out as a particularly promising energy carrier due to its remarkably high gravimetric energy density and its ability to release energy through a carbon-free combustion process. Currently, over 90% of industrial hydrogen is produced through steam reforming of natural gas and other fossil-fuel derivatives, while the remainder is generated via water electrolysis. This hydrogen finds extensive application in numerous industrial sectors, including metallurgical operations, petrochemical refining, pharmaceutical manufacturing, and large-scale chemical synthesis (Ockwig and Nenoff 2007; Wang et al. 2023). However, conventional hydrogen production is invariably accompanied by the

D. M. Y. Mir et al., *Low-Carbon Hydrogen*, Clean Energy Production Technologies,
https://doi.org/10.1007/978-981-95-8142-9_5

formation of various by-products, necessitating subsequent purification and separation processes. These downstream steps are often highly energy-intensive and significantly increase overall production costs, making hydrogen purification a major economic and technical bottleneck that limits the broad adoption of hydrogen as a clean energy vector. Consequently, the development of energy-efficient and cost-effective separation and purification technologies is a critical priority for realizing a hydrogen-based economy. Traditional hydrogen purification techniques include pressure swing adsorption (PSA), vacuum swing adsorption (VSA), temperature swing adsorption (TSA), fractional or cryogenic distillation, and various membrane-based methods (Wang and Zhao 2017; Lin et al. 2019). Among these, adsorption-based processes (PSA, VSA, TSA) are often constrained by high capital expenditure and operational energy demands, and they may not reliably achieve the ultra-high hydrogen purities required for specialized applications. In contrast, membrane separation technologies have gained considerable attention as an alternative due to their relatively low energy requirements, reduced operational costs, continuous operational capability, and capacity to achieve high-purity hydrogen production at scales comparable to conventional techniques (Sholl and Lively 2016). The first commercial application of membrane-based gas separation emerged in 1977 when Monsanto deployed membranes for hydrogen recovery from purge streams in petrochemical and ammonia production facilities. Since then, a wide range of membrane materials—including organic polymers, palladium-based metallic membranes, silica membranes, zeolites, carbon molecular sieves, and mixed tetrahedral–octahedral oxide membranes—has been explored for hydrogen separation (Hazazi et al. 2019; Wu et al. 2025). More recently, porous crystalline metal–organic frameworks (MOFs) have attracted significant interest as next-generation molecular sieve membranes. Comprising metal ions coordinated with organic linkers, MOFs offer exceptional structural tunability, modular synthetic flexibility, and precise control over pore size, topology, and surface chemistry (Moghadam et al. 2017). Unlike conventional porous materials, MOFs allow systematic modification of the pore environment through rational selection of metal nodes and organic ligands, enabling enhanced molecular selectivity and separation efficiency. Gas separation within porous materials is primarily governed by four mechanisms: (a) molecular sieving based on size and shape exclusion, (b) thermodynamic selectivity arising from differential adsorbate–surface interactions, (c) kinetic selectivity driven by differences in diffusion rates, and (d) quantum sieving, which becomes significant for light gases such as hydrogen in ultramicropores (Mondal et al. 2019). Key parameters affecting separation performance include the average pore diameter, the kinetic diameters of the gas molecules, and the chemical nature of the pore surface. Hydrogen, in particular, exhibits relatively weak adsorption compared with most other gases (Kerry 2007); therefore, efficient separation typically requires adsorbents with narrowly distributed micropores that selectively facilitate hydrogen diffusion while excluding larger or more strongly interacting species. During the early development of MOFs, the assessment of selective gas adsorption was largely based on single-component sorption isotherms due to the experimental challenges associated with multicomponent mixtures. These isotherms were generally modeled using adsorption models such as single-site or dual-site Langmuir and Langmuir–Freundlich equations. Binary and multicomponent adsorption equilibria were then estimated using ideal adsorbed solution theory (IAST), circumventing the need for direct mixed-gas experiments (Myers and Prausnitz 1965). In contrast, membrane-based hydrogen separation is evaluated directly by measuring gas permeation under a pressure gradient. For a single gas component, the permeability (P_a) is defined in terms of the steady-state molar flux (N_a), membrane thickness, and the transmembrane pressure difference ($p_2 - p_1$), providing a fundamental metric for assessing membrane performance in hydrogen separation applications.

$$P_a = N_a l / (p_2 - p_1) \tag{5.1}$$

In this context, l represents the membrane thickness, while p_2 and p_1 correspond to the downstream and upstream pressures, respectively. Gas permeability coefficients are conventionally reported in Barrer units, where 1 Barrer is defined as $10^{-10}\,\mathrm{cm^3(STP)\cdot cm\cdot cm^{-2}\cdot s^{-1}\cdot cmHg^{-1}}$. Beyond permeability, the ideal selectivity ($\alpha_{A/B}$)—which quantifies a membrane's ability to separate a binary gas pair (A and B)—is widely considered an intrinsic material property. This parameter is calculated as the ratio of the individual gas permeabilities, providing a convenient measure of the membrane's preferential transport of one gas species over another.

Both permeability and selectivity are fundamentally determined by the structural characteristics of the membrane material. However, these parameters often exhibit a well-known trade-off, in which increased permeability tends to reduce selectivity and vice versa. This limitation is commonly illustrated by the Robeson upper bound, where experimental data for polymeric membranes are plotted on a logarithmic scale of selectivity versus permeability, with the more permeable gas positioned below a characteristic upper-bound line (Robeson 2008). The Robeson plot establishes the practical performance ceiling for homogeneous polymer membranes and serves as a benchmark for evaluating separation efficiency. Nonetheless, this framework does not fully apply to MOF-based membranes, which operate primarily via molecular sieving mechanisms and have demonstrated the ability to exceed the conventional permeability–selectivity trade-off (Joshi et al. 2014; Liao et al. 2025). Beyond conventional hydrogen purification, the separation of hydrogen isotopes—specifically deuterium (D, ^{2}H) and tritium (T, ^{3}H)—poses a distinct technological challenge. These isotopes are critical for applications ranging from nuclear fusion energy systems to advanced industrial and scientific research (Glugla et al. 2003; Kim et al. 2023). Their natural abundances are extremely low (D $\approx$ 0.015% and T $\approx 10^{-18}$%), necessitating highly efficient enrichment methods. The nearly identical physicochemical properties of hydrogen isotopes, however, render their separation particularly difficult using conventional approaches (Fitzgerald et al. 2019). Industrial hydrogen isotope separation is dominated by the Girdler sulfide process and cryogenic hydrogen distillation. The Girdler process relies on isotopic exchange between water and hydrogen sulfide ($H_2O + HDS \rightleftharpoons HDO + H_2S$) in a closed-loop system, followed by electrolysis of heavy water to produce D_2. Cryogenic distillation, in contrast, exploits small differences in boiling points between isotopes (20.3 K for H_2 and 23.7 K for D_2). Both methods, though widely used, are energy-intensive, involve multiple stages, and yield relatively low separation factors ($\approx$2 for the Girdler process and $\approx$1.5 for cryogenic distillation at 24 K). As a more energy-efficient alternative, Beenakker et al. (1995) proposed the kinetic quantum sieving (KQS) mechanism, which leverages confined porous environments to enhance isotope separation. In this mechanism, lighter molecules exhibit shorter de Broglie wavelengths, resulting in higher effective energy barriers at pore entrances and consequently slower diffusion relative to heavier isotopes contrary to classical kinetic theory, where lighter species diffuse more rapidly. When pore sizes approach the de Broglie wavelength of hydrogen isotopes under cryogenic conditions, quantum effects become pronounced, amplifying the diffusion differences between H_2 and D_2 and enabling effective separation. The KQS phenomenon can be further interpreted in terms of zero-point energy (ZPE) and the adsorption potential well depth. When the ZPE of a light isotope exceeds the adsorption potential—typically occurring when pore apertures closely match molecular dimensions—the penetration barrier increases, resulting in lower equilibrium adsorption. Conversely, heavier isotopes with lower ZPE preferentially adsorb under the same conditions. Pore structure and chemical environment also critically influence isotope separation, as stronger-than-van-der-Waals interactions can amplify ZPE differences, promoting selective adsorption. This mechanism, known as chemical affinity quantum sieving (CAQS), provides an additional avenue for efficient isotope separation. In this chapter, we provide a comprehensive discussion of hydrogen gas and isotope separation using both pristine MOFs and MOF-based hybrid membrane systems. Emphasis is placed on classical separation mechanisms, such as kinetic sieving, as well as quantum-driven effects. Special attention is devoted to MOF design strategies aimed at tailoring structural flexibility, pore size distribution, and surface functionality through rational ligand selection, post-synthetic modification (PSM), and hybrid material integration, supported by representative studies from the literature.

5.2 Strategies for MOF-Based Materials for H_2 Separation

5.2.1 MOF-Based Membranes

To date, a wide variety of membranes composed of organic and inorganic building blocks have been developed for the separation of species ranging from ions and gases to organic dyes and even biomolecules (Li et al. 2015). Compared with conventional purification techniques, membrane-based separation offers several advantages, including environmental sustainability, lower energy consumption, and operational simplicity. However, conventionally employed polymeric membranes often exhibit poor reproducibility and limited long-term stability, particularly under elevated temperatures and chemically aggressive conditions. These membranes are susceptible to plasticization effects in the presence of gases such as CO_2, H_2S, and hydrocarbons at high pressures, which compromises their separation performance. Moreover, polymeric membranes are constrained by an inherent trade-off between permeability and selectivity, resulting in relatively low separation efficiency. Consequently, achieving target separation levels typically requires larger membrane areas and multistage processing, thereby increasing capital and operational costs (Koros and Zhang 2017; Raggam et al. 2023). In response to these limitations, substantial research efforts have focused on the development of porous inorganic membranes as viable alternatives, owing to their superior thermal and chemical stability, as well as their enhanced permeability and selectivity. In particular, metal–organic frameworks (MOFs) effectively mitigate the plasticization issues associated with flexible polymer matrices. Since the first report of Cu-based MOFs in 1989 (Hoskins and Robson 1989), MOFs also referred to as porous coordination polymers (PCPs) have emerged as a promising class of membrane materials. Their robustness, high porosity, large specific surface area, tunable functionality, and excellent ion-exchange capabilities make them attractive candidates for molecular separation applications, even under ambient conditions (Ding and Jiang 2020). Notably, the ability to precisely tailor pore size and surface chemistry enables enhanced selectivity and permeability through strengthened framework–adsorbate interactions. MOF membranes are generally classified into two principal categories. The first category comprises polycrystalline MOF membranes, in which pure MOFs are directly grown on a supporting substrate. The second category consists of mixed-matrix membranes (MMMs), where MOF particles are dispersed within a polymer matrix to form a hybrid composite membrane. For polycrystalline MOF membranes, substrate pretreatment or seeding is typically required prior to MOF growth on supports such as metal meshes, polymeric films, or modified and unmodified oxide substrates (e.g., alumina, silica, TiO_2, and ZnO). However, achieving strong interfacial adhesion between the MOF layer and the substrate over large areas remains a significant challenge. To address this issue, various seeding and growth techniques have been developed, including contra-diffusion, electrospinning, reactive seeding, gel–vapor deposition (GVD), and layer-by-layer assembly (Ma et al. 2019). Despite these advances, polycrystalline MOF membranes often suffer from structural defects, grain boundaries, and random crystal orientations, which adversely affect permselectivity and permeation flux. Furthermore, scaling up these membranes for industrial separation processes remains difficult (Hou et al. 2019). In contrast, MMMs alleviate many of these limitations by incorporating pre-synthesized MOF crystals into polymer matrices, forming defect-tolerant composite membranes (Cheng et al. 2018). A critical requirement for MMM performance is the formation of well-integrated interfaces between the MOF fillers and the polymer matrix without particle agglomeration. Additionally, membrane morphology, MOF particle size, and filler loading are key design parameters that govern MMM performance. Guo et al. (2009) demonstrated the fabrication of defect-free HKUST-1 [$Cu_3(BTC)_2$, BTC = 1,3,5-benzenetricarboxylic acid] membranes using a "twin copper source" strategy, wherein copper ions were supplied both from the reaction solution and a

Cu mesh substrate. Prior to membrane synthesis, the Cu mesh substrate was oxidized by heating at 100 °C, as indicated by a visible color transition from yellow to green. Secondary growth of the MOF was then achieved by immersing the oxidized mesh in a 1:1 water–ethanol solution containing $Cu(NO_3)_2{\cdot}3H_2O$ and 1,3,5-benzenetricarboxylic acid (BTC) within a Teflon-lined autoclave at 120 °C for 72 h. This procedure yielded octahedral HKUST-1 crystals, ranging from 5 to 10 μm in size, which occupied the Cu mesh pores to form a continuous membrane approximately 60 μm thick. The resulting HKUST-1 membrane exhibited a high H_2 permeation flux of 1.3×10^{-1} mol m^{-2} s^{-1} at 343 K and an H_2/N_2 separation factor of 7.04 at 298 K. Performance remained stable over 24 h, demonstrating the membrane's robustness and recyclability. Building upon these developments, (Li et al. 2017a, b) introduced a scalable gel–vapor deposition (GVD) strategy for fabricating nanometer-thick ZIF-8 molecular sieving membranes. This method obviates the need for substrate pretreatment and allows precise control over membrane thickness through adjustment of sol concentration and coating parameters. The process involves preparing a zinc-based sol by heat-treating zinc acetate dihydrate, followed by aging to form a viscous gel. This gel is coated onto ammoniated polyvinylidene fluoride (PVDF) hollow fibers, which are then sealed with 2-methylimidazole powder in an autoclave and heated at 150 °C. Under these conditions, ligand vapor reacts with the Zn-based gel to produce a crystalline ZIF-8 membrane. X-ray diffraction confirmed complete conversion to ZIF-8, with no detectable ZnO impurities. Coronas and coworkers further advanced MOF membrane fabrication using microfluidic synthesis to grow ZIF-7 (Zn^{2+}–benzimidazole) and ZIF-8 (Zn^{2+}–2-methylimidazole) membranes on the inner surfaces of hollow polysulfone fibers (Cacho-Bailo et al. 2015; Wu et al. 2024). Unlike conventional exposure-based growth techniques, this one-step microfluidic approach enables rapid, scalable fabrication while minimizing reagent consumption. The polymeric hollow fiber acts as a protective sheath, enhancing the thermal stability of the embedded MOFs. The hollow fiber architecture, with exceptionally high surface-area-to-volume ratios (≈7700 m^2 m^{-3}), is particularly advantageous for membrane applications. The fabricated membranes contained 19.0 wt% ZIF-7 (thickness 2.4 ± 0.4 μm) and 20.8 wt% ZIF-8 (thickness 3.9 ± 0.9 μm). In binary gas separation tests, ZIF-8 exhibited molecular sieving behavior with H_2/N_2 and H_2/CH_4 separation factors of 18.3 and 17.2, respectively, whereas ZIF-7 showed enhanced selectivity (35.1 and 34.6) due to its smaller pore aperture (3.0 Å for ZIF-7 vs. 3.4 Å for ZIF-8). Interestingly, despite CO_2 possessing a kinetic diameter larger than the ZIF-7 pore size, significant CO_2 permeation was observed, attributable to the gate-opening effect even at low relative pressures, suggesting a predominantly size-exclusion-driven separation mechanism. Notably, the H_2/CO_2 separation factor increased to 5.7 at 150 °C, surpassing the Knudsen selectivity limit and indicating nonrestricted CO_2 transport. This fabrication strategy demonstrates an environmentally benign route for producing selective MOF membranes. Nevertheless, challenges remain in synthesizing continuous, defect-free, and well-intergrown MOF membranes without grain boundaries or microcracks. Furthermore, the structural flexibility and limited stability of many MOFs in aqueous environments restrict their practical deployment. To overcome these limitations, recent computational and data-driven screening approaches have been employed to identify promising MOF candidates for specific separation applications prior to experimental synthesis, thereby optimizing resource allocation and accelerating membrane development (Qian et al. 2020; Hu et al. 2025).

5.2.2 *Flexible MOFs*

The ability to modulate the pore environment through reversible structural transformations has positioned flexible metal–organic frameworks (MOFs) as highly promising materials for advanced gas separation and storage applications. Unlike rigid MOFs, which largely retain their framework structure during gas adsorption and desorption and typically exhibit type I sorption isotherms, flexible MOFs undergo significant structural rearrangements in response to external stimuli such as gas pressure. Consequently, their gas sorption behavior differs markedly from that of rigid frameworks. Flexible MOFs commonly display type V or VI adsorption isotherms, which arise from pressure-induced pore opening or contraction during gas uptake. The stepwise adsorption–desorption characteristics of flexible MOFs enable high usable gas capacities while potentially reducing the energy required for gas storage and release. This feature is particularly advantageous for hydrogen (H_2) storage, given the low volumetric energy density of hydrogen, which complicates the achievement of high storage capacities under practical operating conditions. By employing flexible MOFs that allow near-complete adsorption and desorption of hydrogen within a narrow pressure window, the efficiency of hydrogen storage and delivery systems can be substantially enhanced. Beyond hydrogen storage, flexible MOFs also offer new possibilities for hydrogen isotope separation. Traditional isotope separation typically relies on cryogenic operation due to the nearly identical physicochemical properties of hydrogen isotopes. However, the gate-opening behavior inherent to flexible MOFs provides the potential to achieve isotope separation at elevated temperatures, addressing a key limitation of conventional methods. Flexible MOFs exhibit distinctive sorption phenomena, including S-shaped adsorption isotherms, pronounced hysteresis, and multistep adsorption–desorption processes (Choi et al. 2008). For example, Long and co-workers reported a flexible Co(bdp) framework (bdp = 1,4-benzenedi(pyrazolate)) that exhibited a five-step N_2 adsorption profile at 77 K, while showing a single-step hydrogen sorption with significant hysteresis under high-pressure conditions at the same temperature. More recently, the concept of negative gas adsorption (NGA) has emerged, in which the quantity of adsorbed gas decreases as pressure increases within specific temperature and pressure ranges (Shi et al. 2018). This seemingly counterintuitive behavior results from sudden structural transitions within the nanoporous framework, leading to pressure amplification and spontaneous gas release. Such unconventional adsorption profiles highlight the unique application opportunities of flexible MOFs that are unattainable with rigid frameworks. In this section, we provide an overview of the defining features and underlying mechanisms of flexible MOFs, focusing on their dynamic responses to specific gas molecules and emphasizing their potential roles in hydrogen storage and separation technologies.

5.2.3 *Molecular Sieving Effect in Flexible MOFs*

Molecular sieving has emerged as a highly effective gas separation strategy, offering low energy consumption and high selectivity. By precisely controlling pore dimensions—particularly through the use of flexible MOFs—separation can be achieved by selectively allowing or restricting the diffusion of target molecules. Fernandez et al. (2012) investigated two partially fluorinated MOFs, FMOFCu and FMOFZn, as model systems for xenon (Xe) and krypton (Kr) separation. Among these, FMOFCu features a three-dimensional interpenetrated framework constructed from Cu_2 paddle-wheel units, forming internal cavities of approximately 5.1 Å × 5.1 Å connected by narrow bottleneck apertures of roughly 3.5 Å × 3.2 Å. These apertures are slightly smaller than the kinetic diameter of Xe (3.96 Å) but closely match that of Kr (3.60 Å), thereby selectively restricting Xe diffusion into the cavities. Temperature-dependent adsorption studies revealed markedly different behaviors for Kr and Xe in FMOFCu. As temperature decreased, Kr uptake increased, while Xe adsorption decreased. This phenomenon was attributed to reduced framework flexibility at lower temperatures, which elevates kinetic barriers to Xe diffusion and suppresses its adsorption despite favorable thermodynamic interactions. In contrast, Kr diffusion remains largely unaffected by kinetic constraints, and its adsorption is predominantly thermodynamically driven, resulting in enhanced uptake at lower temperatures. Accordingly, FMOFCu acts as a temperature-gated molecular sieve for Xe, achieving an exceptional Xe/Kr selectivity of 36 at 203 K and 0.1 bar. This study marked the first demonstration of Xe/Kr separation using MOFs and illustrated the potential of exploiting framework flexibility for noble gas separations. Building on this concept, Zhou et al. (2019) reported MAF-41, a MOF exhibiting limited yet effective flexibility for the selective accommodation of medium-sized organic molecules. In MAF-41, cavity dimensions lie between the sizes of the target guest molecules, enabling pronounced molecular sieving. Single-component adsorption measurements showed selective uptake of styrene, whereas larger molecules such as ethylbenzene were excluded due to steric constraints. Smaller molecules, including toluene and benzene, failed to induce pore opening because their adsorption energies were insufficient. Grand canonical Monte Carlo (GCMC) simulations further revealed that adsorption enthalpy increases with guest molecule size, explaining the inability of smaller molecules to trigger structural rearrangements. Mixed-gas adsorption experiments demonstrated remarkable separation performance: ethylbenzene/styrene mixtures exhibited a styrene selectivity of 1250, while quaternary mixtures of ethylbenzene/styrene/toluene/benzene achieved selectivities as high as 3300. These findings highlight the potential of pore-opening phenomena in flexible MOFs to enable highly selective separation of intermediate-sized molecules, which are difficult to separate efficiently using rigid porous materials.

5.2.4 *Flexibility of MOFs Induced by Various External Stimuli*

External stimuli, including light (photoresponsive), temperature (thermoresponsive), and pressure, can induce structural flexibility in metal–organic frameworks (MOFs) and porous coordination polymers (PCPs), allowing dynamic control over pore accessibility. By finely tuning framework flexibility in response to these stimuli, selective gas separation can be achieved under precisely tailored operating conditions. Gu et al. (2019) reported a Cu-based PCP incorporating a butterfly-shaped ligand that functions as a kinetic gate through reversible flip–flop molecular motion. Within the intralayer channels, isophthalic acid and phenothiazine-5,5-dioxide (OPTz) units form a narrow gate of approximately 3 Å, which undergoes localized motion with relatively low rotational and flipping energy barriers. Gas diffusion experiments revealed that the diffusion rate of C_2H_4 increased steadily with temperature and pressure, reaching a maximum adsorption capacity at 320 K. This behavior was attributed to the combined effects of strong framework–adsorbate affinity and diffusion-controlled transport. Density functional theory (DFT) calculations indicated that the energy barrier for interlayer diffusion exceeded that for intracage diffusion, identifying interlayer transport as the rate-limiting step governing overall adsorption. As a result of this thermally activated gating mechanism, the PCP exhibited exceptional separation performance, achieving selectivity values of approximately 350 for O_2/Ar and 75 for C_2H_4/C_2H_6. The gradual thermal flipping of the OPTz units effectively expanded the gate aperture at elevated temperatures, permitting gas permeation, while maintaining a closed configuration at lower temperatures, thereby enabling temperature-controlled molecular separation. Expanding on this concept, Dong et al. (2020) demonstrated the separation of ternary gas mixtures by exploiting temperature-dependent gate-opening behavior in combination with pressure control. Adsorption isotherms for CO_2, C_2H_2, and C_2H_4 measured at 195, 263, and 298 K revealed distinct gate-opening responses as a function of temperature. Notably, CO_2 and C_2H_4 exhibited negligible adsorption at 298 K, whereas C_2H_2 showed pronounced uptake. DFT calculations attributed this selective pore opening to favorable hydrogen bonding and van der Waals interactions between C_2H_2 molecules and the SiF_6^{2-} moieties within the framework. Consequently, high-purity C_2H_4 was successfully obtained from $C_2H_4/C_2H_2/CO_2$ ternary mixtures under optimized temperature and pressure conditions. These findings underscore the potential of stimulus-responsive gate-opening mechanisms to finely tune separation performance. They demonstrate that flexible frameworks can be rationally engineered to achieve selective separation of complex multicomponent mixtures by carefully controlling gate-opening pressures and operating temperatures.

5.2.5 Guest-Selective Pore Opening in Flexible MOFs

The interaction between flexible MOF frameworks and specific guest molecules can induce a guest-selective pore-opening response, enabling highly selective molecular discrimination. Foo et al. (2016) demonstrated such a gate-opening mechanism using the framework [Mn(bdc)(dpe)] (H_2bdc = 1,4-benzenedicarboxylic acid; dpe = 1,2-di(4-pyridyl)ethylene). This structure contains isolated zero-dimensional (0D) pores formed by the inclined arrangement of bdc and dpe ligands. Rotation of the aromatic rings in the bdc ligands further constricts the pore volume, generating a kidney-shaped, spatially isolated cavity. While CO_2 adsorption did not trigger any observable gate-opening behavior across the studied temperature range, C_2H_2 adsorption induced a pronounced response. This selective behavior arises from differences in directional electrostatic interactions between the guest molecules and the phenylene rings of the bdc ligand. For C_2H_2, the most energetically favorable configuration occurs when the phenylene rings adopt a perpendicular orientation, maximizing electrostatic interactions and requiring pore expansion. In contrast, CO_2 fails to elicit a structural response. Further validation of this guest-discriminating gate mechanism was achieved by photodimerizing the phenyl rings via a [2 + 2] reaction, which eliminated the inverse CO_2/C_2H_2 selectivity. These results highlight selective gate opening as a powerful strategy for separating molecules with distinct electrostatic properties. Building on this principle, Li et al. (2017a, b) reported efficient C_3H_4/C_3H_6 separation using the flexible and structurally robust MOF ELM-12. This framework incorporates dynamic pendant trifluoromethanesulfonate (OTf^-) groups and contains two types of cavities with dimensions of approximately 3.2 Å × 4.3 Å and 6.1 Å × 4.3 Å × 4.3 Å, closely matching the size and shape of C_3H_4 molecules (6.2 Å × 3.8 Å × 3.8 Å). Neutron powder diffraction (NPD) identified two distinct adsorption sites for C_3H_4, stabilized predominantly via hydrogen bonding interactions. Upon C_3H_4 adsorption, the dynamic OTf^- groups adjust their positions and orientations, facilitating selective capture. The isosteric heat of adsorption (Q_{st}) for C_3H_4 in ELM-12 was 60.6 kJ mol^{-1}, substantially higher than that for C_3H_6 (15.8 kJ mol^{-1}), indicating much stronger host–guest interactions. Ideal adsorbed solution theory (IAST) calculations for a 1:99 C_3H_4/C_3H_6 mixture at 298 K predicted a maximum selectivity of 84, with a C_3H_4 uptake of 0.881 mmol g^{-1} from the mixture. These results demonstrate the effectiveness of combining tailored pore environments, strong adsorption sites, and framework flexibility for selective recognition and separation of small gas molecules. More recently, Sun et al. (2022) introduced ZNU-3, a MOF in which flexibility is modulated by intraframework hydrogen-bonding interactions. Unlike conventional 3,5-pyridinedicarboxylic acid (H_2Pdc)-based frameworks, ZNU-3 retains one protonated carboxylic acid group while the other is deprotonated, forming an intraframework hydrogen bond between –COOH and $–COO^-$ groups. Single-component gas adsorption measurements revealed a C_2H_2 uptake of 81.0 cm^3 (STP) g^{-1}, while adsorption of other gases was negligible, indicating a highly selective gate-opening response toward C_2H_2. Computational modeling suggested that this selective capture arises not only from π–π stacking interactions but also from enhanced hydrogen bonding between C_2H_2 molecules and the carboxylate groups within the framework. Dynamic breakthrough experiments further confirmed the separation capability of ZNU-3, yielding a C_2H_2/CO_2 selectivity of 5.7. These findings underscore the potential of hydrogen-bond-controlled framework flexibility in MOFs for separating gas species with nearly identical molecular sizes.

5.2.6 Functionalization and Post-Synthetic Modifications

Additional or novel chemical functionalities can be incorporated into MOFs through pendant functional groups on organic linkers or via post-synthetic modifications (PSMs) of metal nodes and ligands using substitution or chemical transformation strategies (Kondo et al. 2018). For the separation and purification of small gases such as hydrogen (H_2), precise pore engineering combined with the creation of strong host–guest interaction sites via functionalization or PSMs plays a critical role in enhancing molecular sieving and selective adsorption. A representative example involves Mg-MOF-74, in which open metal sites (OMSs) were modified with coordinatively bound ethylenediamine groups to improve selective hydrogen separation (Wang et al. 2015). The MOF membrane was prepared by seeding MgO nanocrystals onto an Al_2O_3 substrate, followed by sequential solvothermal growth in a solution containing $Mg(NO_3)_2$ and 2,5-dioxido-1,4-benzenedicarboxylic acid (H_4dobdc). Post-synthetic incorporation of ethylenediamine was achieved by refluxing the membrane in an ethylenediamine/toluene solution. The introduced amine functionalities enhanced interactions with CO_2, increasing its adsorption capacity and binding energy. Single-gas (H_2, CO_2, CH_4, N_2) and binary mixture (H_2/CO_2, H_2/N_2, H_2/CH_4) permeation experiments showed that both pristine and amine-functionalized Mg-MOF-74 followed the permeability order $H_2 > N_2 > CH_4 > CO_2$, governed mainly by molecular size and host–guest interactions. While H_2/N_2 and H_2/CH_4 selectivities remained largely unchanged after modification, the H_2/CO_2 selectivity increased substantially—from 11 to 29 in single-gas experiments and from 10.5 to 28 in binary mixtures—due to preferential CO_2 adsorption at the amine-functionalized sites. The improved H_2 permselectivity arises from two synergistic effects: a reduced effective pore volume and strong CO_2 binding at the functionalized sites. Similarly, Zhang et al. (2012) reported an amine-functionalized MIL-53(Al) membrane (NH_2-MIL-53(Al)) designed to enhance hydrogen permeability and selectivity. Continuous membranes were grown on macroporous glass frit discs using colloidal seed-assisted growth. Single-gas permeation experiments demonstrated substantially higher H_2 permeance (26.7×10^{-7} mol m^{-2} s^{-1} Pa^{-1}) compared with CH_4, N_2, and CO_2, resulting in ideal selectivities of 18.5, 19.5, and 27.3 for H_2/CH_4, H_2/N_2, and H_2/CO_2, respectively. Differences in performance were attributed to variations in intrinsic diffusion rates and adsorption affinities. In binary gas experiments, H_2 permeance decreased relative to single-gas measurements due to competitive adsorption effects. Importantly, NH_2-MIL-53(Al) membranes maintained high separation factors (>30) and permeance (>20 × 10^{-7} mol m^{-2} s^{-1} Pa^{-1}) over 5 days of continuous operation, demonstrating excellent operational stability. Ligand-exchange-based PSMs have also been applied to construct core–shell MOF architectures. Sánchez-Laínez et al. (2017) employed partial ligand exchange in ZIF systems, converting ZIF-8 crystals (2-methylimidazole, mIm) into ZIF-7/8 hybrids through immersion in benzimidazole (bIm) solution, with complete conversion to ZIF-7 achieved after approximately 1 week. Partial ligand exchange reduced effective pore window sizes, decreasing N_2 adsorption relative to pristine ZIF-8, while CO_2 adsorption increased by ~20% at 10 wt% conversion. Mixed-matrix membranes (MMMs) incorporating these hybrids into a polybenzimidazole (PBI) matrix were subsequently evaluated. MMMs with 32 wt% ZIF-7/8 (10 wt% conversion) exhibited the highest H_2 permeability (1398 Barrer), whereas optimal H_2/CO_2 selectivity (10.1) was achieved at 20 wt% filler loading. All MMMs surpassed the Robeson upper bound, reflecting superior separation performance. This enhancement was attributed to improved interfacial compatibility between bIm ligands and the PBI matrix, as well as reduced pore apertures that hindered CO_2 diffusion while favoring H_2 transport. These results highlight the versatility of core–shell MOF architectures for tuning membrane properties and suggest their applicability across diverse MOF systems. In a systematic study, Ghalei et al. (2019) investigated hydrogen separation in a series of Zr-based polycrystalline MOF membranes by varying ligand size and functionality. Isoreticular UiO-66-type MOFs were synthesized using terephthalic acid (H_2BDC), 1,4-naphthalenedicarboxylic acid (H_2NDC), 9,10-anthracenedicarboxylic acid (H_2ADC), and functionalized ligands such as 2-aminoterephthalic acid ($H_2BDC–NH_2$) and 2-bromoterephthalic acid ($H_2BDC–Br$). Thin MOF layers were grown solvothermally on 3-aminopropyltriethoxysilane (APTES)-modified Al_2O_3 substrates. Despite similar pore sizes (~6 Å) indicated by N_2 adsorption, H_2 separation performance varied significantly. The UiO-66 membrane with H_2BDC showed relatively low H_2/CO_2 selectivity (~5), whereas membranes with bulkier ligands such as H_2NDC and H_2ADC achieved much higher selectivities of 15.3 and 31.9, respectively. Functional groups ($–NH_2$ and –Br) had minimal influence on H_2 separation. DFT calculations revealed that bulky ligands introduce higher diffusion energy barriers associated with gate opening, mainly affecting larger molecules such as CO_2 while minimally impacting H_2 transport. Enhanced hydrogen selectivity was therefore attributed to narrowed pore apertures created by bulky ligands, which restrict adsorption and diffusion of larger gas molecules while maintaining favorable pathways for hydrogen. Figure 5.1 illustrates the mechanism of hydrogen separation through MOFs. The process involves three key steps: (1) diffusion of a mixed gas stream into the highly porous MOF structure, (2) selective adsorption of H_2 molecules at specific active sites within the framework, and (3) size exclusion, whereby larger molecules such as N_2, CO_2, and CH_4 are blocked from passing through the narrow pores. This combination of high porosity, tunable pore size, and selective adsorption enables MOFs to achieve efficient hydrogen purification with high selectivity and energy efficiency.

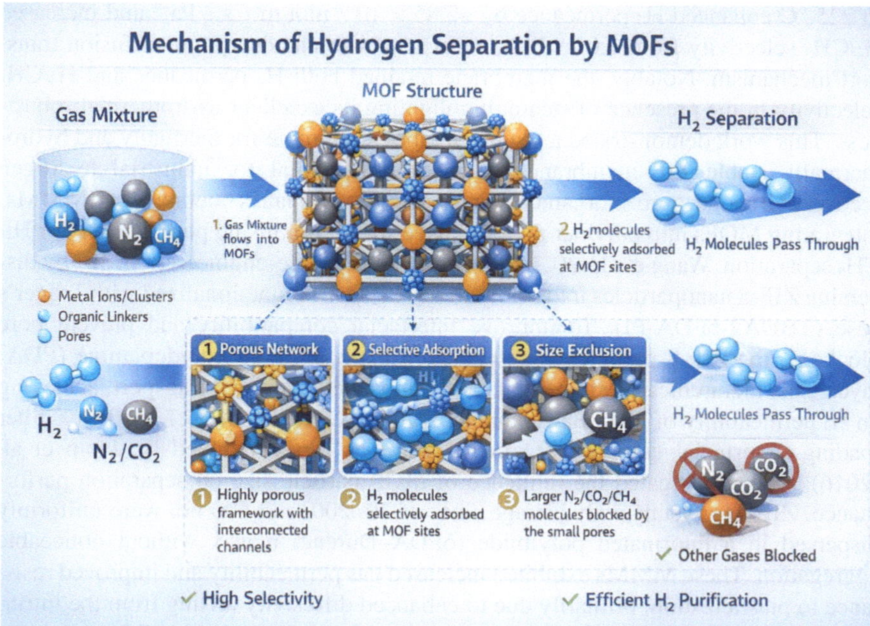

Fig. 5.1 Hydrogen separation pathway using MOF structural networks

5.3 Hydrogen Separation from Other Gases

5.3.1 H_2/CH_4 Separation

The recovery of hydrogen from hydrocarbons during steam methane reforming (SMR), particularly H_2/CH_4 separation, remains a critical challenge for primary gas separation in industrial applications. Hydrocarbon components can reduce permeability in polymeric membranes due to plasticization and surface condensation effects, prompting increasing interest in inorganic membranes and hybrid materials for efficient hydrogen separation. In this context, Huang et al. (2010) reported the fabrication of a ZIF-90 membrane exhibiting exceptional thermal and hydrothermal stability for H_2/CH_4 separation. ZIF-90 is constructed from Zn^{2+} ions and imidazolate-2-carboxyaldehyde (ICA) ligands, forming a framework with a pore aperture of ~3.5 Å. The aldehyde functionality allows imine condensation with 3-aminopropyltriethoxysilane (APTES), significantly improving interfacial adhesion between the ZIF layer and the Al_2O_3 support. A defect-free ZIF-90 membrane activated at 200 °C showed an H_2 permeance of ~2.7×10^{-7} mol m^{-2} s^{-1} Pa^{-1} and an ideal H_2/CH_4 selectivity of 15.2. Single-gas permeation experiments with CO_2, N_2, and C_2H_4 demonstrated a strong dependence on molecular size, with H_2 exhibiting the highest permeance. Binary gas mixture tests at 200 °C yielded selectivities of 7.3 (H_2/CO_2), 11.7 (H_2/N_2), 15.3 (H_2/CH_4), and 62.8 (H_2/C_2H_4). Increasing the temperature from 20 to 225 °C enhanced H_2 permeance by ~2.85 × 10^{-7} mol m^{-2} s^{-1} Pa^{-1} and increased H_2/CH_4 selectivity from 7.0 to 16.4, consistent with an adsorption–diffusion transport mechanism. Notably, the membrane retained both H_2 permeance and H_2/CH_4 selectivity in the presence of steam, highlighting its excellent hydrothermal robustness. This work demonstrated a seed-free fabrication route for thermally and hydrothermally stable ZIF membranes with strong potential for industrial hydrogen recovery. Beyond pure inorganic membranes, mixed-matrix membranes (MMMs) integrating MOFs into polymer matrices have shown promising performance for H_2/CH_4 separation. Wang et al. (2016) developed an interface-engineered MMM by dispersing ZIF-8 nanoparticles into a polyimide (PI) matrix functionalized with Tröger's base (TBDA2-6FDA-PI). To improve interfacial compatibility and prevent pore blockage, the ZIF-8 particles were coated with an ultrathin polydopamine (PDA) layer. This hierarchical design significantly enhanced hydrogen transport, achieving an H_2 permeability of 1858 Barrer and an H_2/CH_4 selectivity of 35.7 at 30 wt% filler loading. Interfacial compatibility in MMMs was further studied by Japip et al. (2016), who investigated the influence of MOF particle size on separation performance. ZIF-71 crystals with average sizes of 30, 200, and 600 nm were uniformly dispersed in a fluorinated polyimide (6FDA–Durene) matrix without noticeable aggregation. These MMMs exhibited increased gas permeability and improved resistance to plasticization, primarily due to enhanced diffusivity arising from the intrinsic microporosity of ZIF-71. Among the tested samples, the MMM containing 200 nm ZIF-71 particles achieved the highest H_2 permeability (2099 Barrer), a threefold increase over the pristine polymer, while maintaining competitive H_2/CH_4 selectivity. Similarly, Ghalei et al. (2017) incorporated functionalized and nanosized UiO-66 MOFs into a polymer of intrinsic microporosity (PIM-1). Optimization of mixing conditions minimized the typical reduction in permeability associated with filler incorporation. The MOF nanoparticles improved polymer chain packing and dispersion, aided by strong interfacial interactions from amine functionalities. MMMs based on UiO-66, UiO-66–NH_2, and UiO-66–Br were evaluated, with the 10 wt% UiO-66–NH_2/PIM-1 membrane delivering the best performance. This membrane exhibited an H_2 permeability of 2641 Barrer and an H_2/CH_4 selectivity of 25.9, while also demonstrating strong CO_2/N_2 separation capability. Collectively, these studies highlight the effectiveness of MOF functionalization, particle size control, and nanoscale dispersion strategies for simultaneously enhancing hydrogen permeability and selectivity in MMMs, demonstrating the potential of hybrid membrane systems for industrial H_2 recovery and purification.

5.3.2 H_2/CO_2 Separation

In the steam methane reforming (SMR) process, hydrogen production involves the water–gas shift (WGS) reaction, where carbon monoxide reacts with steam to generate CO_2 and H_2. The effective removal of CO_2 is critical before downstream utilization or combustion. Conventional polymeric membranes, however, often suffer from CO_2-induced plasticization, resulting in reduced permeability and degraded long-term performance (Ganguli and Bhatt 2023). Typically, increasing membrane thickness improves selectivity but compromises permeance. In a notable departure from this trend, Peng et al. demonstrated the fabrication of ultrathin molecular sieve nanosheet (MSN) membranes (~1 nm thick) derived from a layered Zn-based MOF, $Zn_2(bim)_4$. Layered MOFs are especially advantageous for membrane fabrication, offering large lateral dimensions, high crystallinity, and intrinsic framework flexibility—key factors for molecular sieving. Colloidal suspensions of $Zn_2(bim)_4$ MSNs were deposited onto α-Al_2O_3 supports using a hot-drop coating technique. Single-crystal structural analysis estimated an effective aperture size of ~0.21 nm, which selectively restricts CO_2 molecules (kinetic diameter ≈ 3.3 Å) while allowing H_2 (≈ 2.9 Å) to permeate. The resulting membranes exhibited exceptionally high H_2 permeance and an H_2/CO_2 selectivity exceeding 200. Powder X-ray diffraction (PXRD) confirmed that membranes with fully disordered MSNs demonstrated enhanced H_2/CO_2 selectivity proportional to increased H_2 permeance. Furthermore, these membranes retained excellent thermal and hydrothermal stability when exposed to an equimolar H_2/CO_2 mixture containing ~4 mol% steam at 150 °C. Liu and co-workers developed highly oriented NH_2-MIL-125(Ti) membranes (~0.5 μm thick) with strong CO_2 affinity (Sun et al. 2018). Fabrication involved oriented seeding, in-plane epitaxial growth, and suppression of undesired twin formation, yielding predominantly c-oriented crystals, as confirmed by XRD. The membranes exhibited excellent ideal H_2 selectivity in single- and mixed-gas experiments. Effective H_2/CO_2 separation at temperatures below 90 °C was attributed to strong interactions between CO_2 molecules and amino functional groups, combined with spatial confinement and narrowed diffusion pathways. Although H_2 permeability increased with temperature (30–210 °C), mixed-gas H_2/CO_2 selectivity decreased due to weakened CO_2–NH_2 interactions at elevated temperatures. Nonetheless, the membranes demonstrated outstanding reproducibility, mechanical stability, and thermal robustness, making them suitable for long-term operation (Rahbari et al. 2025). Kang et al. reported Ni-based MOF membranes (JUC-150, $Ni_2(L\text{-}asp)_2(pz)$) with excellent separation performance at temperatures above 200 °C (Kang et al. 2014). JUC-150 is constructed from L-aspartic acid and pyrazine pillaring units, forming a layer-pillared 3D structure with one-dimensional channels (~2.5 × 4.5 Å) along the a-axis. Using a Ni screen as a seed layer during solvothermal growth, uniform coverage was achieved on the support. The ultranarrow pores exhibited strong molecular sieving, allowing predominantly H_2 permeation. Gas transport studies showed a molecular-weight-dependent trend consistent with Knudsen diffusion. H_2 permeance reached 2.96×10^{-7} mol m^{-2} s^{-1} Pa^{-1}, while other gases were largely blocked. Thermal vibrations of methylene groups within the L-asp moiety acted as a dynamic gate, selectively facilitating H_2 transport. Separation factors of 17.1 (H_2/N_2), 26.3 (H_2/CH_4), and 38.7 (H_2/CO_2) were achieved. In a complementary strategy, Sun et al. developed mixed-matrix membranes (MMMs) based on ultrathin graphene oxide (GO)-HKUST-1 composites, leveraging HKUST-1's strong CO_2 affinity [108]. CuO nanosheets and GO were first assembled into CuO-NS@GO films via a layer-by-layer approach, with GO providing flexibility and minimizing interfacial defects. Filtration with a benzene-1,3,5-tricarboxylic acid (BTC) solution converted the films into HKUST-1@GO composites, as verified by XRD and FT-IR. Gas permeation experiments demonstrated that HKUST-1@GO MMMs achieved superior H_2 permeance and H_2/CO_2 selectivity compared to pure GO or CuO-NS@GO membranes. The optimized HKUST-1@GO-4 membrane achieved an H_2/CO_2 selectivity of 73.2 and an H_2 permeance of ~5 × 10^{-7} mol m^{-2} s^{-1} Pa^{-1}, while maintaining excellent mechanical flexibility and stable performance over 120 h. Table 5.1 summarizes recent MOF-based membranes for H_2/CO_2 separation, highlighting their structural features, synthesis strategies, and corresponding hydrogen separation performance.

Table 5.1 Comparative analysis of MOF materials and their hydrogen separation performance

Technology	Separation target	Mechanism	Performance metrics	Reference
Various MOFs	H_2 separation and purification from mixed gas streams (CO_2, CH_4, N2)	Overview of MOF adsorbents and structural design strategies for selective H_2 separation	Selectivity: 2–8, capacity: 1–3 mmol/g	Lim et al. (2021)
MOF-in-COF membranes	H_2/CO_2, H_2/CH_4 separation	Molecular sieving via hybrid channels	H_2/CO_2 selectivity ~180, H_2 permeance 1.2 × 10^3 GPU	Fan et al. (2021)
MOF-based PSA adsorbents	H_2 from steam methane reforming off-gas	Pressure swing adsorption (PSA)	H_2 recovery: >90%	Mao et al. (2021)
ZIF-8 membranes	H_2/CO_2, H_2/N_2 and H_2/CH_4 separation	ZIF-8-based membrane; vapor-phase transformation; improved pore connectivity	H_2 permeance: 3.5 × 10^{-7} mol m^{-2} s^{-1} Pa^{-1}; H_2/CO_2 selectivity: 20	Su et al. (2026)
MOFs under grand canonical Monte Carlo simulations	H_2 purification from syngas under a range of realistic dry and wet conditions	Machine learning-aided screening; hydrophobic frameworks for humid gas separation	H_2 recovery: 91%; H_2 purity: 99.9%	Luo et al. (2026)
MOF-based PSA adsorbents	H_2 separation and storage	AI-guided MOF selection; seasonal storage applications	Storage density: 48 g/L; retention: 98% after 100 cycles	Lee et al. (2026)
Cobalt-based metal–organic framework	H_2/D_2 separation	Co-MOF; ultra-microporous; high density of open metal sites	H_2/D_2 selectivity: 12.3; H_2 uptake: 1.8 wt% at 77 K	Yang et al. (2026)

Table 5.1 (continued)

Technology	Separation target	Mechanism	Performance metrics	Reference
Covalent organic frameworks	Photocatalytic hydrogen evolution using COFs to convert sunlight into sustainable fuel through water splitting	COFs with high crystallinity; stable under HER conditions	HER overpotential: 180 mV at 10 $mA{\cdot}cm^{-2}$	Aslam et al. (2026)
Various MOFs	Separation industrial gas mixtures such as CO_2/N_2, CO_2/CH_4, CO_2/H_2, and CH_4/H_2	Computational screening; top-performing MOFs for CO_2/H_2 separation	CO_2/H_2 selectivity: Up to 50; H_2 recovery: 89%	Mudhulu et al. (2026)

5.4 H_2 Isotope Separation

Currently, the separation of hydrogen isotopes in MOF-based materials is primarily dictated by quantum sieving mechanisms. These mechanisms are generally categorized into two main types. The first, kinetic quantum sieving (KQS), relies on confinement effects arising from pore dimensions that are comparable to the de Broglie wavelengths of the isotopic molecules, leading to differential diffusion rates. The second, chemical affinity quantum sieving (CAQS), exploits isotope-specific interactions with strongly adsorbing sites within the MOF framework, where differences in adsorption energies between isotopes drive selective uptake. Together, KQS and CAQS offer complementary approaches for achieving efficient hydrogen isotope separation under controlled conditions.

5.4.1 *Kinetic Quantum Sieving (KQS)*

Hydrogen isotope separation remains challenging for conventional molecular sieving techniques because of the nearly identical molecular sizes of H_2 and D_2 (Li et al. 2022). Beenakker et al. (1995) first proposed the concept of kinetic quantum sieving (KQS), demonstrating theoretically that gas transport within confined pores becomes restricted when the difference between the pore aperture and the molecular size is comparable to the de Broglie wavelength (λ) of the gas molecules. Under such conditions, the lighter isotope (H_2), which possesses a larger de Broglie wavelength, experiences a higher diffusion energy barrier than the heavier isotope (D_2), thereby enabling selective D_2/H_2 separation. For hydrogen isotopes, the difference between λ_H_2 and λ_D_2 increases with decreasing temperature, making the KQS effect more pronounced under cryogenic conditions. Consequently, the KQS separation mechanism is primarily governed by the pore architecture of the material and the kinetic properties of the isotopes. However, achieving an optimal balance between separation selectivity and adsorption capacity remains difficult, as reducing pore size enhances isotope selectivity but simultaneously limits adsorption capacity. Moreover, the requirement for low operating temperatures restricts the practical applicability of KQS-based separation. Motivated by these limitations, extensive efforts have been directed toward elucidating the relationship between pore aperture and isotope separation efficiency (Kim et al. 2019). Zhang and co-workers introduced the concept of the local optimal operating temperature (LOOT) to describe the interplay between pore size, temperature, and isotope selectivity, demonstrating that materials with smaller pore apertures exhibit higher LOOT values (Cao et al. 2020). Accordingly, an ideal material for efficient D_2/H_2 quantum sieving should feature uniform, hydrophilic pores with a single dominant adsorption site. These insights provide valuable guidance for the rational design of materials and the optimization of operating conditions to achieve enhanced hydrogen isotope separation performance.

5.4.2 Chemical Affinity Quantum Sieving (CAQS)

The chemical affinity quantum sieving (CAQS) effect enables hydrogen isotope separation through thermodynamically driven adsorption differences. FitzGerald et al. (2013) first demonstrated the CAQS mechanism at strong adsorption sites in M-MOF-74 (M = Fe, Co, Ni). Since then, numerous MOF systems exploiting the CAQS effect for hydrogen isotope separation have been reported (Li et al. 2022). Unlike KQS, which relies primarily on pore confinement and kinetic effects, CAQS is governed by the chemical nature of the adsorption sites and the strength of the interaction between the framework and hydrogen isotopes. The fundamental principle of CAQS arises from the mass-dependent zero-point energy (ZPE) of hydrogen isotopes adsorbed at strong binding sites within nanoporous materials. Because heavier isotopes exhibit lower ZPE, they possess higher adsorption enthalpies (Q_{st}), resulting in preferential adsorption. Specifically, D_2 has a lower ZPE and higher Q_{st} than H_2, leading to enhanced D_2 selectivity at chemically active sites (Kim et al. 2019). Consequently, materials suitable for CAQS-based isotope separation must contain metal active sites or other strong adsorption centers, while maintaining sufficiently large pore apertures to allow isotope diffusion into the framework. Importantly, CAQS-driven hydrogen isotope separation can be effective even at relatively elevated temperatures (e.g., 77 K), offering a distinct advantage over purely kinetic sieving mechanisms. Because both quantum sieving mechanisms contribute to isotope separation in MOF materials, increasing attention has been devoted to their synergistic coupling. Studies have shown that the coexistence of CAQS and KQS significantly enhances D_2/H_2 adsorption selectivity compared with either mechanism alone. When both effects operate simultaneously, MOFs can exploit chemical affinity and physical confinement in tandem, thereby improving overall separation efficiency and enabling effective isotope capture even at low concentrations (Kim et al. 2023).

5.5 Conclusions and Future Perspectives

In this chapter, we have highlighted the pivotal role of flexible metal–organic frameworks (MOFs) in the storage and separation of hydrogen and its isotopes. Characterized by dynamically adjustable pore apertures, flexible MOFs exhibit exceptional gas separation performance due to their tunable adsorption selectivity and responsiveness to external stimuli such as pressure and temperature. Advanced experimental techniques, including in situ X-ray diffraction (XRD), single-component gas adsorption, neutron powder diffraction (NPD), and crystallographic analyses combined with theoretical tools such as ideal adsorbed solution theory (IAST), provide critical insights into the structural factors governing separation. These include ligand motifs and functional groups, host–guest and guest–guest interactions, and the energy barriers associated with framework transformations. Despite the inherent challenges of hydrogen and isotope separation, flexible MOFs have demonstrated the ability to overcome limitations of conventional separation technologies. However, the application of flexible MOFs for hydrogen isotope separation remains relatively underexplored. Systematic investigations into their separation mechanisms can yield valuable design principles for next-generation materials. The degree of framework flexibility, which strongly depends on metal–ligand chemistry and crystal size, is a key determinant for tailoring MOFs toward specific hydrogen separation applications. Notably, the synergistic interplay between kinetic quantum sieving (KQS) and chemical affinity quantum sieving (CAQS) enables enhanced hydrogen isotope separation. Incorporating ligands with strong binding sites into open and flexible frameworks offers a promising strategy to exploit CAQS effects while mitigating temperature-related limitations. Furthermore, flexible MOF-based membranes present distinct advantages over bulk crystalline MOFs and thin films, including cost-effectiveness, operational simplicity, and superior separation efficiency. These membranes emerge as viable alternatives to energy-intensive conventional isotope separation processes. Overall, this chapter provides critical guidance for advancing hydrogen separation technologies. By elucidating the principles and applications of flexible MOFs in hydrogen and isotope separation, it contributes to the development of sustainable, low-carbon energy solutions poised to address global energy challenges.

References

Aslam AA, Nazir MS, ul Hassan S, Akhtar MN, Hussain M, Amjad S, Farooq A, Khan AA (2026) A review on recent developments towards hydrogen evolution reaction for covalent organic frameworks. Renew Sust Energ Rev 230:116695

Beenakker JJM, Borman VD, Krylov SY (1995) Molecular transport in subnanometer pores: zero-point energy, reduced dimensionality and quantum sieving. Chem Phys Lett 232:379–382

Cacho-Bailo F, Catalan-Aguirre S, Etxeberria-Benavides M, Karvan O, Sebastian V, Tellez C, Coronas J (2015) Metal-organic framework membranes on the inner-side of a polymeric hollow fiber by microfluidic synthesis. J Membr Sci 476:277–285

Cao D, Ren J, Gong Y, Huang H, Fu X, Chang M, Chen X, Xiao C, Liu D, Yang Q, Zhong C (2020) Quantum sieving of H 2/D 2 in MOFs: a study on the correlation between the separation performance, pore size and temperature. J Mater Chem A 8(13):6319–6327

Cheng Y, Ying Y, Japip S, Jiang SD, Chung TS, Zhang S, Zhao D (2018) Advanced porous materials in mixed matrix membranes. Adv Mater 30:1–20

Choi HJ, Dinca M, Long JR (2008) Broadly hysteretic H2 adsorption in the microporous metal–organic framework Co (1, 4-benzenedipyrazolate). J Am Chem Soc 130(25):7848–7850

Ding M, Jiang H-L (2020) Improving water stability of metal-organic frameworks by a general surface hydrophobic polymerization. CCS Chem 2:2740–2748

Dong Q, Zhang X, Liu S, Lin RB, Guo Y, Ma Y, Yonezu A, Krishna R, Liu G, Duan J, Matsuda R (2020) Tuning gate-opening of a flexible metal–organic framework for ternary gas sieving separation. Angew Chem Int Ed 59(50):22756–22762

Fan H, Peng M, Strauss I, Mundstock A, Meng H, Caro J (2021) MOF-in-COF molecular sieving membrane for selective hydrogen separation. Nat Commun 12(1):38

Fernandez CA, Liu J, Thallapally PK, Strachan DM (2012) Switching Kr/Xe selectivity with temperature in a metal–organic framework. J Am Chem Soc 134(22):9046–9049

FitzGerald SA, Pierce CJ, Rowsell JL, Bloch ED, Mason JA (2013) Highly selective quantum sieving of D2 from H2 by a metal–organic framework as determined by gas manometry and infrared spectroscopy. J Am Chem Soc 135(25):9458–9464

Fitzgerald SA, Mukasa D, Rigdon KH, Zhang N, Barnett BR (2019) Hydrogen isotope separation within the metal-organic framework Cu(i)-MFU-4 I. J Phys Chem C 123:30427–30433

Foo ML, Matsuda R, Hijikata Y, Krishna R, Sato H, Horike S, Hori A, Duan J, Sato Y, Kubota Y, Takata M (2016) An adsorbate discriminatory gate effect in a flexible porous coordination polymer for selective adsorption of CO2 over C2H2. J Am Chem Soc 138(9):3022–3030

Ganguli A, Bhatt V (2023) Hydrogen production using advanced reactors by steam methane reforming: a review. Front Therm Eng 3:1143987

Ghalei B, Sakurai K, Kinoshita Y, Wakimoto K, Isfahani AP, Song Q, Doitomi K, Furukawa S, Hirao H, Kusuda H, Kitagawa S (2017) Enhanced selectivity in mixed matrix membranes for CO2 capture through efficient dispersion of amine-functionalized MOF nanoparticles. Nat Energy 2(7):1–9

Ghalei B, Wakimoto K, Wu CY, Isfahani AP, Yamamoto T, Sakurai K, Higuchi M, Chang BK, Kitagawa S, Sivaniah E (2019) Rational tuning of zirconium metal–organic framework membranes for hydrogen purification. Angew Chem Int Ed 58(52):19034–19040

Glugla M, Lasser R, Dorr L, Murdoch DK, Haange R, Yoshida H (2003) The inner deuterium/tritium fuel cycle of ITER. Fusion Eng Des 69:39–43

Gu C, Hosono N, Zheng JJ, Sato Y, Kusaka S, Sakaki S, Kitagawa S (2019) Design and control of gas diffusion process in a nanoporous soft crystal. Science 363(6425):387–391

Guo H, Zhu G, Hewitt IJ, Qiu S (2009) "Twin copper source" growth of metal–organic framework membrane: Cu3 (BTC) 2 with high permeability and selectivity for recycling H2. J Am Chem Soc 131(5):1646–1647

Hazazi K, Ma X, Wang Y, Ogieglo W, Alhazmi A, Han Y, Pinnau I (2019) Ultra-selective carbon molecular sieve mem-branes for natural gas separations based on a carbon-rich intrinsically microporous polyimide precursor. J Membr Sci 585:1–9

Hoskins BF, Robson R (1989) Infinite polymeric frame-works consisting of three dimensionally linked rod-like segments. J Am Chem Soc 111:5962–5964

Hou J, Zhang H, Simon GP, Wang H (2019) Polycrystalline advanced microporous framework membranes for efficient separation of small molecules and ions. Adv Mater 32:1902009

Hu H, Guo X, Yang L, Wu Y, Yang G, Luo X (2025) Adsorption materials toward highly-efficient lithium extraction from non-conventional lithium sources. Adv Mater 37(38):2506055

Huang A, Dou W, Caro J (2010) Steam-stable zeolitic imidazolate framework ZIF-90 membrane with hydrogen selectivity through covalent functionalization. J Am Chem Soc 132(44):15562–15564

Japip S, Xiao Y, Chung TS (2016) Particle-size effects on gas transport properties of 6FDA-Durene/ZIF-71 mixed matrix membranes. Ind Eng Chem Res 55(35):9507–9517

Joshi RK, Carbone P, Wang FC, Kravets VG, Su Y, Grigorieva IV, Wu HA, Geim AK, Nair RR (2014) Precise and ultrafast molecular sieving through graphene oxide membranes. Science 343(6172):752–754

Kang Z, Xue M, Fan L, Huang L, Guo L, Wei G, Chen B, Qiu S (2014) Highly selective sieving of small gas molecules by using an ultra-microporous metal–organic framework membrane. Energy Environ Sci 7(12):4053–4060

Kerry FG (2007) Industrial gas handbook: gas separation and purification. CRC Press, Boca Raton

Kim JY, Oh H, Moon HR (2019) Hydrogen isotope separation in confined nanospaces: carbons, zeolites, metal–organic frameworks, and covalent organic frameworks. Adv Mater 31(20):1805293

Kim H, Jee S, Park J, Jung M, Muhammad R, Choi K, Oh H (2023) High D2/H2 selectivity performance in MOF-303 under ambient pressure for potential industrial applications. Sep Purif Technol 325:124660

Kondo A, Yashiro T, Okada N, Hiraide S, Ohkubo T, Tanaka H, Maeda K (2018) Selective molecular-gating adsorption in a novel copper-based metal–organic framework. J Mater Chem A 6(14):5910–5918

Koros WJ, Zhang C (2017) Materials for next-generation molecularly selective synthetic membranes. Nat Mater 16:289–297

Lee RW, Patil AS, Zhang Y (2026) Machine learning-derived stage-specific design rules for metal-organic framework selection in seasonal hydrogen storage. Sci Rep 16:4964

Li Y, Wang S, He G, Wu H, Pan F, Jiang Z (2015) Facilitated transport of small molecules and ions for energy-efficient membranes. Chem Soc Rev 44:103–118

Li W, Su P, Li Z, Xu Z, Wang F, Ou H, Zhang J, Zhang G, Zeng E (2017a) Ultrathin metal–organic framework membrane production by gel–vapour deposition. Nat Commun 8(1):406

Li L, Lin RB, Krishna R, Wang X, Li B, Wu H, Li J, Zhou W, Chen B (2017b) Flexible–robust metal–organic framework for efficient removal of propyne from propylene. J Am Chem Soc 139(23):7733–7736

Li L, Ji C, Wang W, Wu F, Tan YX, Yuan D (2022) The effect of pore sizes on D 2/H 2 separation conducted by MOF-74 analogues. Inorg Chem Front 9(8):1674–1680

Liao CL, Wu XQ, Wang M, Sun B, Hai G, Wu YP, Zhao J, Lu JY, Li DS (2025) An ultramicroporous metal-organic framework featuring local-molecule trap for efficient C2H2/CO2 separation. Sep Purif Technol 363:132075

Lim DW, Ha J, Oruganti Y, Moon HR (2021) Hydrogen separation and purification with MOF-based materials. Mater Chem Front 5(11):4022–4041

Lin RB, Xiang S, Xing H, Zhou W, Chen B (2019) Exploration of porous metal-organic frameworks for gas separation and purification. Coord Chem Rev 378:87–103

Luo J, Gibaldi M, Rostom S, Symonds RT, Hughes RW, Woo TK (2026) High-throughput screening of experimental metal-organic frameworks for high water content syngas hydrogen purification under pressure swing adsorption conditions. ACS Appl Mater Interfaces. https://doi.org/10.1021/acsami.5c19342

Ma Y, Dong Z, You M, Zhang Y, Feng X, Ma X, Meng J (2019) Formation of a thin and continuous MOF membrane with 2-D MOF nanosheets as seeds: *via* layer-by-layer growth. Chem Commun 55:10146–10149

Mao D, Griffin JM, Dawson R, Fairhurst A, Bimbo N (2021) Metal organic frameworks for hydrogen purification. Int J Hydrog Energy 46(45):23380–23405

Moghadam PZ, Li A, Wiggin SB, Tao A, Maloney AGP, Wood PA, Ward SC, Fairen-Jimenez D (2017) Development of a Cambridge structural database subset: a collection of metal-organic frameworks for past, present, and future. Chem Mater 29:2618–2625

Mondal SS, Kreuzer A, Behrens K, Schütz G, Holdt HJ, Hirscher M (2019) Systematic experimental study on quantum sieving of hydrogen isotopes in metal-amide-imidazolate frameworks with narrow 1-D channels. ChemPhysChem 20(10):1311–1315

Mudhulu S, Nath J, Kuncharam BVR, Gupta S (2026) High-throughput computational screening of metal organic frameworks (MOFs) for CO_2 selective separations: trends, challenges, and future perspectives. Chem Eng J Green Sustain 1:100025. https://doi.org/10.1016/j.cejgas.2026.100025

Myers AL, Prausnitz JM (1965) Thermodynamics of mixed-gas adsorption. AIChE J 11(1):121–127

Ockwig NW, Nenoff TM (2007) Membranes for hydrogen separation. Chem Rev 107:4078–4110

Qian Q, Asinger PA, Lee MJ, Han G, Mizrahi Rodriguez K, Lin S, Benedetti FM, Wu AX, Chi WS, Smith ZP (2020) MOF-based membranes for gas separations. Chem Rev 120(16):8161–8266

Raggam S, Mohammad M, Choo Y, Naidu G, Zargar M, Shon HK, Razmjou A (2023) Advances in metal organic framework (MOF)–based membranes and adsorbents for lithium-ion extraction. Sep Purif Technol 307:122628

Rahbari A, Chakrapani TH, Shuang F, Krokidas P, Habibi P, Lagerweij VJ, Ramdin M, Vlugt TJH, Hajibeygi H, Dey P, Tsimpanogiannis IN, Moultos OA (2025) Molecular simulation of hydrogen systems: from properties and methods to applications and future directions. Chem Rev 125(24):11878–12029. https://doi.org/10.1021/acs.chemrev.5c00617

Robeson LM (2008) The upper bound revisited. J Membr Sci 320(1-2):390–400

Sánchez-Laínez J, Veiga A, Zornoza B, Balestra SR, Hamad S, Ruiz-Salvador AR, Calero S, Téllez C, Coronas J (2017) Tuning the separation properties of zeolitic imidazolate framework core–shell structures via post-synthetic modification. J Mater Chem A 5(48):25601–25608

Shi YX, Li WX, Zhang WH, Lang JP (2018) Guest-induced switchable breathing behavior in a flexible metal–organic framework with pronounced negative gas pressure. Inorg Chem 57(14):8627–8633

Sholl DS, Lively RP (2016) Seven chemical separations to change the world. Nature 532:435–437

Su P, Zhang J, Wu W, Xiao Y, Xu T, Wu S, Wang Y, Li W (2026) Vapor-phase crystalline transformation of metal-organic framework membranes for efficient hydrogen separation. J Membr Sci 742:125168

Sun Y, Liu Y, Caro J, Guo X, Song C, Liu Y (2018) In-plane epitaxial growth of highly c-oriented NH2-MIL-125 (Ti) membranes with superior H2/CO2 selectivity. Angew Chem Int Ed 57(49):16088–16093

Sun WQ, Hu JB, Jiang YJ, Xu N, Wang LY, Li JH, Hu YQ, Duttwyler S, Zhang YB (2022) Flexible molecular sieving of C2H2 from CO2 by a new cost-effective metal organic framework with intrinsic hydrogen bonds. Chem Eng J 439:135745

Wang Y, Zhao D (2017) Beyond equilibrium: metal-organic frameworks for molecular sieving and kinetic gas separation. Cryst Growth Des 17:2291–2308

Wang N, Mundstock A, Liu Y, Huang A, Caro J (2015) Amine-modified Mg-MOF-74/CPO-27-Mg membrane with enhanced H2/CO2 separation. Chem Eng Sci 124:27–36

Wang Z, Wang D, Zhang S, Hu L, Jin J (2016) Interfacial design of mixed matrix membranes for improved gas separation performance. Adv Mater 28(17):3399–3405

Wang S, Cheng M, Luo L, Ji X, Liu C, Bi K, Zhou L (2023) High-throughput screening of metal–organic frameworks for hydrogen purification. Chem Eng J 451:138436

Wu X, Zhang H, Zhang X, Guan Q, Tang X, Wu H, Feng M, Wang H, Ou R (2024) Sustainable lithium extraction enabled by responsive metal-organic frameworks with ion-sieving adsorption effects. Proc Natl Acad Sci 121(6):e2309852121

Wu Q, He X, Cui C, Qi B, Wei J (2025) Recent developments in metal–organic framework-based mixed matrix membranes for hydrogen separation. Sep Purif Technol 354:128946

Yang Y, Ji X, Ping E, Kong L, Sun L, Zhang L, Zhou Y (2026) A cobalt-based metal-organic framework material featuring ultra-micropores and an exceptionally high density of open metal sites enables efficient H2/D2 separation: molecular simulation and experimental study. Int J Hydrog Energy 205:153264

Zhang F, Zou X, Gao X, Fan S, Sun F, Ren H, Zhu G (2012) Hydrogen selective NH2-MIL-53 (Al) MOF membranes with high permeability. Adv Funct Mater 22(17):3583–3590

Zhou DD, Chen P, Wang C, Wang SS, Du Y, Yan H, Ye ZM, He CT, Huang RK, Mo ZW, Huang NY (2019) Intermediate-sized molecular sieving of styrene from larger and smaller analogues. Nat Mater 18(9):994–998

Chapter 6
Hydrogen Storage and Transportation: Advanced Materials, Technologies, and Future Perspectives

Abstract Hydrogen is widely regarded as a promising energy carrier due to its exceptionally high gravimetric energy density (~120 MJ kg^{-1}) and its ability to enable electricity generation without greenhouse gas emissions when used in fuel cell systems. Despite these advantages, the absence of safe, efficient, and economically viable storage and transportation solutions remains a major obstacle to its large-scale deployment. This chapter provides a concise overview of the evolution of hydrogen energy development, followed by an in-depth examination of contemporary hydrogen storage and transportation pathways, including compressed and liquefied hydrogen, solid-state storage through physical adsorption, and chemical conversion-based approaches. The performance, advantages, limitations, safety considerations, and environmental impacts of each method are critically assessed. Furthermore, recent progress in materials science aimed at improving hydrogen storage capacity, stability, and transport efficiency is highlighted. This chapter also reviews current research and development trends, market dynamics, and representative examples of emerging commercial and practical systems. Key technical challenges, system-level requirements, and desirable characteristics for future hydrogen storage and transportation infrastructures are discussed in the context of energy security, environmental sustainability, and ecological protection. By synthesizing technological advances and future perspectives, this chapter aims to stimulate innovative strategies for efficient hydrogen storage and transportation, thereby supporting the global transition toward a sustainable, low-carbon energy system.

Keywords Hydrogen · Compressed hydrogen gas storage · Liquid hydrogen storage · Organic hydrogen carriers · Metal hydride · Zeolites

6.1 Introduction

Global energy consumption has been rising consistently, driven by population growth, rapid urbanization, and expanding economic activity. Projections indicate that the global population may reach approximately 9.7 billion by 2050, which will

D. M. Y. Mir et al., *Low-Carbon Hydrogen*, Clean Energy Production Technologies,
https://doi.org/10.1007/978-981-95-8142-9_6

further intensify energy demand. At present, fossil fuels, namely, coal, oil, and natural gas, remain the dominant energy sources, supplying nearly 80% of global energy needs (Hassan et al. 2021; Zhang et al. 2024). The extensive reliance on these fuels has serious environmental consequences, as their combustion releases large quantities of greenhouse gases, particularly carbon dioxide (CO_2). These emissions are a major contributor to global warming and climate change, leading to adverse impacts such as extreme weather events, sea-level rise, ecosystem degradation, and biodiversity loss (Jones et al. 2023; Kabir et al. 2023). In addition, fossil fuel utilization emits harmful air pollutants, including nitrogen oxides (NO_x), sulfur oxides (SO_x), and particulate matter (PM), which degrade air quality and pose significant risks to human health. The uneven geographical distribution of fossil fuel resources has also heightened geopolitical tensions and raised concerns regarding long-term energy security (Asghar et al. 2021). To address these environmental and security challenges, countries worldwide are increasingly investing in renewable energy technologies such as solar, wind, and hydropower (Xiaosan et al. 2021; Obaideen et al. 2023). These renewable sources generate little to no greenhouse gas emissions during operation, thereby substantially lowering the carbon footprint of the energy sector and supporting climate mitigation goals (Pandit et al. 2023). Within this transition, hydrogen has emerged as a transformative component of the future energy system. As a clean, renewable, and highly adaptable energy carrier, hydrogen can be deployed across multiple sectors, including transportation, electricity generation, and industrial applications (Osman et al. 2022). Hydrogen's significance arises from its clean energy attributes, renewable nature, capacity for energy storage, and broad applicability across sectors. In addition, its potential for domestic production strengthens energy security, while the technological innovations required for its deployment stimulate economic growth and generate employment opportunities (Hassan et al. 2023). The increasing penetration of intermittent renewable energy sources has intensified the need for effective energy storage solutions to balance supply and demand and minimize energy curtailment. Various storage technologies such as batteries, compressed air energy storage, pumped hydroelectric storage, flywheels, and thermal energy storage have been proposed. Nevertheless, these systems often face limitations related to efficiency, scalability, safety, and high capital costs. In this context, hydrogen storage represents a compelling alternative, enabling surplus renewable electricity to be converted into hydrogen via low-carbon processes and subsequently used in combustion engines or fuel cells. Owing to its exceptionally high gravimetric energy density, hydrogen holds considerable promise for deep decarbonization of the global energy mix (Jahanbakhsh et al. 2024). Despite its potential, the widespread adoption of hydrogen is hindered by challenges associated with its safe, efficient, and economical storage. Consequently, the development of advanced hydrogen storage materials has become a critical research priority. Designing materials capable of high-capacity hydrogen uptake and controlled release requires careful consideration of factors such as structural characteristics, porosity, surface area, thermodynamic stability, kinetics, and safety. Addressing these parameters is essential for advancing hydrogen storage technologies and realizing their role in clean energy systems (Mondal et al. 2024). Hydrogen storage strategies are generally categorized into physical-based and material-based approaches. Physical storage involves storing hydrogen as a compressed gas, cryogenic liquid, or in cryo-compressed form by manipulating pressure and temperature conditions. In contrast, material-based or solid-state storage employs substances such as metal hydrides, complex hydrides, and carbon-based materials including activated carbon, graphene, and carbon nanotubes. These materials store hydrogen through absorption or adsorption mechanisms, with hydrogen release typically facilitated by thermal input or catalytic processes. This area has prompted extensive research into both homogeneous and heterogeneous catalysts to improve hydrogen release efficiency and kinetics (Navlani-García et al. 2018). Although experimental efforts have driven much of the progress in hydrogen storage research, development has often been slow due to the iterative and empirical nature of traditional experimentation. Recently, however, advances in computational and theoretical tools—such as density functional theory (DFT), high-throughput screening, crystal structure prediction, and emerging machine learning techniques—have begun to transform the field. These data-driven and computational approaches enable faster materials discovery, improved performance prediction, and accelerated innovation, thereby supporting more rapid advancement of energy technologies and associated industries (Lu 2021).

6.2 Hydrogen Storing Methods

6.2.1 Compressed Hydrogen Gas Storage

Compressed hydrogen gas (CAG) storage is one of the most widely adopted methods for hydrogen transportation. In this approach, hydrogen is stored by compressing the gas to very high pressures, typically in the range of 3.5×10^7 to 7×10^7 Pa, in order to attain a sufficiently high energy density. One of the key advantages of CAG storage is the rapid refueling and discharge capability, which is particularly beneficial for transportation applications. However, the compression process is energy intensive and leads to a considerable loss of approximately 13%–18% of the hydrogen's heating value. The storage vessels used in this method are generally cylindrical and must satisfy several critical requirements, including low cost, low weight, high mechanical strength, and resistance to hydrogen permeation and diffusion (Omran et al. 2022). Hydrogen gas is commonly stored using four types of pressure vessels, classified as Type I through Type IV. Type I vessels are fully metallic pressure tanks designed to store hydrogen at pressures between 1.5×10^7 and 3×10^7 Pa. These tanks are relatively inexpensive and are among the most commonly used high-pressure storage options; however, their substantial weight makes them unsuitable for onboard vehicular applications. Type II vessels consist of a thick metallic liner reinforced with a partial wrapping of carbon-fiber composite, offering improved strength-to-weight performance compared to Type I tanks (Elberry et al. 2021). Type III vessels further reduce weight by employing an aluminum liner fully wrapped with carbon fibers and can withstand hydrogen pressures ranging from 7×10^7 to 8.5×10^7 Pa. Type IV vessels represent the most advanced design, utilizing a fully composite structure with a nonmetallic polymer liner and an outer carbon-fiber-reinforced shell. These tanks can achieve weight reductions of up to 70% compared to Type I vessels, making them particularly attractive for mobile, automotive, and aeronautical applications. Owing to their high pressure tolerance and low mass, Type IV tanks are increasingly favored for hydrogen energy storage systems. The selection of a suitable hydrogen storage vessel depends on several factors, including the target storage pressure, volumetric and gravimetric storage density, and overall system cost. Importantly, cost considerations must be evaluated in conjunction with performance and safety requirements, ensuring that the chosen vessel is appropriate for the intended application. In addition to conventional pressure vessels, glass microsphere-based storage has been investigated as an alternative method for hydrogen containment. However, this approach suffers from high material density and gradual hydrogen leakage at ambient temperatures, leading to hydrogen losses during handling and transportation (Le Duigou et al. 2017). Safety remains a critical concern for CAG storage systems, as high-pressure vessels surrounded by flammable hydrogen gas can pose explosion risks if subjected to mechanical shock or structural failure. Refueling operations at pressures exceeding 7×10^7 Pa demand meticulous planning and specialized infrastructure at hydrogen filling stations. Consequently, the development of robust transportation and distribution infrastructure is essential for the large-scale deployment of hydrogen as an energy carrier. Recent advancements in CAG technology have focused on improving storage efficiency and operational safety through the use of advanced materials, optimized tank geometries, and enhanced liner and coating technologies. These developments aim to improve hydrogen containment, reduce permeation, and increase overall system reliability. Although CAG storage is a technologically mature and comparatively economical option, it is accompanied by several intrinsic constraints. The use of high-pressure storage vessels necessitates strict handling procedures and periodic maintenance, while the relatively low volumetric energy density of compressed hydrogen can limit vehicle driving range and overall system performance. In addition, safety concerns remain significant, as the combination of extremely high operating pressures and hydrogen's inherent flammability introduces risks associated with tank rupture, leakage, or explosion. To address these challenges, a range of mitigation strategies has been proposed, including the optimization of tank geometries, the development of cryogenic pressure-vessel concepts, enhanced thermal insulation schemes, and the incorporation of advanced composite materials. Ultimately, safe and reliable operation relies on a comprehensive framework encompassing robust engineering design, adherence to stringent manufacturing standards, well-defined operational protocols, and regular inspection and maintenance. A holistic strategy that integrates novel materials, innovative vessel architectures, alternative storage concepts, and rigorous safety measures has been shown to substantially reduce the risks associated with CAG storage systems (Yue et al. 2021). Within the domain of fuel-cell electric vehicles (FCEVs), research has predominantly focused on the advancement of Type IV hydrogen storage tanks featuring polymer liners, driven by the dual objectives of maximizing gravimetric and volumetric storage density while minimizing overall system weight (Wang et al. 2022). Key challenges associated with these liners—such as hydrogen permeation, thermal aging, and long-term mechanical degradation—have motivated extensive investigations into material modification strategies and advanced molding and processing techniques aimed at improving liner durability and barrier performance (Okonkwo et al. 2023). Parallel efforts have sought to enhance pressure-vessel efficiency through structural innovations, including box-shaped or conformable tank designs reinforced with internal tension struts and carbon-fiber-reinforced polymer composites (Zhang et al. 2023). Notably, researchers at the Technical University of Munich have demonstrated substantial progress in the development of conformable, cuboidal hydrogen storage tanks for FCEVs. Their modular design approach combines thermoplastic frameworks with composite tension struts and carbon-fiber-reinforced epoxy materials, enabling improved packaging efficiency and structural integrity (Oztas et al. 2022). Beyond automotive applications, hybrid hydrogen storage systems and large-scale subsurface storage solutions—such as salt caverns and engineered rock caverns—are also under active investigation as viable options for accommodating large hydrogen volumes in stationary energy systems (Ancona et al. 2023). Concurrent advancements in FCEV technology have facilitated the commercialization of several hydrogen-powered vehicle models, including the Toyota Mirai, Hyundai Nexo, and Honda Clarity Fuel Cell. Among these, the Toyota Mirai has become a global reference for FCEVs, achieving widespread market adoption. Mercedes-Benz further demonstrated the maturity of compressed hydrogen storage with the GLC F-Cell which achieved a driving range of approximately 478 km per hydrogen fill. The Hyundai Nexo extends this performance, offering a driving range of about 611 km, reflecting ongoing improvements in both storage system design and fuel cell efficiency. Similarly, the Honda Clarity Fuel Cell exhibits enhanced energy efficiency, with an EPA-rated combined fuel economy of 68 mi kg^{-1}. Concept vehicles such as Audi's A7 h-tron quattro and BMW's i Hydrogen NEXT further exemplify the evolving design philosophies and performance targets of hydrogen-powered mobility, with storage tank configurations adapted to different developmental stages. Overall, ongoing research continues to focus on improving the performance, safety, and adaptability of compressed hydrogen storage systems, with applications extending beyond transportation to include stationary energy storage and grid-level stabilization. Ultimately, the selection of an appropriate hydrogen storage solution requires a careful balance among factors such as weight, cost, safety, application-specific requirements, transportation logistics, and infrastructure availability, as different materials and vessel designs offer distinct trade-offs in performance and practicality.

6.2.2 Liquid Hydrogen Storage

Liquid hydrogen (LH_2) storage is based on maintaining hydrogen in a highly condensed liquid phase by cooling it to cryogenic temperatures below −252.8 °C. Achieving and sustaining such extreme conditions necessitates the use of specialized cryogenic storage vessels and involves a substantial energy penalty during the liquefaction process. This challenge is further exacerbated by unavoidable boil-off losses that occur due to heat ingress during storage and handling. In practical implementations, LH_2 is commonly stored in insulated, pressurized cryogenic tanks operating at pressures typically in the range of 250–350 atm (Langmi et al. 2022). Owing to its markedly higher volumetric energy density, LH_2 enables more compact storage compared to compressed gaseous hydrogen or cryogenic gaseous hydrogen, making it particularly attractive for applications where space efficiency is critical. Dedicated LH_2 storage containers are engineered to safely maintain hydrogen at temperatures below −240 °C, close to its critical phase-transition limits. Historically, liquid hydrogen storage has been successfully demonstrated in aerospace and industrial applications, offering advantages such as hydrogen discharge rates comparable to those of compressed gas systems while requiring lower adiabatic compression energy. Compared with compressed hydrogen, LH_2 possesses a significantly higher density, allowing for smaller tank volumes and potentially lower storage-system footprints and associated costs. These attributes position LH_2 as a promising option for high-density hydrogen storage requirements. However, the necessity for ultra-low operating temperatures and stringent safety and handling protocols introduces considerable technical and operational complexity. Although cryogenic pressure vessels with reduced insulation thickness can achieve comparable or even extended dormancy times with lower operational costs, LH_2 storage remains energy-intensive due to liquefaction demands and poses safety risks arising from hydrogen's broad flammability limits. Moreover, despite the high density of LH_2, the combined mass of cryogenic tanks and insulation materials presents challenges related to gravimetric efficiency and overall system weight. Cryogenic hydrogen storage, in its conventional form, maintains hydrogen in the liquid state at temperatures near −253 °C under approximately atmospheric pressure (Sundén 2019) and is widely employed for large-scale storage and long-distance transportation. Storage technologies such as Dewar vessels, vacuum-insulated cryogenic tanks, and cryocoolers are utilized to minimize heat ingress and suppress hydrogen evaporation. One of the principal advantages of cryogenic storage particularly in the form of cryo-compressed hydrogen is its exceptionally high hydrogen storage capacity (Madadi Avargani et al. 2022). Cryo-compressed hydrogen storage integrates moderate pressurization with cryogenic temperatures, enabling hydrogen densities comparable to liquid hydrogen while retaining hydrogen in the gaseous or supercritical phase. This hybrid approach significantly mitigates boil-off losses commonly associated with conventional LH_2 storage, thereby addressing one of the most critical limitations of liquid hydrogen systems (Chalk and Miller 2006). Consequently, cryogenic and cryo-compressed storage concepts are regarded as technically viable solutions for the efficient storage and transport of large hydrogen quantities, particularly in long-duration or large-scale applications where loss minimization is essential. Nevertheless, cryogenic hydrogen storage continues to face several challenges, including residual boil-off, complex insulation requirements, and the high energy demand associated with liquefaction and refrigeration processes. Cryogenic storage vessels are typically designed for operating pressures up to approximately 850 kPa and are manufactured in compliance with established safety regulations, such as the ASME Boiler and Pressure Vessel Code (BPVC) (Ulucan et al. 2023). Recent research efforts have focused on the development of advanced cryogenic materials, innovative tank geometries, and next-generation liquid hydrogen carriers to address issues such as low-temperature material embrittlement and hydrogen-induced degradation, while simultaneously enhancing storage efficiency, durability, and transport reliability (Bosu and Rajamohan 2023). Increasing attention has also been directed toward vehicle-scale cryogenic hydrogen storage, with studies evaluating both LH_2 and cryo-compressed hydrogen as onboard storage solutions. These investigations examine the thermodynamic and transient behavior of hydrogen within insulated pressure vessels, including refueling, discharge, and dormancy periods. Key parameters such as temperature and pressure evolution, phase equilibrium, and transitions between supercritical, liquid, and gaseous states during operation have been analyzed in detail (Ahluwalia and Peng 2008). The primary challenges associated with cryogenic hydrogen storage include the energy-intensive nature of hydrogen liquefaction, evaporative losses during storage, and the requirement for controlled venting to prevent excessive pressure build-up. Hydrogen behavior during storage and release is governed by thermodynamic properties, mass and energy balances, and phase equilibria, all of which are strongly dependent on temperature, pressure, and hydrogen composition. Heat transfer processes—namely, conduction, convection, and radiation—play a decisive role in system performance and must be carefully managed through effective insulation and thermal control strategies. Venting systems are essential for pressure regulation, and their impact on storage efficiency, hydrogen losses, and operational safety must be systematically evaluated. As a result, ongoing research emphasizes advanced storage architectures, improved thermal management solutions, optimized venting strategies, and seamless integration with fuel-cell systems, supported by detailed numerical modeling and simulation studies aimed at enhancing both performance and safety. Beyond ground-based applications, cryogenic hydrogen is increasingly being explored as a sustainable aviation fuel, offering significant potential for reducing carbon emissions in the aviation sector. Continued advancements in LH_2 and cryo-compressed hydrogen storage technologies are therefore expected to play a critical role in enabling low-carbon transportation systems and supporting the broader transition toward sustainable and hydrogen-based energy infrastructures.

6.2.3 Liquid Organic Hydrogen Carriers

Liquid organic hydrogen carriers (LOHCs) have emerged as promising and potentially transformative technologies for hydrogen storage and transportation. These systems store hydrogen through reversible chemical reactions, enabling hydrogen uptake and release under near-ambient conditions. As a result, LOHCs eliminate the need for high-pressure compression or cryogenic temperatures, offering inherent safety advantages and simplified handling. In addition, LOHCs possess higher volumetric energy densities than compressed hydrogen gas, which facilitates more efficient storage and long-distance transportation (Preuster et al. 2017). In LOHC systems, hydrogen is chemically bound to organic molecules during the hydrogenation process and later released through catalytic dehydrogenation when needed. Several carrier compounds, including methanol, dibenzyltoluene, and toluene, have demonstrated economic and logistical benefits compared with conventional compressed hydrogen storage (Makepeace et al. 2019). Owing to their operational safety, recyclability, and compatibility with existing fuel infrastructure, LOHCs offer an environmentally benign and efficient pathway for hydrogen storage and transport, thereby supporting the broader development of a hydrogen-based energy economy (Modisha et al. 2019). A wide variety of organic compounds—such as benzene, cyclohexane, toluene, and methylcyclohexane—have been investigated as LOHCs, highlighting the versatility and adaptability of this technology (Valentini et al. 2022). From an economic perspective, LOHCs have proven competitive, particularly for long-term storage and large-scale transportation of hydrogen. Their cost advantages and scalability have attracted significant industrial interest and investment from major stakeholders, including Umicore and Hyundai Motors (Niermann et al. 2019). At the system level, recent studies have combined experimental and modeling approaches to optimize LOHC performance. For instance, Rao et al. (2022) developed a detailed simulation of hydrogen production via dehydrogenation of perhydrodibenzyltoluene, integrating process modeling using DW-Sim and Python. Their one-dimensional reactor model accounted for both internal and external mass transfer resistances and examined the influence of key operational parameters, such as feed temperature (240–250 °C), wall temperature (350–380 °C), hydrogen burner efficiency, and reactor pressure (1.2 atm), on hydrogen yield and system performance. Further exploration of LOHC chemistry has been conducted by Rüde et al. (2022), who investigated the benzyltoluene/perhydrobenzyltoluene (H_0-BT/H_{12}-BT) system as a safe and efficient hydrogen storage medium. This LOHC pair demonstrated high volumetric hydrogen storage density, excellent stability over repeated hydrogenation–dehydrogenation cycles, and low viscosity, making it suitable for operation under cold conditions. Their study systematically analyzed reaction equilibria, physicochemical properties, and operating conditions, reporting hydrogenation–dehydrogenation equilibrium behavior over a temperature range of 220–360 °C at total pressures of 1×10^5 and 9×10^5 Pa. Innovative reactor concepts have also been proposed to enhance the practicality of LOHC systems. Badakhsh et al. (2022) introduced a thermally self-sustained reactor designed for hydrogen release from methylcyclohexane (MCH), integrating an H_2 burner, a heat-pipe-based dehydrogenator, and an advanced thermal management module. This configuration enabled carbon oxide–free hydrogen production suitable for onboard and mobile applications, achieving high reforming efficiency and MCH conversion. Such designs demonstrate the feasibility of compact and efficient LOHC-based hydrogen supply systems. Fundamentally, LOHC technology relies on catalytic hydrogenation and dehydrogenation reactions of hydrocarbon molecules, with system efficiency strongly influenced by the availability and quality of heat for dehydrogenation (Chu et al. 2023). Utilizing waste heat—such as that generated from fuel cells or industrial processes—has been identified as one of the most effective strategies for improving overall energy efficiency. By leveraging otherwise unused thermal energy, LOHC systems can significantly reduce operational energy penalties, reinforcing their technological and economic attractiveness. Consequently, LOHCs represent a robust and scalable solution for hydrogen storage and transport, with strong potential to support a sustainable and resilient hydrogen economy.

6.2.4 Chemisorption-Based Solid Hydrogen Storage

High-entropy alloys (HEAs) and hydrides belong to the class of chemisorption-based hydrogen storage materials, in which hydrogen is stored through the formation of chemical bonds with host elements. Depending on the nature of the hydrogen–host interaction, hydrides are traditionally categorized into three principal types: ionic (saline) hydrides, metallic hydrides, and covalent hydrides (Britannica 2024). The formation of hydrides follows a sequence of distinct physicochemical processes. Initially, molecular hydrogen (H_2) is adsorbed onto the material surface, where it dissociates into atomic hydrogen. These hydrogen atoms then diffuse from the surface into the bulk of the host lattice, where they ultimately form stable chemical bonds through chemisorption (Barthélémy et al. 2017). The incorporation of hydrogen into the host lattice induces a pronounced lattice expansion, typically on the order of 20%–30% compared to the original crystal volume (Møller et al. 2017). Among these steps, the dissociation of molecular hydrogen is particularly energy-intensive, requiring approximately 432 kJ mol^{-1}, and therefore plays a dominant role in governing the overall hydrogenation kinetics. Surface oxide or hydroxide layers further hinder hydrogen dissociation and subsequent diffusion, thereby slowing the chemisorption process. However, structural defects such as cracks, grain boundaries, and surface imperfections can act as preferential pathways and active sites for hydrogen uptake, partially offsetting these kinetic limitations (Aguey-Zinsou and Ares-Fernández 2010). The rate-limiting step in hydride formation is highly material-dependent, but it can often be alleviated through targeted material modification strategies. The incorporation of catalysts and elemental substitution have proven effective in reducing activation barriers, weakening metal–hydrogen bonds, and destabilizing hydride phases, thereby lowering the enthalpy required for hydrogen absorption and desorption. Despite their inherently high gravimetric and volumetric hydrogen storage capacities, hydrides frequently suffer from slow reaction kinetics, high hydrogen release temperatures, incomplete reversibility, and vulnerability to air poisoning. These limitations collectively constrain their large-scale and practical implementation in hydrogen storage systems. To address these challenges, a range of advanced material engineering approaches has been explored. Nanostructuring techniques and severe plastic deformation can generate a high density of defects, refined grain structures, and fresh surfaces, effectively increasing the number of active sites while shortening hydrogen diffusion paths (Klopčič et al. 2023). In parallel, air sensitivity can be mitigated through surface-engineering strategies such as nanoconfinement and polymer encapsulation. These protective coatings enable selective hydrogen transport while preventing the ingress of oxygen and moisture, thereby preserving structural integrity and sustaining performance over repeated hydrogenation–dehydrogenation cycles (Saeid et al. 2025). Hydride formation generally occurs either through the direct interaction of hydrogen gas with metals or via electrochemical hydrogen generation from water using appropriate metal electrodes (Usman 2022). The chemisorption process involved in metal hydride formation is typically exothermic (Graetz 2009). As a result, the release of hydrogen from hydrides requires an energy input comparable to that released during absorption, making hydrogen desorption thermodynamically favorable at elevated temperatures and reduced pressures, where hydrogen diffusivity within the material is enhanced (Usman 2022). In summary, chemisorption-based hydrogen storage materials can be broadly classified into metal hydrides, complex hydrides, and high-entropy alloys (HEAs). Each category presents distinct advantages in terms of storage capacity and thermodynamic stability, as well as specific challenges related to kinetics, reversibility, and operational conditions, underscoring the need for continued materials innovation to enable their practical deployment in hydrogen energy systems.

6.2.5 Metal Hydride

Metal hydrides are compounds formed through the direct bonding of hydrogen with metals or metal alloys, where the resulting metal–hydrogen bonds are primarily ionic or polar covalent in nature. Hydrogen storage in these materials occurs via chemical reactions and bond formation, which impart enhanced thermodynamic stability and intrinsic safety compared with physical storage methods (Aceves et al. 2006, 2010). Metal hydrides encompass a wide range of materials, including pure metals, intermetallic compounds, and metallic alloys, many of which exhibit good reversibility under moderate temperature and pressure conditions, along with high volumetric hydrogen storage capacities (Blackman et al. 2006). In recent years, metal hydrides have gained substantial industrial interest and acceptance as viable hydrogen storage media. Among them, intermetallic compounds have played a particularly important role, as they typically demonstrate higher hydrogen absorption capacities and more favorable sorption characteristics. The extensive availability of intermetallic hydrides with diverse compositions, coupled with the relative ease of forming solid solutions, has further accelerated the development and optimization of metal hydride-based storage systems (Rusman and Dahari 2016). Metal hydrides are commonly classified according to the type of intermetallic compound and its interaction with hydrogen atoms, with each class defined by a distinct stoichiometry and crystal structure. These compounds are generally represented by the empirical formula $A_mB_nH_x$, where elements A and B serve different functional roles. Element A typically exhibits a strong affinity for hydrogen and provides structural stability, while element B acts as a destabilizing component with comparatively lower hydrogen affinity, thereby facilitating reversible hydrogen absorption and desorption (Almeida Neto et al. 2022). A well-known example is $LaNi_5H_6$, an AB_5-type metal hydride that exhibits a gravimetric hydrogen storage capacity of approximately 1.37 wt% at 22 °C and 1 bar pressure (Almeida Neto et al. 2022). From the standpoint of solid-state hydrogen storage, metal hydrides are further categorized into several structural families based on their compositional ratios and crystal frameworks. These include A-type, AB-type, AB_2-type, AB_3-type, AB_5-type, and A_2B-type hydrides. Each category exhibits distinct thermodynamic, kinetic, and structural characteristics, which influence hydrogen storage capacity, operating conditions, and suitability for specific applications. Consequently, understanding these classifications is essential for tailoring metal hydride systems to meet the performance requirements of advanced hydrogen storage technologies.

6.2.5.1 A-Type of Metal Hydride

This category of materials is commonly referred to as binary hydrides, which primarily include light-metal hydride systems when the constituent metal element (A) is magnesium (Mg), lithium (Li), sodium (Na), calcium (Ca), or aluminum (Al). In these materials, hydrogen forms strong chemical bonds with the host metal atoms, resulting in highly stable hydride phases in which hydrogen is incorporated directly into the crystal lattice. Binary hydrides are particularly attractive for hydrogen storage owing to their low density, high hydrogen content, and generally reversible hydrogen absorption–desorption behavior. However, because these light metals lack d-electrons, the resulting metal–hydrogen interactions are relatively strong. As a consequence, hydrogen uptake and release typically require elevated temperatures and pressures to overcome the associated thermodynamic stability (Liu et al. 2023). Among binary hydrides, magnesium- and vanadium-based systems have been the subject of extensive investigation. Magnesium hydride (MgH_2) is especially prominent due to its high reversible gravimetric hydrogen storage capacity of approximately 7.6 wt% under atmospheric pressure conditions. Despite this advantage, its practical application is limited by a high hydrogen desorption temperature of around 300 °C, which poses challenges for integration into low-temperature energy systems (Kumar and Soren 2023). In contrast, vanadium-based hydrides operate under significantly milder conditions. Pure vanadium can reversibly store hydrogen with a gravimetric capacity of about 1 wt% at a desorption plateau pressure of 3–4 bar and temperatures near 40 °C, highlighting its favorable thermodynamic properties despite its lower storage capacity (Kumar et al. 2017). Other light-metal hydrides, such as lithium hydride (LiH) and sodium hydride (NaH), exhibit exceptionally high thermodynamic stability due to strong ionic bonding. As a result, these materials require decomposition temperatures exceeding 500 °C to release hydrogen, which severely restricts their practical utility for reversible hydrogen storage applications, despite their inherently high hydrogen content (Schüth et al. 2004). In addition to binary hydrides, AB-type intermetallic hydrides constitute another important class of hydrogen storage materials. These compounds are generally low-cost, lightweight, and capable of storing hydrogen at moderate gravimetric capacities of approximately 2 wt% under near-ambient operating conditions (Ha et al. 2023). AB-type hydrides are particularly valued for their good reversibility, rapid hydrogen absorption and desorption kinetics, and excellent cycling stability. Their storage performance can be further improved through compositional optimization, elemental doping, and crystal structure engineering, which allow fine-tuning of both thermodynamic stability and kinetic behavior (Bishnoi et al. 2024). Representative examples of AB-type hydrides include TiNi and TiFe intermetallic compounds. TiNi-based hydrides exhibit hydrogen storage capacities of up to approximately 1.2 wt% at temperatures around 127 °C and hydrogen pressures of 10 bar (Rusman and Dahari 2016). TiFe hydrides demonstrate comparatively higher performance, with a gravimetric hydrogen storage capacity of about 1.9 wt% at a desorption plateau pressure of 8.6 bar and near-ambient temperatures of approximately 30 °C. Owing to the natural abundance, low cost, and favorable mechanical properties of titanium and iron, TiFe-based hydrides have received considerable research attention and are widely regarded as promising candidates for scalable, economically viable hydrogen storage systems (Chen et al. 2021a, b).

6.2.5.2 AB_2-Type of Metal Hydride

AB_2-type hydrides are typically derived from combinations of rare-earth or early transition metals with nonmagnetic transition metals, resulting in materials that exhibit high structural stability alongside moderate hydrogen storage capacities. These intermetallic compounds readily undergo hydride phase formation under elevated hydrogen pressures and display a strong dependence of their thermodynamic behavior and absorption–desorption kinetics on compositional variations and impurity levels (Principi et al. 2009). In most AB_2 systems, the A-site element is titanium (Ti) or zirconium (Zr), while the B-site is occupied by transition metals such as vanadium (V), chromium (Cr), manganese (Mn), or iron (Fe) (Nagar et al. 2023). Common examples of AB_2-type hydrides include $TiMn_2$ and $ZrMn_2$. $TiMn_2$ exhibits a reversible gravimetric hydrogen storage capacity of approximately 1.9 wt% at a desorption pressure of about 8.5 bar and ambient temperature (25 °C). In contrast, $ZrMn_2$ demonstrates a gravimetric hydrogen density of roughly 1.77 wt% at a pressure of 1 bar and an elevated temperature of 167 °C (Rusman and

Dahari 2016). These materials are particularly attractive due to their relatively rapid hydrogen absorption and desorption kinetics, good cycling stability, and mechanical robustness. However, their overall hydrogen storage capacities remain moderate, which limits their suitability for applications requiring high gravimetric or volumetric energy densities.

6.2.5.3 AB_3-Type of Metal Hydride

In AB_3-type hydrides, element **A** is generally a rare-earth metal such as lanthanum, cerium, or yttrium, while element **B** consists of d-block transition metals. This class has attracted increasing interest, particularly for electrochemical and energy storage applications, due to their favorable hydrogen absorption characteristics and structural versatility (Denys et al. 2007). Emerging compounds within this group include $CaNi_3$, which exhibits a gravimetric hydrogen storage capacity of approximately 1.8 wt% at 0.4 bar and 20 °C. Although less extensively studied than AB_5-type systems, AB_3 hydrides show promise for applications requiring moderate operating conditions.

6.2.5.4 AB_5-Type of Metal Hydride

AB_5-type hydrides are among the most extensively investigated intermetallic hydrides and are typically composed of rare-earth f-block elements as A and d-block transition metals as **B** (Bishnoi et al. 2024). These materials generally crystallize in a hexagonal structure, which has been shown to be particularly favorable for hydrogen absorption. AB_5 intermetallic compounds are inherently brittle and tend to pulverize into fine powders during repeated hydrogenation-induced volumetric expansion (Nagar et al. 2023). This pulverization increases the available surface area, thereby facilitating faster hydrogen absorption and desorption kinetics. In addition, AB_5 hydrides exhibit excellent cyclic stability, retaining their performance over numerous hydrogenation–dehydrogenation cycles. Their low equilibrium pressures and rapid reaction rates further enhance their suitability for practical hydrogen storage applications (Rusman and Dahari 2016). $LaNi_5$-based hydrides are the most prominent examples in this class, exhibiting maximum gravimetric hydrogen storage capacities of approximately 1.5 wt% at operating temperatures around 70 °C and desorption pressures near 8.5 bar.

6.2.5.5 A_2B-Type of Metal Hydride

A_2B-type hydrides are intermetallic compounds formed between alkaline earth metals (element A) and transition metals (element B). Although this class of hydrides has received comparatively less attention than other hydride families, A_2B-type systems display several attractive characteristics, including relatively high gravimetric hydrogen storage capacities and generally good reversibility. Nevertheless, the formation of thermodynamically stable hydride phases often leads to elevated hydrogen desorption temperatures, which significantly constrains their practical applicability in low-temperature hydrogen storage systems (Bishnoi et al. 2024). Among the most extensively investigated A_2B-type hydrides are Mg_2Ni- and Ti_2Ni-based compounds, which have attracted considerable interest owing to their favorable crystal structures and hydrogen storage behavior (Rusman and Dahari 2016; Bishnoi et al. 2024). Mg_2Ni exhibits a gravimetric hydrogen storage capacity of approximately 3.59 wt% at a hydrogen pressure of 1 bar and a temperature of −18 °C, while Ti_2Ni demonstrates a gravimetric hydrogen density of about 1.93 wt% at 70 bar and 30 °C (Balcerzak et al. 2015). Despite these promising attributes, the relatively high operating pressures and temperatures required for hydrogen desorption remain a major barrier to widespread deployment. More broadly, although a wide range of metal hydrides exhibit diverse and potentially attractive hydrogen storage properties, their gravimetric storage capacities generally remain below the targets set by the U.S. Department of Energy (DOE) (Simanullang and Prost 2022). Most technologically mature metal hydride systems display gravimetric capacities below 3 wt%, which severely limits their suitability for practical hydrogen storage, particularly in mobile and onboard vehicular applications. For example, TiFe, $LaNi_5$, and $TiMn_{1.5}$ store approximately 1.89 wt%, 1.4 wt%, and 1.86 wt% hydrogen, respectively (Gasiorowski et al. 2004). In addition to limited gravimetric density, many metal hydrides suffer from complex activation procedures, high sensitivity to surface oxidation, and sluggish hydrogen absorption–desorption kinetics. These challenges, coupled with limited hydrogen diffusivity within the bulk material, render conventional metal hydrides currently impractical for onboard vehicular hydrogen storage systems (Rusman and Dahari 2016). Consequently, continued research is required to optimize their chemical composition, microstructure, and synthesis routes. Advanced strategies such as alloying, nanostructuring, surface modification, and catalyst incorporation are essential to overcome existing limitations and to fully realize the potential of metal hydrides for next-generation hydrogen storage applications.

6.3 Physisorption-Based Materials

In this class of hydrogen storage materials, hydrogen interacts with the host surface predominantly through physisorption, which is governed by weak van der Waals forces, electrostatic interactions, and limited orbital overlap. The typical binding energy between hydrogen molecules and the adsorbent surface ranges from approximately 4 to 10 kJ mol^{-1}, enabling fast and fully reversible adsorption–desorption kinetics. However, the weak nature of these interactions also results in poor hydrogen retention at elevated temperatures, thereby restricting practical storage under ambient conditions. Physisorption-based hydrogen storage relies heavily on the presence of porous architectures to maximize hydrogen uptake. High specific surface area and large pore volume increase the number of accessible adsorption sites available for hydrogen molecules. While higher surface area is often associated with improved hydrogen storage capacity, experimental studies have demonstrated that storage performance is also critically influenced by pore size and pore size distribution, rather than surface area alone. This phenomenon is clearly illustrated by investigations of chitosan-derived porous carbon materials possessing similar specific surface areas of approximately 3300 $m^2 g^{-1}$, yet exhibiting distinct pore size distributions. These materials were classified according to dominant pore types, such as type A pores (~2.5 nm). At 1 bar and −196 °C, all samples displayed comparable hydrogen storage capacities. However, under elevated pressures of 20 bar, materials featuring optimized combinations of type A and type B pores achieved a gravimetric hydrogen storage capacity of 6.9 wt%, whereas samples dominated by ultramicropores (type A) reached only 6.2 wt%, and those containing predominantly type C pores exhibited a lower capacity of 5.6 wt% (Huang et al. 2018). These results clearly demonstrate that even for materials with identical surface areas, hydrogen uptake can vary substantially depending on the optimization of pore size distribution. Porous adsorbents are conventionally classified based on pore size into micropores (<2 nm), mesopores (2–50 nm), and macropores (>50 nm). In the context of hydrogen adsorption, micropores play a critical role in providing high surface area and strong confinement effects that enhance storage capacity, particularly at cryogenic temperatures. Mesopores facilitate rapid hydrogen diffusion and improve adsorption–desorption kinetics, while macropores, although contributing less directly to storage capacity, improve structural stability and mass transport during repeated hydrogenation and dehydrogenation cycles. To capitalize on the advantages of each pore regime, advanced adsorbents have been engineered with interconnected micro-, meso-, and macroporous networks, commonly referred to as hierarchically porous materials. Such architectures consistently exhibit superior hydrogen storage performance compared to single-scale porous materials of identical chemical composition under comparable operating conditions (Wu et al. 2020). Recent studies on hierarchically porous activated carbons synthesized via CO_2 activation with carefully optimized activation durations to generate ultramicroporous structures have demonstrated significant improvements in hydrogen storage performance. For instance, a gravimetric hydrogen storage capacity of 2.36 wt% at 1 bar and −196 °C has been reported, surpassing that of commercial activated carbons, which typically store less than 2 wt% hydrogen under similar conditions (Conte et al. 2023). These findings underscore the considerable potential of hierarchically porous materials for physisorption-based hydrogen storage applications. Despite these advances, the intrinsically low binding energy between hydrogen molecules and the adsorbent surface remains a fundamental limitation, particularly for achieving efficient hydrogen storage at near-ambient temperatures. In this section, various classes of physisorption-based hydrogen storage materials—including metal–organic frameworks (MOFs), zeolites, carbon-based materials, silica-based adsorbents, and polymeric materials—have been discussed in detail, along with their

Table 6.1 Comprehensive summary of hydrogen storage technologies and their key characteristics

No.	Hydrogen storage technology	Principle/ mechanism	Advantages	Challenges/ limitations	Reference
1	Compressed hydrogen storage	Physical storage of hydrogen via high-pressure compression in composite cylinders	Commercially mature; fast refueling; simple operation	High compression energy; safety concerns; bulky tanks	Rivard et al. (2019)
2	Compressed hydrogen storage	Hydrogen is stored under high pressure (350–700 bar) in specially designed tanks	Mature technology; fast refueling; suitable for transport applications	Low volumetric density; high energy cost for compression; safety concerns	Alam et al. (2025)
3	Liquid hydrogen storage	Hydrogen is cooled to cryogenic temperatures (−253 °C) and stored as a liquid	High energy density; applicable for aerospace and long-range transport	High liquefaction cost; boil-off losses; insulation challenges	Singh et al. (2025)
4	Metal hydride storage	Hydrogen reacts reversibly with metals/alloys forming metal hydrides	Safe and compact; moderate storage pressure	Slow kinetics; heavy materials; temperature control required	Bhatnagar (2025)
5	Chemical hydrogen storage	Hydrogen is stored in chemical compounds (e.g., ammonia, formic acid)	High volumetric density; liquid-phase storage possible	Requires complex regeneration and catalysts; limited reversibility	Farghaly et al. (2025)
6	Adsorption storage (MOFs, activated carbon)	Hydrogen molecules adsorb on porous materials at low temperature/ pressure	Lightweight; tunable surface area; potential for high gravimetric capacity	Low adsorption capacity at ambient conditions; costly synthesis of materials	La Monaca et al. (2025)

respective advantages and limitations. A comparative overview of these materials is provided in Table 6.1, summarizing their hydrogen storage capacities, operating conditions, key benefits, and associated challenges.

6.3.1 Zeolites

Zeolites are crystalline aluminosilicate materials with a general chemical formula of$MX/Z[(AlO_2)X(SiO_2)Y]\cdot mH_2O$, where *M* denotes an exchangeable alkali or alkaline earth metal cation. Their well-defined porous framework and high crystallinity make them potential candidates for hydrogen storage applications (Prabhukhot et al. 2016). However, similar to other physisorption-based materials, zeolites exhibit extremely low hydrogen storage capacities at ambient conditions, typically below 0.1 wt%. Since hydrogen adsorption in zeolites is strongly temperature dependent, lowering the operating temperature substantially enhances storage performance. Experimental studies have demonstrated that cooling from room temperature to −196 °C leads to a significant increase in hydrogen uptake. For instance, NaY and CaX zeolites have been reported to achieve gravimetric hydrogen storage capacities of 1.81 wt% and 2.19 wt%, respectively, at −196 °C under a pressure of 15 bar. Despite this improvement, the maximum theoretical hydrogen storage capacity of zeolites remains limited to approximately 2.86 wt% (Vitillo et al. 2005). This intrinsically low gravimetric capacity represents a major limitation for zeolite-based hydrogen storage systems, as their performance remains far below the targets set by the U.S. Department of Energy. Furthermore, the dependence on cryogenic temperatures severely restricts their practical applicability, particularly for onboard and large-scale energy storage systems.

6.3.2 Carbonaceous Materials

Carbon is a unique element in the periodic table, characterized by the electronic configuration $1s^2\ 2s^2\ 2p^2$, which allows it to form a diverse range of covalent bonds. Its low atomic mass, combined with the ability to exist in multiple allotropic forms, provides exceptional flexibility for designing advanced carbon-based materials with tailored properties. From the perspective of hydrogen storage, carbon materials are particularly appealing owing to their high specific surface area, adjustable pore structures, and large surface-to-volume ratios, all of which are highly favorable for adsorption-driven hydrogen storage mechanisms (Soni et al. 2024). Consequently, a wide variety of carbonaceous materials including carbon nanotubes (CNTs), graphene, and fullerenes have been extensively explored as potential solid-state hydrogen storage media. Beyond their intrinsic hydrogen adsorption capabilities, carbon-based materials are also commonly employed as functional additives in metal hydride systems, where they play a crucial role in improving hydrogen storage performance. Specifically, the incorporation of carbon materials can enhance hydrogen absorption–desorption kinetics, improve reversibility, and mitigate issues related to particle agglomeration and surface oxidation (Desai et al. 2023).

(a) *Carbon nanotubes (CNTs)*

Carbon nanotubes (CNTs) are among the most extensively studied carbonaceous materials for hydrogen storage. Their cylindrical nanostructure endows them with exceptional mechanical strength, high thermal stability, low density, and large specific surface area. Similar to other physisorption-based materials, CNTs exhibit improved hydrogen uptake at cryogenic temperatures and elevated pressures. In practical applications, CNTs are often incorporated as additives to improve hydrogen storage capacity and reaction kinetics in composite systems. Despite these advantages, the hydrogen adsorption mechanism in CNTs remains unclear, and reported storage capacities often suffer from poor reproducibility, which limits their independent and reliable application (Simanullang and Prost 2022).

(b) *Graphene*

Graphene is a two-dimensional allotrope of carbon renowned for its exceptional physicochemical properties, including high thermal conductivity, large specific surface area, mechanical flexibility, superior strength, chemical inertness, and non-toxicity. Structurally, graphene consists of a single atomic layer of sp^2-hybridized carbon atoms arranged in a hexagonal honeycomb lattice, with each carbon atom covalently bonded to three neighboring atoms (Simanullang and Prost 2022). Owing to these characteristics, graphene has attracted considerable interest as a potential hydrogen storage material, and its storage behavior has been extensively investigated through theoretical and experimental studies. A number of theoretical investigations have predicted promising hydrogen storage capacities for graphene-based systems. For example, a theoretical gravimetric hydrogen storage capacity as high as 8.3 wt% has been proposed for hydrogen chemisorption on graphene surfaces (Tozzini and Pellegrini 2013). Other studies have suggested that bilayer graphene can store hydrogen predominantly through physisorption, achieving capacities of approximately 2–3 wt% at 50 bar and 3–4 wt% at 100 bar (Patchkovskii et al. 2005). Hydrogen adsorption on graphene is strongly temperature-dependent; one theoretical analysis predicted hydrogen coverages exceeding 40% at 100 bar and −173 °C, whereas increasing the temperature resulted in a sharp decline in adsorption due to enhanced molecular mobility and weakened physisorption interactions (Simanullang and Prost 2022). Further computational studies have indicated that chemical modification can substantially enhance graphene's hydrogen storage performance. In particular, lithium-decorated graphene has been theoretically predicted to achieve gravimetric hydrogen storage capacities as high as 12.8 wt%, owing to enhanced polarization-induced interactions between hydrogen molecules and the metal-decorated surface (Ataca et al. 2008). Experimental efforts have also demonstrated the potential of graphene-based materials. For instance, nickel-catalyzed porous graphene has exhibited a hydrogen storage capacity of 4.22 wt% at 5 bar and −196 °C, although this capacity decreased significantly to 1.95 wt% at room temperature, underscoring the persistent challenge of hydrogen retention under ambient conditions. Notably, experimental work conducted at IIT Madras on Pd_3Co-decorated boron- and nitrogen-doped graphene composites reported a gravimetric hydrogen storage capacity of 4.6 wt% at 30 bar under ambient temperature conditions, demonstrating the effectiveness of synergistic metal decoration and heteroatom doping in improving room-temperature hydrogen storage performance (Simanullang and Prost 2022). Beyond its role as a standalone storage medium, graphene is also extensively employed as an additive in metal hydride systems, where it contributes to hydride destabilization, improved hydrogen diffusion, and enhanced absorption–desorption kinetics.

(c) *Fullerenes*

Fullerenes are carbon allotropes composed of closed, three-dimensional cage-like structures, commonly forming hollow spheres or ellipsoids (Simanullang and Prost 2022). Theoretical studies suggest that hydrogen can be encapsulated within these cage structures. For example, C_{60} fullerenes are predicted to accommodate up to 58 hydrogen atoms, while larger fullerenes such as C_{720} could theoretically encapsulate up to 800 hydrogen atoms (Pupysheva et al. 2008). Experimental investigations have shown that fullerenes exhibit very slow hydrogen sorption kinetics. In one study, C_{60} initially demonstrated a hydrogen storage capacity of only 0.6 wt% at 200 °C and 120 bar after 10 h. However, after prolonged exposure of approximately 4500 h, the storage capacity increased significantly to 8.2 wt%. Although this indicates high storage potential, the extremely long absorption time severely limits practical applicability (Simanullang and Prost 2022). Doping has been shown to improve the hydrogen storage performance of fullerenes. For instance, lithium-doped fullerenes (fulleranes) can reversibly desorb up to 5 wt% hydrogen at a substantially lower temperature of around 270 °C, compared to 500–600 °C for conventional hydrogenated fullerenes (Teprovich et al. 2012). Additionally, composites formed by mixing $LiBH_4$ with C_{60} in a 70:30 wt% ratio have demonstrated improved hydrogen adsorption conditions, operating at 10 bar and 150 °C (Ward et al. 2013). Consequently, fullerenes such as C_{60} are primarily employed as additives to enhance desorption kinetics, reversibility, and operating conditions rather than as standalone hydrogen storage materials.

6.3.3 Silica

Silica-based mesoporous materials belong to the class of physisorption-based hydrogen storage media and are distinguished by their well-defined pore structures, high thermal and chemical stability, and structural tunability, which make them attractive for hydrogen storage applications (Gebretatios et al. 2024). These materials are relatively easy to synthesize, cost-effective, and environmentally benign (Conte et al. 2023). At an industrial scale, silica precursors are commonly derived from abundant quartz rock minerals, further supporting their economic feasibility (Errington et al. 2023). Despite these advantages, the hydrogen storage capacity of pristine silica-based materials is inherently low. For example, hexagonal mesoporous silica (HMS) exhibits a gravimetric hydrogen storage capacity of approximately 0.5 wt% at −196 °C and 1.1 bar. Surface functionalization has been shown to enhance hydrogen uptake; incorporation of vinyl trimethoxysilane into HMS increased the gravimetric capacity to 0.72 wt%, attributed to stronger interactions between hydrogen molecules and the modified silica surface (Owens et al. 2015). Similarly, Mobil Composition of Matter-41 (MCM-41) demonstrates a hydrogen storage capacity of 0.44 wt% at −196 °C, which increases to 0.53 wt% upon nickel doping, highlighting the role of metal additives in improving adsorption performance. In contrast, silica synthesized using 3-aminopropyl triethoxysilane as a precursor exhibited a comparatively lower hydrogen uptake of 0.23 wt% (Conte et al. 2023). Hydrogen adsorption in silica-based mesoporous materials is strongly temperature dependent, with acceptable storage capacities achieved only under cryogenic conditions. At ambient temperatures, their gravimetric hydrogen storage capacity remains far below the targets established by the U.S. Department of Energy, limiting their practical application. To address this challenge, several studies have explored the incorporation of transition metals such as Ni, Pd, Co, and Fe to promote the hydrogen spillover effect (Dündar-Tekkaya and Yürüm 2016). Spillover involves the dissociation of hydrogen molecules on metal sites followed by migration of atomic hydrogen onto the silica surface, thereby enhancing overall storage capacity and improving adsorption–desorption kinetics. Nevertheless, further research is required to significantly improve hydrogen storage performance at room temperature, which is essential for the sustainable deployment of silica-based materials in practical hydrogen storage systems.

6.3.4 Polymers

The hydrogen storage performance of polymers is largely governed by their specific surface area and pore structure, as hydrogen is stored primarily through physisorption. Consequently, the enthalpy of adsorption in polymer-based materials is relatively low, typically in the range of 4–7 kJ mol^{-1}. For effective and reversible hydrogen adsorption at room temperature, an adsorption enthalpy in the range of approximately 13–20 kJ mol^{-1} is generally required. As a result, most polymer-based hydrogen storage systems exhibit meaningful hydrogen uptake only under cryogenic conditions. Polymers of intrinsic microporosity (PIMs) are among the most extensively studied polymeric materials for hydrogen storage. For example, PIM-1 has been reported to exhibit a reversible gravimetric hydrogen storage capacity of approximately 0.95–1.04 wt% at −196 °C and 1 bar, while PIM-7 demonstrates a gravimetric capacity of around 1 wt% under similar conditions. In general, most PIM-based materials show gravimetric hydrogen storage capacities below 2 wt% at cryogenic temperatures (Ramimoghadam et al. 2016). This dependence on extremely low temperatures presents a major barrier to their practical deployment, as maintaining such conditions is energy-intensive and often impractical for real-world applications. When evaluated against the hydrogen storage targets set by the U.S. Department of Energy, polymer-based materials fall significantly short in terms of gravimetric capacity. The primary limitations of polymers include their low hydrogen uptake, reliance on cryogenic operating temperatures, and comparatively low accessible porous surface area. Despite these shortcomings, polymers play an important complementary role in hydrogen storage systems. When combined with metal hydrides, polymers can enhance dimensional stability, improve resistance to

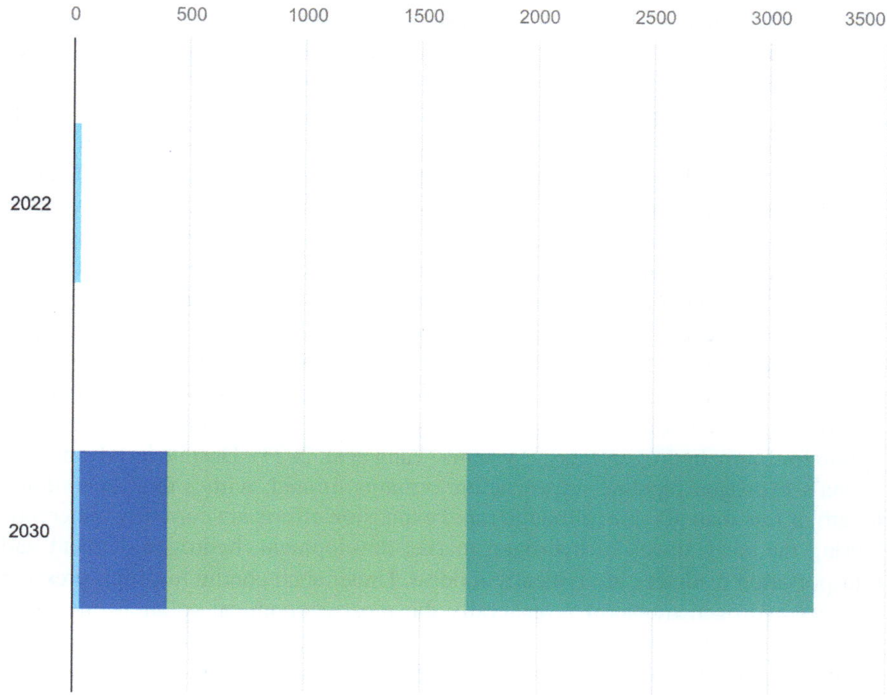

Fig. 6.1 Potential underground hydrogen storage capacity in Europe by 2030 based on announced projects (GWh). (IEA 2024)

air and moisture, and increase overall processability of the storage system (Almeida Neto et al. 2022). As such, polymer–hydride composites represent a promising strategy to leverage the structural and protective advantages of polymers while mitigating the intrinsic limitations of polymer-only hydrogen storage materials. The potential underground hydrogen storage capacity in Europe by 2030, expressed in GWh, is summarized in Fig. 6.1 based on announced projects. The figure indicates a significant expansion of large-scale subsurface hydrogen storage, reflecting growing interest in seasonal energy balancing and hydrogen-based energy systems across Europe (IEA 2024).

6.4 Hydrogen Transportation Technologies

Hydrogen transportation strategies are strongly influenced by both the physical storage state of hydrogen and the quantity to be delivered. At present, hydrogen is primarily transported in three forms: gaseous hydrogen, liquid hydrogen, and solid-state hydrogen storage systems (Ma et al. 2008). Among these options, the most technically mature and widely implemented methods for large-scale hydrogen transportation include high-pressure gaseous hydrogen trailers, liquid hydrogen tankers, and pipeline-based hydrogen transport. High-pressure gaseous hydrogen transportation using tube trailers currently represents the dominant hydrogen delivery mode in China, particularly for short- to medium-distance distribution. This approach benefits from relatively low initial infrastructure requirements and operational simplicity. Liquid hydrogen (LH_2) tankers, in contrast, offer significantly higher volumetric energy density and high hydrogen purity, making them suitable for transporting larger quantities over longer distances. However, these advantages are counterbalanced by substantial drawbacks, including the high energy demand associated with hydrogen liquefaction, significant capital investment in cryogenic infrastructure, and elevated operational and maintenance costs. Pipeline-based hydrogen transportation—including both dedicated pure hydrogen pipelines and pipelines that transport hydrogen blended with natural gas—enables continuous, large-scale, long-distance hydrogen delivery at comparatively low unit costs, making it an attractive option for future hydrogen energy systems (Yang et al. 2023). Despite its advantages, China's hydrogen pipeline infrastructure remains limited, with a total operational length of less than 400 km, although rapid expansion efforts are currently underway. During the early stages of hydrogen market development, hydrogen demand and transportation distances are typically modest. Under such conditions, high-pressure gaseous hydrogen transport is generally the most economical solution due to its lower conversion requirements and minimal infrastructure investment, and it is commonly implemented using onboard hydrogen storage cylinders. As the hydrogen economy transitions into a more mature phase, transportation distances increase and demand scales up, leading to the growing adoption of both compressed gaseous hydrogen and cryogenic liquid hydrogen transport options, often supported by advancements in low-temperature liquefaction technologies. In the long term, continued progress in high-density, high-safety hydrogen storage technologies is expected to facilitate the development of an extensive hydrogen pipeline network. This future infrastructure will be supported by comprehensive technical standards governing hydrogen storage and transportation, encompassing solid-state hydrogen storage systems, liquid organic hydrogen carriers (LOHCs), and pipeline distribution regulations. Figure 6.2 presents projected hydrogen production and transportation costs to Germany in 2035 (USD kg^{-1} H_2). This analysis compares levelized hydrogen supply costs from different production regions and transport pathways, incorporating cost components associated with hydrogen production, conversion processes, long-distance transport via pipelines or maritime shipping, reconversion losses, and terminal handling. All values are expressed in constant 2035 USD per kilogram of delivered hydrogen and account for expected technology learning effects and infrastructure utilization assumptions (IEA 2025).

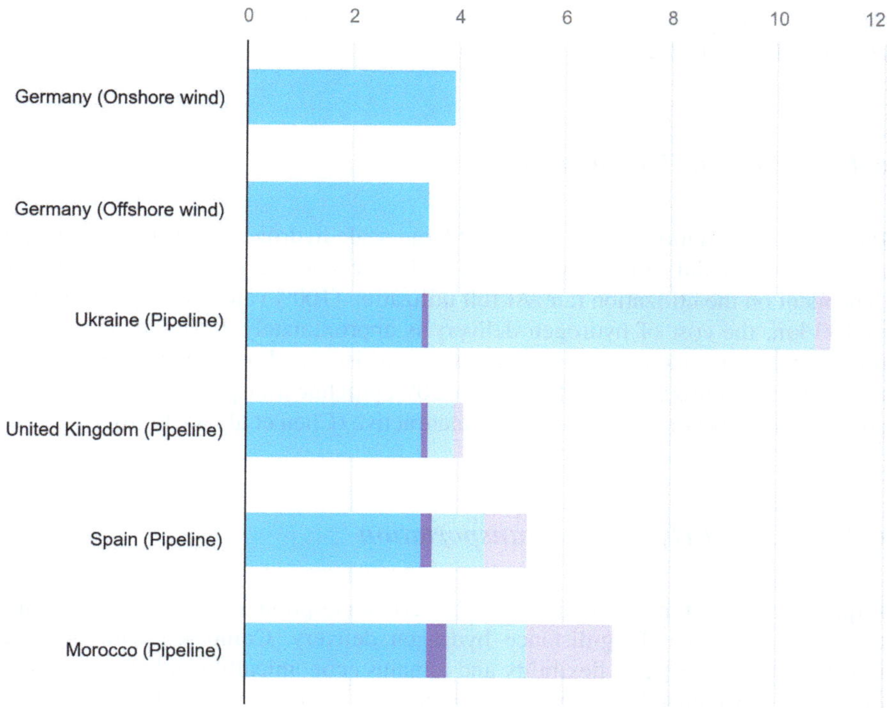

Fig. 6.2 Cost implications of international hydrogen supply to Germany in 2035 (USD kg^{-1} H_2). (IEA 2025)

6.4.1 Long Pipe Trailer Transportation

Hydrogen transport using long-tube trailers operating at 20 MPa is primarily suitable for short-distance delivery, typically within a radius of about 200 km. An analysis of the cost structure shows that disassembly operations, labor, and fuel consumption are the dominant contributors to overall transportation costs. Consequently, this mode of transport is economically viable mainly for applications involving limited transport distances and relatively small hydrogen volumes. At present, domestic regulations restrict the maximum operating pressure of long-tube trailers to 20 MPa, whereas internationally, trailers designed to operate at pressures up to 50 MPa have already been deployed. If domestic standards governing storage and transportation pressures were relaxed, a greater amount of hydrogen could be transported within the same trailer volume, leading to a substantial reduction in unit transportation costs. For example, at a transport distance of 100 km, the cost of hydrogen delivery using 20 MPa and 50 MPa long-tube trailers is approximately 8.7 yuan/kg and 5.6 yuan/kg, respectively, with the 50 MPa option reducing costs to about 65% of those associated with 20 MPa operation. Accordingly, it is anticipated that 50 MPa long-tube trailers will be widely adopted in China by 2025 (Chen et al. 2021a, b; Riera et al. 2023).

6.4.2 *Pipeline Transportation*

Pipeline transportation is well suited for large-scale hydrogen delivery; however, it requires substantial upfront investment and its economic performance is highly dependent on the utilization rate. At full utilization (100%) and a transport distance of 100 km, the cost of hydrogen delivery is approximately 1.20 yuan/kg. As the transport distance increases to 500 km, the cost rises to about 3.0 yuan/kg. In contrast, when the utilization rate falls below 20%, pipeline transportation loses its cost advantage and becomes economically unattractive (Chen et al. 2021a, b).

6.4.3 *Liquid Hydrogen Transportation*

Liquid hydrogen transportation is a highly efficient option and is particularly suitable for large-scale, long-distance hydrogen delivery. Compared with pipeline transport, it offers greater flexibility and remains economically viable for distances of up to approximately 2000 km. Transporting liquid hydrogen by tank trucks allows a higher amount of hydrogen to be delivered per trip compared to gaseous transport, thereby improving cost-effectiveness. However, even large-capacity liquid hydrogen tank trucks typically carry only a few tons of hydrogen per journey. In contrast, maritime transportation using liquid hydrogen carrier ships can deliver hundreds of tons or more in a single trip, making sea transport a significantly more efficient option than road-based tank trucks for long-distance, large-volume hydrogen transportation (Chen et al. 2021a, b).

6.5 Conclusion

In conclusion, this chapter highlights the significant progress made in hydrogen storage technologies and materials, which are fundamental to the large-scale deployment of hydrogen as a clean and sustainable energy carrier. With the global population projected to reach 9.7 billion by 2050, the demand for reliable and environmentally benign energy solutions is becoming increasingly critical. Although investments in renewable energy sources such as solar, wind, and hydropower continue to grow, the intermittency of these resources necessitates the development of efficient and scalable energy storage systems, particularly to reduce reliance on fossil fuels and their associated environmental impacts. Hydrogen storage stands out as a promising pathway due to hydrogen's high energy density and its potential role in deep decarbonization across multiple sectors. Physical storage methods, including compressed gaseous hydrogen and liquid hydrogen storage, have reached varying levels of technological maturity. Compressed hydrogen storage is already well established, especially in transportation applications such as fuel cell electric vehicles. Liquid hydrogen storage offers significantly higher energy density; however, it poses substantial challenges related to cryogenic operation, safety, energy consumption, and infrastructure requirements. Material-based hydrogen storage strategies, encompassing metal hydrides, complex hydrides, and carbon-based materials, present attractive alternatives to conventional physical storage. These systems store hydrogen through absorption or adsorption mechanisms and release it via thermal or catalytic processes. Ongoing research is focused on optimizing material properties such as structure, porosity, surface area, storage capacity, stability, and safety as well as developing efficient catalysts to improve hydrogen uptake and release kinetics. In this context, computational chemistry, high-throughput screening, and machine learning have emerged as powerful tools for accelerating material discovery and optimization, significantly advancing hydrogen storage research. Despite these advances, several challenges remain, including limited volumetric and gravimetric energy densities, safety concerns, high costs, and the need for robust storage and transportation infrastructure. Addressing these issues is essential to fully harness hydrogen's potential in transportation, power generation, and industrial applications. Future research directions are expected to emphasize advanced material design, nanostructuring, integrated and hybrid storage systems, improved characterization techniques, sustainable hydrogen production pathways, and enhanced safety strategies. The integration of data-driven and computational approaches with experimental research will further accelerate innovation. Collectively, these efforts will play a crucial role in improving the efficiency, safety, and practicality of hydrogen storage technologies, thereby enabling the widespread adoption of hydrogen as a cornerstone of a sustainable global energy system.

References

Aceves SM, Berry GD, Martinez-Frias J, Espinosa-Loza F (2006) Vehicular storage of hydrogen in insulated pressure vessels. Int J Hydrog Energy 31:2274–2283

Aceves SM, Espinosa-Loza F, Ledesma-Orozco E, Ross TO, Weisberg AH, Brunner TC, Kircher O (2010) High-density automotive hydrogen storage with cryogenic capable pressure vessels. Int J Hydrog Energy 35:1219–1226

Aguey-Zinsou KF, Ares-Fernández JR (2010) Hydrogen in magnesium: new perspectives toward functional stores. Energy Environ Sci 3:526–543. https://doi.org/10.1039/b921

Ahluwalia RK, Peng JK (2008) Dynamics of cryogenic hydrogen storage in insulated pressure vessels for automotive applications. Int J Hydrog Energy 33(17):4622–4633. https://doi.org/10.1016/j.ijhydene.2008.05.090

Alam T, Haque I, Ali N (2025) Editorial: Assessment of renewable energy systems for energy conversion and storage. Front Energy Res 13:1663365. https://doi.org/10.3389/fenrg.2025.1663365

Almeida Neto GRD, Matheus FH, Gonçalves Beatrice CA, Leiva DR, Pessan LA (2022) Fundamentals and recent advances in polymer composites with hydrideforming metals for hydrogen storage applications. Int J Hydrog Energy 47:34139–34164. https://doi.org/10.1016/j.ijhydene.2022.08.004

Ancona MA, Catena F, Ferrari F (2023) Optimal design and management for hydrogen and renewables based hybrid storage micro-grids. Int J Hydrog Energy 48(54):20844–20860. https://doi.org/10.1016/j.ijhydene.2022.10.204

Asghar U, Rafiq S, Anwar A, Iqbal T, Ahmed A, Jamil F et al (2021) Review on the progress in emission control technologies for the abatement of CO, SO and NO from fuel combustion. J Environ Chem Eng 9:106064

Ataca C, Aktürk E, Ciraci S, Ustunel H (2008) High-capacity hydrogen storage by metallized graphene. Appl Phys Lett 93:043123

Badakhsh A, Song D, Moon S, Jeong H, Sohn H, Woo Nam S, Soon Kim P, Hui Seo J, Kim Y, Lee J, Woo Choung J, Kim Y (2022) COX-free LOHC dehydrogenation in a heatpipe reformer highly integrated with a hydrogen burner. Chem Eng J 449:137679. https://doi.org/10.1016/j.cej.2022.137679

Balcerzak M, Jakubowicz J, Kachlicki T, Jurczyk M (2015) Hydrogenation properties of nanostructured Ti2Ni-based alloys and nanocomposites. J Power Sources 280:435–445. https://doi.org/10.1016/j.jpowsour.2015.01.135

Barthélémy H, Weber M, Barbier F (2017) Hydrogen storage: recent improvements and industrial perspectives. Int J Hydrog Energy 42:7254–7262

Bhatnagar A (2025) ICEAM 2023 and ICHEAM-2024. Int J Mater Res. https://doi.org/10.1515/ijmr-2025-3005

Bishnoi A, Pati S, Sharma P (2024) Architectural design of metal hydrides to improve the hydrogen storage characteristics. J Power Sources 608:234609. https://doi.org/10.1016/j.jpowsour.2024.234

Blackman JM, Patrick JW, Arenillas A, Shi W, Snape CE (2006) Activation of carbon nanofibres for hydrogen storage. Carbon N Y 44:1376–1385

Bosu S, Rajamohan N (2024) Recent advancements in hydrogen storage—comparative review on methods, operating conditions and challenges. Int J Hydrog Energy 52:352–370. https://doi.org/10.1016/j.ijhydene.2023.01.344

Britannica (2024) https://www.britannica.com/science/hydride. Accessed 24 Nov 2024

Chalk SG, Miller JF (2006) Key challenges and recent progress in batteries, fuel cells, and hydrogen storage for clean energy systems. J Power Sources 159(1):73–80. https://doi.org/10.1016/j.jpowsour.2006.04.058

Chen Z, Ma Z, Zheng J, Li X, Akiba E, Li H-W (2021a) Perspectives and challenges of hydrogen storage in solid-state hydrides. Chin J Chem Eng 29:1–12

Chen Y, Zhang X, Zhang R (2021b) Current status and future prospects of China's hydrogen energy industry chain. New Ind 11(4):176–180

Chu C, Wu K, Luo B, Cao Q, Zhang H (2023) Hydrogen storage by liquid organic hydrogen carriers: catalyst, renewable carrier, and technology—a review. Carbon Resour Convers 6(4):334–351. https://doi.org/10.1016/j.crcon.2023.03.007

Conte G, Simari C, Putz A-M, Bonaventura CP, Porto M, Desiderio G, Agostino RG, Policicchio A (2023) Hydrogen storage and mobility determined by NMR to various organically functionalized porous silica synthetized by using the post-grafting method. Int J Hydrog Energy 48:27319–27329. https://doi.org/10.1016/j.ijhydene.2023.03.350

Denys RV, Riabov B, Yartys VA, Delaplane RG, Sato M (2007) Hydrogen storage properties and structure of La1− xMgx (Ni1− yMny) 3 intermetallics and their hydrides. J Alloys Compd 446:166–172

Desai FJ, Uddin MN, Rahman MM, Asmatulu R (2023) A critical review on improving hydrogen storage properties of metal hydride via nanostructuring and integrating carbonaceous materials. Int J Hydrog Energy 48:29256–29294. https://doi.org/10.1016/j.ijhydene.2023.04.029

Dündar-Tekkaya E, Yürüm Y (2016) Mesoporous MCM-41 material for hydrogen storage: a short review. Int J Hydrog Energy 41:9789–9795. https://doi.org/10.1016/j.ijhydene.2016.03.050

Elberry AM, Thakur J, Santasalo-Aarnio A, Larmi M (2021) Large-scale compressed hydrogen storage as part of renewable electricity storage systems. Int J Hydrog Energy 46(29):15671–15690. https://doi.org/10.1016/j.ijhydene.2021

Errington E, Guo M, Heng JYY (2023) Synthetic amorphous silica: environmental impacts of current industry and the benefit of biomass-derived silica. Green Chem 25:4244–4259

Farghaly AA, Onajah S, Lalley M, Kort-Kamp W, Kreider ME, Alia SM, Hamlyn R, Yu H, Zelenay P, Myers DJ (2025) High-throughput experimentation for accelerated discovery of PGM-free catalysts for anion-exchange membrane water electrolyzers. In: Electrochemical society meeting abstracts 248 (No. 39). The Electrochemical Society, Inc, pp 1892–1892

Gasiorowski A, Iwasieczko W, Skoryna D, Drulis H, Jurczyk M (2004) Hydriding properties of nanocrystalline Mg2-xMxNi alloys synthesized by mechanical alloying (M = Mn, Al). J Alloys Compd 364:283–288. https://doi.org/10.1016/S0925-8388(03)00544-9

Gebretatios AG, Banat F, Witoon T, Cheng CK (2024) Synthesis of sustainable rice husk ash-derived nickel-decorated MCM-41 and SBA-15 mesoporous silica materials for hydrogen storage. Int J Hydrog Energy 51:255–266. https://doi.org/10.1016/j.ijhydene.2023.11.154

Graetz J (2009) New approaches to hydrogen storage. Chem Soc Rev 38:73–82

Ha T, Kim JH, Sun C, Lee YS, Kim DI, Suh JY, Jang JI, Lee J, Kim Y, Shim JH (2023) Crucial role of Ce particles during initial hydrogen absorption of AB-type hydrogen storage alloys. Nano Energy 112. https://doi.org/10.1016/j.nanoen.202

Hassan A, Ilyas SZ, Jalil A, Ullah Z (2021) Monetization of the environmental damage caused by fossil fuels. Environ Sci Pollut Res Int 28:21204–21211

Hassan Q, Sameen AZ, Salman HM, Jaszczur M, Al-Jiboory AK (2023) Hydrogen energy future: advancements in storage technologies and implications for sustainability. J Energy Storage 72:108404

Huang J, Liang Y, Dong H, Hu H, Yu P, Peng L, Zheng M, Xiao Y, Liu Y (2018) Revealing contribution of pore size to high hydrogen storage capacity. Int J Hydrog Energy 43:18077–18082

IEA (2024) Potential underground hydrogen storage capacity based on project announcements in Europe, 2030. IEA, Paris. https://www.iea.org/data-and-statistics/charts/potential-underground-hydrogen-storage-capacity-based-on-project-announcements-in-europe-2030

IEA (2025) Hydrogen production and transport costs to Germany, 2035. IEA, Paris. https://www.iea.org/data-and-statistics/charts/hydrogen-production-and-transport-costs-to-Germany-2035

Jahanbakhsh A, Louis Potapov-Crighton A, Mosallanezhad A, Tohidi Kaloorazi N, Maroto-Valer MM (2024) Underground hydrogen storage: a UK perspective. Renew Sust Energ Rev 189:114001

Jones MW, Peters GP, Gasser T, Andrew RM, Schwingshackl C, Gütschow J et al (2023) National contributions to climate change due to historical emissions of carbon dioxide, methane, and nitrous oxide since 1850. Sci Data 10:155

Kabir M, Habiba UE, Khan W, Shah A, Rahim S, Rios-Escalante PRD et al (2023) Climate change due to increasing concentration of carbon dioxide and its impacts on environment in 21st century; a mini review. J King Saud Univ Sci 35:102693

Klopčič N, Grimmer I, Winkler F, Sartory M, Trattner A (2023) A review on metal hydride materials for hydrogen storage. J Energy Storage 72:108456. https://doi.org/10.1016/j.e

Kumar N, Soren S (2023) Magnesium metal nano composites—a solid state hydrogen storage material. Mater Today Proc. https://doi.org/10.1016/j.matpr.2023.03.434

Kumar S, Jain A, Ichikawa T, Kojima Y, Dey GK (2017) Development of vanadium based hydrogen storage material: a review. Renew Sust Energ Rev 72:791–800. https://doi.org/10.1016/j.rser.2017.01.063

La Monaca A, Pava V, Welburn A, Yekani R, Feldmann T, Bélanger F, Demopoulos GP (2025) BiVO4 photoanodes for photo-assisted electrochemical water splitting. In: Electrochemical society meeting abstracts 247 (No. 39). The Electrochemical Society, Inc, p 2090

Langmi HW, Engelbrecht N, Modisha PM, Bessarabov D (2022) Chapter 13 - Hydrogen storage. In: Smolinka T, Garche J (eds) Electrochemical power sources: fundamentals, systems, and applications. Elsevier, pp 455–486. https://doi.org/10.1016/B978-0-12-819424-9.00006-9

Le Duigou A, Bader A-G, Lanoix J-C, Nadau L (2017) Relevance and costs of large scale underground hydrogen storage in France. Int J Hydrog Energy 42(36):22987–23003. https://doi.org/10.1016/j.ijhydene.2017.06.239

Liu Y, Zhang W, Zhang X, Yang L, Huang Z, Fang F, Sun W, Gao M, Pan H (2023) Nanostructured light metal hydride: fabrication strategies and hydrogen storage performance. Renew Sust Energ Rev 184:113560. https://doi.org/10.1016/j.rser

Lu Z (2021) Computational discovery of energy materials in the era of big data and machine learning: a critical review. Mater Rep: Energy 1:100047

Ma JX, Liu SJ, Zhou W, Pan X (2008) Comparison of hydrogen transportation methods for hydrogen refueling station. J Tongji Univ Nat Sci 36(5):615–619

Madadi Avargani V, Zendehboudi S, Cata Saady NM, Dusseault MB (2022) A comprehensive review on hydrogen production and utilization in North America: prospects and challenges. Energy Convers Manag 269:115927. https://doi.org/10.1016/j.enconman.2022.115927

Makepeace JW, He T, Weidenthaler C, Jensen TR, Chang F, Vegge T, Ngene P, Kojima Y, de Jongh PE, Chen P, David WIF (2019) Reversible ammoniabased and liquid organic hydrogen carriers for high-density hydrogen storage: recent progress. Int J Hydrog Energy 44(15):7746–7767. https://doi.org/10.1016/j.ijhydene.2019.01.144

Modisha PM, Ouma CNM, Garidzirai R, Wasserscheid P, Bessarabov D (2019) The prospect of hydrogen storage using liquid organic hydrogen carriers. Energy Fuel 33(4):2778–2796. https://doi.org/10.1021/acs.energyfuels.9b00296

Møller KT, Jensen TR, Akiba E, Li H (2017) Hydrogen—a sustainable energy carrier. Prog Nat Sci: Mater Dermatol Int 27:34–40. https://doi.org/10.1016/j.pnsc.2016

Mondal K, Malode SJ, Shetti NP, Alqarni SA, Pandiaraj S, Alodhayb A (2024) Porous nanostructures for hydrogen generation and storage. J Energy Storage 76:109719

Nagar R, Srivastava S, Hudson SL, Amaya SL, Tanna A, Sharma M, Achayalingam R, Sonkaria S, Khare V, Srinivasan SS (2023) Recent developments in state-of-the-art hydrogen energy technologies–review of hydrogen storage materials. Sol Compass 5:100033

Navlani-García M, Mori K, Kuwahara Y, Yamashita H (2018) Recent strategies targeting efficient hydrogen production from chemical hydrogen storage materials over carbon-supported catalysts. NPG Asia Mater 10:277–292

Niermann M, Drünert S, Kaltschmitt M, Bonhoff K (2019) Liquid organic hydrogen carriers (LOHCs)—techno-economic analysis of LOHCs in a defined process chain. Energy Environ Sci 12(1):290–307. https://doi.org/10.1039/C8E

Obaideen K, Olabi AG, Al Swailmeen Y, Shehata N, Abdelkareem MA, Alami AH et al (2023) Solar energy: applications, trends analysis, bibliometric analysis and research contribution to sustainable development goals (SDGs). Sustainability 15:1418

Okonkwo PC, Barhoumi EM, Ben Belgacem I, Mansir IB, Aliyu M, Emori W, Uzoma PC, Beitelmal WH, Akyüz E, Radwan AB, Shakoor RA (2023) A focused review of the hydrogen storage tank embrittlement mechanism process. Int J Hydrog Energy 48(35):12935–12948. https://doi.org/10.1016/j.ijhydene.2022.12.252

Omran A, Nesterenko N, Valtchev V (2022) Ab initio mechanistic insights into the stability, diffusion and storage capacity of sI clathrate hydrate containing hydrogen. Int J Hydrog Energy 47(13):8419–8433. https://doi.org/10.1016/j.ijhydene.2021.12.186

Osman AI, Mehta N, Elgarahy AM, Hefny M, Al-Hinai A, Al-Muhtaseb AH et al (2022) Hydrogen production, storage, utilisation and environmental impacts: a review. Environ Chem Lett 20:153

Owens D, Han A, Sun L, Mao Y (2015) Synthesis of VTMS(X)-HMS-3 mesoporous ordered silica for hydrogen storage. Int J Hydrog Energy 40:2736–2741. https://doi.org/10.1016/j.ijhydene.2014.12.100

Oztas KA, Kunze K, Jois K, Sackmann J, Zaremba S, Ruf MG (2022) A numerical approach to design a type II box-shaped pressure vessel with inner tension struts. Int J Hydrog Energy 47(6):3927–3938. https://doi.org/10.1016/j.ijhydene.2021.11.005

Pandit K, Jeffrey C, Keogh J, Tiwari MS, Artioli N, Manyar HG (2023) Techno-economic assessment and sensitivity analysis of glycerol valorization to biofuel additives via esterification. Ind Eng Chem Res 62:9201–9210

Patchkovskii S, Tse JS, Yurchenko SN, Zhechkov L, Heine T, Seifert G (2005) Graphene nanostructures as tunable storage media for molecular hydrogen. Proc Natl Acad Sci USA 102:10439–10444. https://doi.org/10.1073/pnas.0501030102

Prabhukhot PR, Wagh MM, Gangal AC (2016) A review on solid state hydrogen storage material. Adv Energy Power 4(2):11–22

Preuster P, Papp C, Wasserscheid P (2017) Liquid organic hydrogen carriers (LOHCs): toward a hydrogen-free hydrogen economy. Acc Chem Res 50(1):74–85. https://doi.org/10.1021/acs.accounts.6b00474

Principi G, Agresti F, Maddalena A, Russo SL (2009) The problem of solid state hydrogen storage. Energy 34:2087–2091

Pupysheva OV, Farajian AA, Yakobson BI (2008) Fullerene nanocage capacity for hydrogen storage. Nano Lett 8:767–774. https://doi.org/10.1021/nl071436g

Ramimoghadam D, Gray EM, Webb CJ (2016) Review of polymers of intrinsic microporosity for hydrogen storage applications. Int J Hydrog Energy 41:16944–16965. https://doi.org/10.1016/j.ijhydene.2016.07.134

Rao N, Lele AK, Patwardhan AW (2022) Optimization of liquid organic hydrogen carrier (LOHC) dehydrogenation system. Int J Hydrog Energy 47(66):28530–28547. https://doi.org/10.1016/j.ijhydene.2022.06.197

Riera JA, Lima RM, Knio OM (2023) A review of hydrogen production and supply chain modeling and optimization. Int J Hydrog Energy 48(37):13731–13755

Rivard E, Trudeau M, Zaghib K (2019) Hydrogen storage for mobility: a review. Materials 12(12):1973. https://doi.org/10.3390/ma12121973

Rüde T, Dürr S, Preuster P, Wolf M, Wasserscheid P (2022) Benzyltoluene/perhydro benzyltoluene—pushing the performance limits of pure hydrocarbon liquid organic hydrogen carrier (LOHC) systems. Sustain Energy Fuels 6(6):1541–1553. https://doi.org/10.1039/D1SE01767E

Rusman NAA, Dahari M (2016) A review on the current progress of metal hydrides material for solid-state hydrogen storage applications. Int J Hydrog Energy 41:12108–12126

Saeid MF, Abdulkadir BA, Abidin SZ, Setiabudi HD (2025) Nanoconfinement of magnesium hydride in porous scaffolds for hydrogen storage: kinetics, thermodynamics, and future prospects. Mater Sci Semicond Process 188:109225. https://doi.org/10.1016/j.mssp.2024.109225

Schüth F, Bogdanović B, Felderhoff M (2004) Light metal hydrides and complex hydrides for hydrogen storage. Chem Commun 36:2249–2258

Simanullang M, Prost L (2022) Nanomaterials for on-board solid-state hydrogen storage applications. Int J Hydrog Energy 47:29808–29846. https://doi.org/10.1016/j.ijhydene.2022.06.301

Singh A, Mishra S, Gautam A (2025) Architecture, modeling and simulation of fuel cell-based vehicles. Arab J Sci Eng 50:1–45

Soni K, Panwar NL, Lanjekar PR (2024) Emergence of carbonaceous material for hydrogen storage: an overview. Clean Energy 8:147–168. https://doi.org/10.1093/ce/zkae041

Sundén B (2019) Hydrogen, batteries and fuel cells. Academic Press, pp 37–55. https://doi.org/10.1016/B978-0-12-816950-6.00003-8

Teprovich JA, Wellons MS, Lascola R, Hwang SJ, Ward PA, Compton RN, Zidan R (2012) Synthesis and characterization of a lithium-doped fullerane (Li x C60-Hy) for reversible hydrogen storage. Nano Lett 12:582–589. https://doi.org/10.1021/NL203045V/SUPPL_FILE/NL203045V_SI_001.PDF

Tozzini V, Pellegrini V (2013) Prospects for hydrogen storage in graphene. Phys Chem Chem Phys 15:80–89

Ulucan TH, Akhade SA, Ambalakatte A, Autrey T, Cairns A, Chen P, Cho YW, Gallucci F, Gao W, Grinderslev JB, Grubel K, Jensen TR, de Jongh PE, Kothandaraman J, Lamb KE, Lee Y-S, Makhloufi C, Ngene P, Olivier P, Webb CJ, Wegman B, Wood BC, Weidenthaler C (2023)

Hydrogen storage in liquid hydrogen carriers: recent activities and new trends. Prog Energy 5(1):012004. https://doi.org/10.1088/2516-1083/acac5c

Usman MR (2022) Hydrogen storage methods: review and current status. Renew Sust Energ Rev 167:112743. https://doi.org/10.1016/j.rser.2022.112743

Valentini F, Marrocchi A, Vaccaro L (2022) Liquid organic hydrogen carriers (LOHCs) as H-source for bio-derived fuels and additives production. Adv Energy Mater 12(13):2103362. https://doi.org/10.1002/aenm.202103362

Vitillo JG, Ricchiardi G, Spoto G, Zecchina A (2005) Theoretical maximal storage of hydrogen in zeolitic frameworks. Phys Chem Chem Phys 7:3948–3954

Wang X, Tian M, Chen X, Xie P, Yang J, Chen J, Yang W (2022) Advances on materials design and manufacture technology of plastic liner of type IV hydrogen storage vessel. Int J Hydrog Energy 47(13):8382–8408. https://doi.org/10.1016/j.ijhydene.2021.12.198

Ward PA, Teprovich JA, Peters B, Wheeler J, Compton RN, Zidan R (2013) Reversible hydrogen storage in a LiBH4-C60 nanocomposite. J Phys Chem C 117:22569–22575. https://doi.org/10.1021/JP4079103/SUPPL_FILE/JP4079103_SI_001.PDF

Wu L, Li Y, Fu Z, Su BL (2020) Hierarchically structured porous materials: synthesis strategies and applications in energy storage. Natl Sci Rev 7:1667–1701. https://doi.org/10.1093/nsr/nwaa183

Xiaosan Z, Qingquan J, Shoukat Iqbal K, Manzoor A, Zia UR (2021) Achieving sustainability and energy efficiency goals: assessing the impact of hydroelectric and renewable electricity generation on carbon dioxide emission in China. Energy Policy 155:112332

Yang L, Wang S, Zhang Z, Lin K, Zheng M (2023) Current development status, policy support and promotion path of China's green hydrogen industries under the target of carbon emission peaking and carbon neutrality. Sustainability 15:10118. https://doi.org/10.3390/su151310118

Yue M, Lambert H, Pahon E, Roche R, Jemei S, Hissel D (2021) Hydrogen energy systems: a critical review of technologies, applications, trends and challenges. Renew Sust Energ Rev 146:111180. https://doi.org/10.1016/j.rser.2021.1111

Zhang J, Lei L, Zhou W, Li G, Yan Y, Ni Z (2023) Cryogenic mechanical and hydrogen-barrier properties of carbon fiber composites for type V cryocompressed hydrogen storage vessels. Compos Commun 43:101733. https://doi.org/10.1016/j.coco.2023.101733

Zhang L, Jia C, Bai F, Wang W, An S, Zhao K, Li Z, Li J, Sun H (2024) A comprehensive review of the promising clean energy carrier: hydrogen production, transportation, storage, and utilization (HPTSU) technologies. Fuel 355:129455

Chapter 7
Hydrogen-Based Energy Storage in Power Grids: A Pathway to Renewable Integration and Decarbonization

Abstract The rapid expansion of variable renewable energy sources such as wind and solar has introduced significant challenges to power grid stability, flexibility, and reliability. Energy storage technologies are therefore essential to enable large-scale renewable integration and deep decarbonization of electricity systems. Among the available options, hydrogen-based energy storage has emerged as a promising long-duration and seasonal storage solution due to its high energy density, scalability, and cross-sector coupling potential. This chapter examines the role of hydrogen as an energy carrier within modern power grids, focusing on power-to-hydrogen-to-power pathways enabled by electrolysis, hydrogen storage, and reconversion through fuel cells or hydrogen-fueled turbines. Key technological components, system architectures, and integration strategies are critically reviewed, along with operational, economic, and infrastructure-related challenges. The chapter further explores the contribution of hydrogen-based storage to grid flexibility, renewable curtailment reduction, and emissions mitigation, highlighting its synergies with transportation, industry, and heating sectors. Policy frameworks, market mechanisms, and future research directions necessary for large-scale deployment are also discussed. Overall, hydrogen-based energy storage is presented as a pivotal enabler for achieving resilient, low-carbon power systems and advancing the global transition toward sustainable energy futures.

Keywords Hydrogen energy · Hydrogen electrification · Fuel cell systems · Hydrogen coupling · Local equilibrium

7.1 Introduction

Nations worldwide are undergoing a profound transition in the global energy landscape as they strive to mitigate climate change, reduce greenhouse gas emissions, and move toward more sustainable and resilient energy systems (Sikiru et al. 2024a; Ur Rehman et al. 2024). In this context, hydrogen has emerged as a strong candidate among the various alternative energy options under consideration, largely due to its

D. M. Y. Mir et al., *Low-Carbon Hydrogen*, Clean Energy Production Technologies,
https://doi.org/10.1007/978-981-95-8142-9_7

versatility, high energy content, and environmentally benign nature when produced from low-carbon pathways (Sharma et al. 2021; He et al. 2024). However, integrating hydrogen into existing energy and electricity networks presents both significant opportunities and notable challenges, particularly in terms of infrastructure development and overall economic feasibility (Maestre et al. 2021; Wang et al. 2024). Hydrogen, the most abundant element in the universe, can be produced from a wide range of feedstocks, including natural gas, coal, biomass, and water, using processes such as steam methane reforming (SMR), gasification, and electrolysis (Sikiru et al. 2024a). When generated using renewable energy sources such as wind, solar, and hydropower, hydrogen is widely regarded as a clean, sustainable, and long-term energy carrier capable of supporting deep decarbonization (Mir 2025; Ishaq et al. 2022). Its applications span multiple sectors, including transportation, industrial processes, and residential and commercial heating, thereby positioning hydrogen as a central pillar in global decarbonization strategies aimed at addressing climate change (Ghorbani et al. 2024; Sikiru et al. 2024b). One of the hydrogen's most compelling advantages lies in its ability to store and transport energy efficiently. Unlike electricity, which must be consumed instantaneously or stored in batteries with limited capacity and duration, hydrogen can be stored in large quantities and transmitted over long distances through pipelines, in a manner comparable to natural gas infrastructure (Acar and Dincer 2018). This characteristic makes hydrogen particularly attractive for balancing the intermittency of renewable energy sources and providing long-term and seasonal energy storage solutions. Nevertheless, the large-scale integration of hydrogen into current power systems necessitates substantial upgrades and modifications to existing infrastructure (Kovač et al. 2021). The infrastructural requirements for hydrogen integration are multifaceted and encompass production, storage, transportation, and distribution systems (SaberiKamarposhti et al. 2024). The establishment of hydrogen production facilities whether through renewable-powered electrolysis or alternative production routes is a critical step in the value chain (Ishaq et al. 2022). Strategic siting of these facilities is essential to ensure resource availability, minimize losses, and optimize distribution logistics (Ishaq et al. 2022; Sikiru et al. 2024a). At the same time, continuous technological innovation is required to improve efficiency and reduce production costs, which remain a key barrier to widespread adoption. Hydrogen storage infrastructure represents another vital component of system integration. Hydrogen can be stored as a compressed gas, in liquid form, or through chemical carriers such as ammonia and metal hydrides, each option presenting distinct technical, safety, and economic considerations (Sikiru et al. 2024a). Compressed hydrogen storage is technologically mature and relatively simple, but it requires high-pressure vessels and suffers from low volumetric energy density. In contrast, liquid hydrogen storage offers higher energy density but requires cryogenic temperatures, leading to increased energy consumption, insulation requirements, and overall system costs (Aziz 2021). While hydrogen integration offers substantial potential to accelerate the global transition toward a clean and sustainable energy future, transportation and distribution infrastructure pose additional challenges related to material compatibility, hydrogen embrittlement, leakage, and stringent safety standards (Zainal et al. 2024). In many cases, the construction of dedicated hydrogen pipelines may be necessary, demanding significant capital investment. Consequently, the economic viability of hydrogen deployment depends on multiple factors, including production costs, market demand, regulatory frameworks, and the pace of technological advancement. At present, the cost of green hydrogen produced via electrolysis remains considerably higher than that of conventional fossil fuels. However, ongoing research and development efforts aim to reduce these costs through economies of scale, efficiency improvements, and technological innovation (Mohammed et al. 2023). Policy support and industrial participation play a decisive role in shaping hydrogen markets. Governments and industry stakeholders can stimulate demand by introducing favorable policies, financial incentives, and the development of hydrogen hubs and demonstration projects. Parallel to these efforts, advances in digitalization and artificial intelligence (AI) are increasingly recognized as transformative enablers of sustainable energy systems. Mohammed et al. (2023) highlight that the intelligent design and deployment of AI solutions aligned with environmental objectives can fundamentally reshape sustainability planning and accelerate large-scale energy transitions (Mohammed et al. 2023). Collaboration among researchers, industry, and policymakers is therefore essential to realize intelligent circular economies, resilient energy infrastructures, and long-term sustainable development. In this regard, the integration of machine learning models with big data analytics significantly enhances the forecasting of energy demand, hydrogen production, and grid stability. Advanced digital tools and analytics are transforming the management and integration of hydrogen in energy systems. These technologies enhance the optimization of hydrogen storage and distribution networks, improve predictive capabilities for the variability of renewable energy generation, and support more informed decision-making for the deployment and utilization of hydrogen within existing power grids. Beyond planning, big data and digital monitoring enable real-time oversight of hydrogen infrastructure, facilitating predictive maintenance, strengthening safety management, and enhancing operational reliability. Looking forward, the large-scale integration of hydrogen into energy systems will increasingly emphasize infrastructure requirements, which must be evaluated through comprehensive techno-economic assessments. The widespread deployment of pipelines, storage facilities, and refueling stations is essential to achieve scalability and operational efficiency. Concurrently, machine learning and artificial intelligence (AI) are expected to play pivotal roles in optimizing system performance, improving operational reliability, and enhancing the efficient allocation of resources. These tools are particularly valuable for managing the intermittency of renewable energy sources such as wind and solar, improving grid flexibility, and maximizing the utilization of generated renewable power. Hydrogen is positioned as a cornerstone of future sustainable energy systems, acting as a versatile energy carrier that enables effective coupling between renewable generation and variable energy demand. This chapter contributes significantly to the literature by presenting a detailed economic and techno-economic assessment of hybrid hydrogen production systems that integrate Steam Methane Reforming (SMR), water electrolysis, and CO_2 capture. Unlike previous studies that often examine these technologies in isolation, this work evaluates the combined economic feasibility, efficiency, and environmental sustainability of their integrated operation, thereby providing a more comprehensive perspective. Through quantitative analysis of cost structures, process efficiencies, and integration strategies, the study reveals how the synergistic combination of SMR, electrolysis, and carbon capture can enhance overall performance. The integration of SMR with electrolysis and CO_2 capture emerges as a promising pathway to achieve an optimal balance among cost-effectiveness, energy efficiency, and environmental impact. By considering these technologies collectively, the study addresses significant knowledge gaps in the literature, offering concrete insights into system-level trade-offs and performance metrics. For example, incorporating CO_2 capture within SMR-based systems can reduce carbon emissions by up to 90% while only marginally increasing hydrogen production costs, highlighting a practical route to sustainable hydrogen generation in the near- to medium-term. Cost indicators, such as a reported 13% increase in production expenses due to CO_2 capture, provide tangible benchmarks for evaluating the economic viability of integrated hydrogen systems. Despite substantial research into hydrogen production, storage, and utilization technologies, several critical areas remain underexplored. Large-scale integration of hydrogen into power grids, associated infrastructure requirements (pipelines, storage facilities, refueling stations), and system-level techno-economic implications have received limited attention. Many studies focus narrowly on individual technologies without evaluating grid compatibility, system integration, and comprehensive cost analysis. As the global energy transition accelerates, addressing these gaps is essential to understand the infrastructural, technical, and economic dimensions of hydrogen adoption. This review seeks to fill these gaps by providing a structured and holistic analysis, making it a valuable resource for both researchers and policymakers. More broadly, the transition to a low-carbon energy system represents a fundamental shift toward cleaner, more resilient, and sustainable energy supply chains. As countries strive for carbon neutrality, diversify their energy sources, and reduce greenhouse gas emissions, renewable energy technologies such as wind, solar, and hydropower are increasingly central. However, the intermittent nature of these resources limits their ability to reliably meet rising global energy demand on their own. Hydrogen emerges as a versatile and clean energy carrier capable of bridging these gaps. It can be produced through multiple pathways, including renewable-based water electrolysis, and deployed across diverse sectors such as industry, transportation, and power generation. Hydrogen plays a particularly critical role in decarbonizing hard-to-electrify sectors, such as steel production and long-distance transport, while simultaneously serving as a long-duration energy storage medium to manage renewable energy variability. Within the broader context of the global energy transition, hydrogen supports carbon emission reduction, energy security enhancement, and economic growth, facilitated by the development of advanced production technologies and supportive infrastructure. Consequently, hydrogen is poised to become a pivotal enabler of resilient, low-carbon, and sustainable energy systems in the coming decades.

7.2 Hydrogen Energy Storage in Renewable Energy System

According to forecasts by the International Energy Agency (IEA), nearly one-third of global electricity generation is expected to come from intermittent renewable energy sources such as wind and solar by 2040. This increasing penetration of variable renewables will necessitate effective solutions for large-scale, long-term electricity storage (Fig. 7.1).

Among the available options, hydrogen production and storage has emerged as a particularly promising approach for addressing these challenges (Moore and Shabani 2016). Hydrogen enables surplus renewable electricity to be converted into a storable energy carrier that can be utilized when the generation from wind and solar is insufficient. Hydrogen can be stored in several different forms depending on the application and system requirements. The most widely adopted method to date is storage as compressed gas. An alternative approach involves liquefying hydrogen at cryogenic temperatures, while other options include physisorption where hydrogen is physically adsorbed on the surface of solid materials and chemisorption, in which hydrogen is stored within metal hydrides. Comprehensive reviews of these hydrogen storage technologies are available in the literature (von Colbe et al. 2019)

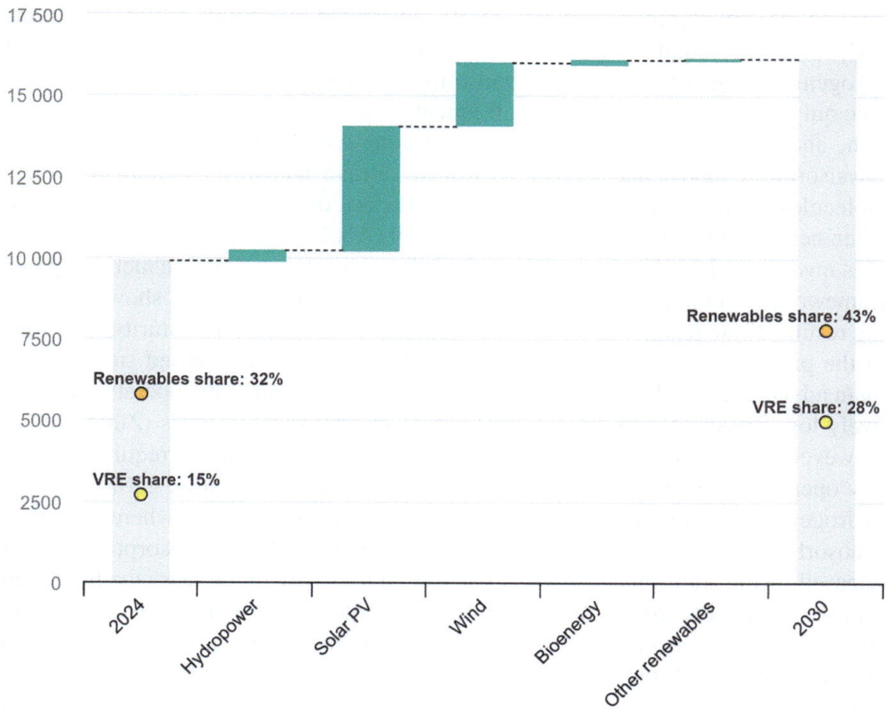

Fig. 7.1 Technology-specific forecasts of global renewable power generation for the years 2024 and 2030, highlighting projected contributions from key renewable technologies (IEA 2025)

and are beyond the scope of the present study. For stationary power system applications, compressed gas storage and metal hydride storage are generally considered the most relevant, and these are the only two storage technologies employed in the projects reviewed in this chapter. Among the analyzed projects, seven rely exclusively on compressed gas storage, six utilize only metal hydride storage, and two combine both storage approaches. Compressed hydrogen storage is currently the most mature and commonly used method. Commercial storage tanks, such as those used in Toyota's Mirai fuel cell vehicle, can store hydrogen at pressures of up to 700 bar. However, the compression process typically consumes approximately 20% of the hydrogen's energy content (Abdin et al. 2015). The primary advantages of compressed gas storage include technical simplicity, relatively low costs, and established commercial availability. Conversely, its drawbacks include comparatively low volumetric energy density relative to fossil fuel-based systems and safety concerns associated with high-pressure operation. Hydrogen can also be stored in liquid form, which significantly increases its volumetric energy density compared to compressed gas storage. Liquid hydrogen has a volumetric energy density of approximately 2.2 kWh/L, whereas compressed hydrogen provides around 1.3 kWh/L at 700 bar and 0.8 kWh/L at 350 bar (Hirscher 2010). Despite this improvement, liquid hydrogen still offers less than 40% of the volumetric energy density of liquefied natural gas (LNG), which is about 5.8 kWh/L (Hirscher 2010). Furthermore, liquid hydrogen storage is associated with substantial disadvantages, including the high energy demand for liquefaction, hydrogen boil-off losses, and the high cost of cryogenic storage systems. Due to boil-off issues, liquid hydrogen is generally suitable only for applications involving frequent usage or refueling, such as transportation, and is therefore not well suited for long-term stationary energy storage. Physisorption represents another hydrogen storage pathway, wherein hydrogen molecules are physically adsorbed onto the surface of solid materials and released when needed, for instance, to supply a fuel cell (Abdin et al. 2015). Common materials investigated for this purpose include carbon-based materials and metal–organic frameworks (Abdin et al. 2015). Although research in this area has shown promising results, these technologies have not yet reached commercial maturity, and none of the projects reviewed in this chapter employ physisorption-based storage. The main advantages of physisorption include low system complexity, operation at relatively low pressures, and the potential use of inexpensive materials (Züttel 2003). However, limitations such as low hydrogen storage density and the requirement for low operating temperatures restrict its current applicability (Züttel 2003). Finally, hydrogen can be stored through chemisorption in metal hydrides, where hydrogen is absorbed into a metal or metal alloy in powdered form. During absorption, heat is released, while additional heat input is required to desorb and release the hydrogen for use (Abdin et al. 2015). A key limitation of certain metal hydride materials is the strong bonding between hydrogen and the host metal, which necessitates high desorption temperatures—exceeding 650 °C in the case of lithium-based materials (Abdin et al. 2015). Nevertheless, some metal hydrides offer exceptionally high gravimetric hydrogen storage capacities, such as 18 wt% for $LiBH_4$ (Abdin et al. 2015). Alternative materials, including intermetallic compounds, have also been explored. For example, the TiFe-based alloy studied by Endo et al. (2019) exhibits a lower storage capacity of approximately 1.4 wt% but operates at relatively mild temperatures (hydrogen absorption at 30 °C and desorption at 45 °C) and moderate pressures. These characteristics can significantly reduce system costs and improve safety. Overall, ongoing research in hydrogen storage is focused on identifying optimal metal hydride materials and storage concepts that can combine low operating pressures and moderate absorption–desorption temperatures with high gravimetric and volumetric energy densities. Achieving this balance is critical for advancing hydrogen storage technologies suitable for large-scale, stationary power system applications in a future dominated by renewable energy.

7.3 Emerging Technologies for Hydrogen Electrification

Hydrogen energy is typically integrated into the power system through electrification processes, where it functions as an energy carrier rather than a primary energy source. This electrification is most commonly achieved via gas-to-electricity conversion, in which the chemical energy stored in hydrogen is converted into electrical energy for grid supply. In this context, fuel-cell-based technologies represent the principal pathway for hydrogen electrification, offering high conversion efficiencies and low environmental impacts. Accordingly, this section presents an overview of hydrogen electrification technologies with a focus on fuel cell systems, followed by a discussion of medium- and long-term hydrogen storage options and the range of ancillary services that hydrogen energy can provide to modern power systems.

7.3.1 Hydrogen Electrification Technology

Fuel cells are capable of efficiently utilizing the energy stored in hydrogen by directly converting its chemical energy into electrical energy, achieving conversion efficiencies in the range of 60–80%. This process occurs through electrochemical reactions rather than combustion, resulting in minimal energy losses and producing only water as the sole by-product, thereby making fuel cells a highly efficient and environmentally benign power generation technology.

7.3.1.1 Principle of Operation

A hydrogen fuel cell supplies hydrogen to the anode, where it undergoes ionization, releasing electrons and H^+ ions, while air is fed to the cathode, where oxygen is reduced to form negatively charged oxygen species (Ragab et al. 2023). Similar to electrolyzers, different types of fuel cells are distinguished by the direction of charge transport and the nature of the ionic charge carriers within the electrolyte, which determines whether water is produced at the anode or the cathode. Most fuel cells employ platinum-based catalysts supported on carbon materials, although the specific catalyst–electrolyte interactions vary among fuel cell technologies. These variations lead to differences in operating temperature, efficiency, durability, and application suitability. Subsequently, the operating principles and performance of a polymer electrolyte membrane fuel cell (PEMFC) are discussed as a representative example.

7.3.1.2 Structure of Fuel Cell Systems

A typical fuel cell system comprises a fuel cell stack along with a range of auxiliary components, including hydrogen storage tanks, pumps, air compressors, power electronics, and thermal management subsystems. An individual fuel cell generally produces a rated voltage in the range of 0.6–0.8 V under nominal operating conditions (Boretti 2024). The overall stack voltage can be increased by connecting multiple cells in series, while the output current—and consequently the power capacity—can be enhanced by operating fuel cell stacks in parallel, similar to the configuration strategies used in electrolyzers. Additionally, increasing the effective active area of the cells provides another means of boosting current output. To ensure stable and efficient operation, fuel cell systems incorporate several balance-of-plant components. Air compressors are used to supply oxygen-rich air to the reaction chamber, while hydrogen storage tanks deliver hydrogen at controlled flow rates and pressures. Thermal and water management are also critical aspects of system design. Condensers are typically employed to cool the compressed intake air, and humidification units are used to maintain adequate membrane hydration, preventing the proton exchange membrane from drying out and thereby preserving ionic conductivity (Zhang et al. 2024). Power conditioning equipment, such as inverters, is commonly integrated to convert the direct current (DC) electricity generated by the fuel cell into alternating current (AC) suitable for grid connection or end-use applications. From an operational perspective, fuel cells generally achieve their highest efficiencies when operating near their optimal load conditions. Operating at current densities below the peak power density can reduce voltage losses associated with activation, ohmic, and concentration overpotentials, thereby improving overall cell efficiency (Yu et al. 2011). Through careful system design and advanced control strategies, fuel cell systems can be engineered to consistently operate within this optimal efficiency range, maximizing performance, durability, and energy conversion efficiency (Da Silva et al. 2024).

7.3.1.3 Hydrogen Ancillary Service Technologies

Hydrogen energy systems, encompassing both electrolyzers and fuel cells, can provide a wide range of ancillary services that enhance the flexibility, reliability, and resilience of modern power grids. These services include peak load shaving, frequency regulation, mitigation of negative electricity pricing events, voltage support, and emergency power restoration or black-start capability (Gu et al. 2023). Similar to other energy storage technologies, hydrogen systems can also alleviate power supply limitations in transmission and distribution networks that arise from insufficient line capacity or network congestion (Yan et al. 2024). Addressing such congestion through conventional redispatch or curtailment strategies can be costly; in particular, managing power flow congestion in systems with high penetration of variable renewable generation often leads to significantly increased dispatch costs. In this context, the flexible coupling of hydrogen electrolysis and electrification processes offers an efficient and comparatively low-cost means of regulating power flows and relieving grid congestion. Frequency regulation is a fundamental ancillary service aimed at maintaining grid stability by keeping system frequency and voltage magnitude close to their nominal reference values. This is achieved through the coordinated injection or absorption of active power to continuously balance electricity supply and demand. In most contemporary power systems, frequency control is organized into two hierarchical levels. The first level, known as primary frequency control, provides rapid and continuous response to short-term frequency deviations (Deng et al. 2023). When frequency deviations persist beyond approximately 30 seconds, secondary or frequency restoration control is activated, offering greater regulation capacity over longer durations. Fuel cells and electrolyzers are well suited to participate in both levels of frequency control, as they can dynamically adjust their power output or consumption in response to frequency signals (Deng et al. 2023). During periods of frequency decline, fuel cells can increase electricity generation, while electrolyzers can reduce hydrogen production by lowering their power consumption. This bidirectional flexibility enables hydrogen systems to function as cost-effective demand response resources. Another important advantage of electrified hydrogen technologies is their ability to reduce the occurrence and severity of negative electricity prices in wholesale markets. Negative prices typically arise due to limited flexibility on the generation side, especially in power systems with high shares of inflexible or must-run renewable generation. In electricity markets with significant hydrogen deployment, electrolyzers can absorb excess electricity by increasing hydrogen production during periods of oversupply, while fuel cells can decrease output or operate strategically, thereby contributing to more flexible and stable spot price formation. In this way, hydrogen systems can act similarly to conventional generators or controllable loads, providing an effective mechanism for market-based grid balancing. Hydrogen-based devices also offer valuable voltage support capabilities (Fu et al. 2023). When connected to the grid through power electronic converters, both fuel cells and electrolyzers can regulate reactive power and power factor, adjusting their operating points to support voltage levels at the point of common coupling. This capability is particularly important in weak or renewable-rich grids where voltage fluctuations are more pronounced. Furthermore, in the event of a widespread power outage, fuel cell systems can provide fast, quiet, and reliable black-start functionality, restoring power more rapidly and with lower noise and emissions compared to conventional diesel or gas generators (Yang et al. 2023). This Table 7.1 summarizes key hydrogen technologies that provide ancillary services in power grids, including electrolyzers, fuel cells, and integrated power-to-gas systems.

Table 7.1 Hydrogen technologies enabling ancillary services in renewable integrated grids

Technology	Ancillary/grid service provided	Mechanism	References
PEM electrolyzers + fuel cells (PtG)	Frequency and voltage support; ramp-rate control	Power-to-gas systems using PEM electrolyzers provide voltage support and fast frequency response via controlled power modulation and hydrogen storage	Jain et al. (2023)
PEM electrolyzers (frequency reserve)	Primary frequency reserves	Dynamic models show PEM electrolyzers can contribute faster frequency reserve than conventional generators	Alshehri et al. (2019)
PEM electrolyzers	Frequency regulation, voltage support, congestion management, black start	Electrolyzers can modulate power draw to help balance supply/demand; fast dynamic response suitable for ancillary service markets	Cozzolino and Bella (2024)
Hydrogen storage + fuel cell backup	Black start, emergency power, peak load support	Stored hydrogen discharged via fuel cells provides grid support during outages or peak demand	Dai et al. (2024)
Integrated electrolyzer systems (market value)	Frequency containment and restoration reserves, economic ancillary participation	Optimization studies show providing ancillary services increases electrolyzer profitability and grid service value	Johnsen et al. (2025)

7.3.1.4 Hydrogen–Electric Power Systems

Similar to conventional gasoline engines, certain internal combustion engines and gas turbines can operate directly on hydrogen to generate mechanical and electrical power. However, due to hydrogen's relatively low volumetric energy density, hydrogen-fueled internal combustion engines generally exhibit lower thermodynamic efficiencies, typically in the range of 20–25%, which is inferior to that of conventional gasoline engines (Teoh et al. 2023). Moreover, although hydrogen combustion does not emit carbon dioxide, it can lead to the formation of nitrogen oxides (NO_x), particularly nitrogen dioxide, which remains a harmful air pollutant and poses challenges for air quality management (Park and Kim 2023). In contrast, utilizing fuel cells as the primary power source in hydrogen–electric power systems effectively avoids combustion-related air pollutants, as electricity is generated through electrochemical reactions rather than high-temperature combustion. In the transportation sector, vehicles powered by fuel cells differ fundamentally from battery-based new energy vehicles. While battery electric vehicles often face limitations in driving range due to battery capacity constraints, fuel cell electric vehicles (FCEVs) can achieve significantly longer driving ranges because hydrogen has a higher gravimetric energy density and can be refueled rapidly. It is projected that hydrogen-powered vehicles will account for approximately 3% of global car sales by 2030, with this share potentially increasing to around 36% by 2050 as technologies mature and infrastructure expands (Albatayneh et al. 2023). In response to this growth potential, numerous automotive manufacturers are actively advancing fuel cell powertrain technologies, with a strong emphasis on improving system reliability, durability, and operational safety. A prominent example is Toyota's Mirai fuel cell vehicle, which employs a commercially available polymer electrolyte membrane fuel cell (PEMFC) system. The Mirai features a volumetric power density of approximately 3.1 km/L and delivers a peak power output of 144 kW, demonstrating the practical viability of fuel cell technology for passenger vehicles. Ongoing and future developments aim to further enhance hydrogen storage safety, reduce system size, and improve overall vehicle integration. Currently, most commercially available fuel cell vehicles rely on high-pressure compressed hydrogen tanks for onboard storage, reflecting a balance between technological maturity, safety, and performance. Beyond road transportation, fuel cell technology is gaining momentum in maritime and rail applications. Maritime shipping alone accounts for roughly 2.5% of global greenhouse gas emissions, highlighting the urgent need for cleaner propulsion alternatives. Fuel cells offer sufficient power density and energy capacity to support long-distance marine operations and can meet the auxiliary power demands of large vessels, where battery-only solutions remain impractical due to weight and range limitations. Similarly, fuel cell-powered trains are emerging as a viable solution for non-electrified rail lines. Hydrogen-powered regional trains are already in operation across several European countries and are expected to capture more than 30% of the regional rail market in the future, underscoring the growing role of hydrogen fuel cells in decarbonizing heavy-duty and long-range transportation sectors (Dall'Armi et al. 2023). Hydrogen mobility infrastructure is steadily expanding in the United States, driven by the deployment of fuel cell electric vehicles (FCEVs), and supporting hydrogen refueling stations (HRSs). As shown in Fig. 7.2, the percentage share of FCEVs and HRSs in the USA in 2022 highlights the current adoption levels and spatial distribution of hydrogen-based transportation technologies. Despite being in the early stages, these trends indicate a growing investment in hydrogen infrastructure, which is expected to accelerate with ongoing technological improvements and policy support.

7.4 Hydrogen–Electric Coupling

Electric–hydrogen coupling systems have attracted increasing attention in recent research, with studies demonstrating that incorporating both generation and transmission planning within an integrated electric–hydrogen framework can significantly reduce overall system costs (Gonzalez-Romero et al. 2020). By jointly optimizing electricity and hydrogen infrastructures, such integrated planning enables more efficient utilization of resources, enhances system flexibility, and mitigates the need for costly grid reinforcements. Broadly, electric–hydrogen coupling can be classified based on the point of integration within the energy system, namely, load-side electric–hydrogen coupling and power-side electric–hydrogen coupling.

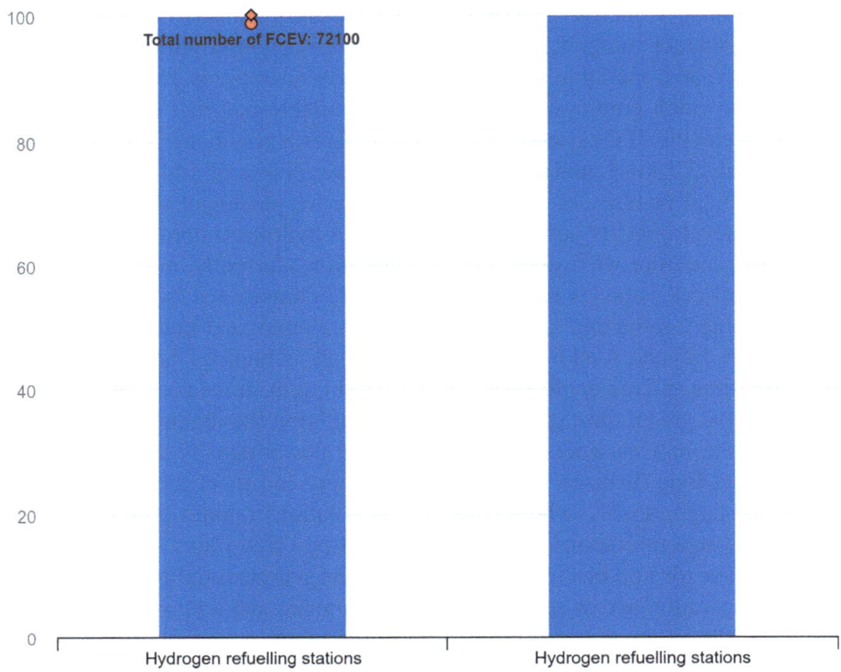

Fig. 7.2 Share of fuel cell electric vehicle and hydrogen refueling station stock in the USA in 2022 (%) (IEA 2023)

7.4.1 Load-Side Electric–Hydrogen Coupling

Using China as a representative example, electricity produced by large-scale, centralized renewable energy plants in the northwestern regions can be transmitted over long distances to the central and eastern load-intensive areas via ultra-high-voltage (UHV) transmission lines. Once delivered, this electricity can be utilized to produce hydrogen through electrolysis, enabling the deployment of hydrogen-based technologies—such as fuel cells or hydrogen gas turbines—that provide essential grid services, including voltage support and capacity reserves. This load-side electricity–hydrogen coupling approach offers several significant advantages (Zhang et al. 2023). Firstly, hydrogen production facilities benefit from considerable locational flexibility. Because electrolyzers draw electricity directly from the grid, hydrogen plants can be strategically located near major hydrogen demand centers. This proximity reduces costs associated with hydrogen conversion, storage, and long-distance transportation, while also mitigating safety risks inherent to extensive hydrogen logistics networks. Secondly, UHV transmission infrastructure can be leveraged to optimize large-scale allocation of renewable electricity. Comparative studies indicate that the cost of transporting hydrogen via pipelines increases substantially with distance from approximately USD 4.5/kg for 400 km to USD 10/kg for 1000 km (Khaleel et al. 2021; Qiu et al. 2023). In contrast, the cost of electricity transmission via UHV lines rises more modestly over the same distance range, from about USD 0.05/kWh to USD 0.09/kWh. When converted to an energy-equivalent basis, the unit cost of electricity transmission is roughly one-quarter to one-fifth that of hydrogen transport, demonstrating the superior economic competitiveness of UHV electricity transmission for long-distance energy delivery. A further advantage of the load-side coupling model is the pronounced difference between peak and off-peak electricity prices in load centers. These price differentials can help offset efficiency losses inherent in the electricity–hydrogen–electricity conversion cycle, thereby enhancing the overall economic attractiveness of the system. Despite these benefits, the load-side electricity–hydrogen coupling model has certain limitations. Although UHV transmission lines may entail relatively low upfront capital costs, this does not necessarily result in lower overall energy delivery expenses. The inherent intermittency of renewable generation requires supplementary support from flexible power sources such as coal-fired plants, hydropower, electrochemical storage, or pumped hydro to maintain reliable and stable power delivery. The integration of these complementary resources can add significant investment costs, increasing the total system expenditure. Additionally, the operational flexibility of electrolyzers may not be fully exploited under this model. Since variability in renewable generation is largely managed on the supply side through flexible resources and storage, the electricity reaching the load side tends to be relatively stable. This stability limits the ability of electrolyzers to actively participate in grid balancing or to absorb fluctuations from renewable generation, thereby constraining their potential contribution to overall system flexibility (Yan 2021).

7.4.2 Power-Side Electric–Hydrogen Coupling

Electricity–hydrogen coupling on the power supply side can generally be classified into three distinct modes: the electricity transmission mode, the hydrogen transmission mode, and the local balancing mode (Salama et al. 2022). These modes differ in terms of energy conversion pathways, infrastructure requirements, and their roles in integrating renewable energy sources into the power system.

7.4.3 Electricity-Transmission Method

In this configuration, electrolyzers, hydrogen storage facilities, and hydrogen gas turbines are strategically located at the power generation side to mitigate curtailment and minimize losses associated with the variability of renewable energy sources. Electrolyzers in this setup are characterized by high operational flexibility and rapid response capabilities, allowing them to efficiently absorb surplus electricity generated by renewable plants (Marquez et al. 2024). The hydrogen produced through this process can either be used to satisfy local energy demands or stored for later re-electrification as required. This generation-side coupling model offers significant benefits by simultaneously enhancing the integration of renewable energy and improving the utilization efficiency of transmission infrastructure. As traditional coal- and natural-gas-fired power plants are gradually retired, the combination of electrolyzers, hydrogen storage, and hydrogen gas turbines assumes a critical role in grid management, including peak shaving, frequency regulation, and load balancing. During the periods of high renewable output, electrolyzers can operate at or near full capacity, converting excess electricity into hydrogen and storing it in dedicated tanks. When renewable generation declines, stored hydrogen can be dispatched through hydrogen gas turbines to supply electricity, stabilizing transmission lines and ensuring reliable power delivery. This dynamic operation supports a higher penetration of renewable energy while maintaining efficient system performance. However, similar to the load-side electricity–hydrogen coupling approach, this power-side configuration relies on an electricity–hydrogen–electricity conversion chain, which is inherently inefficient and leads to relatively high energy losses. These conversion losses remain a significant limitation and highlight the need for technological advancements, improved system design, and optimized operational strategies to enhance the overall viability and economic feasibility of power-side electricity hydrogen coupling systems.

7.4.4 Hydrogen-Delivery Mode

In this configuration, hydrogen produced from renewable electricity at the power generation site is transported over long distances to centrally located load centers via dedicated hydrogen pipelines. The main advantage of this model is its ability to relieve stress on the electrical grid by shifting the burden of long-distance energy transmission from power lines to hydrogen infrastructure. By converting surplus renewable electricity into hydrogen at the generation site, this approach facilitates the large-scale deployment and utilization of renewable resources, particularly in regions geographically distant from major consumption centers (Rocha et al. 2024). Unlike electricity, hydrogen production and consumption do not require instantaneous balancing, which makes hydrogen systems inherently more tolerant of temporal fluctuations between supply and demand. Furthermore, hydrogen pipelines can provide a limited degree of line-pack storage, offering additional operational flexibility and the ability to buffer short-term mismatches between generation and usage. Despite these benefits, several challenges constrain the widespread adoption of long-distance hydrogen transport. The construction of dedicated hydrogen pipelines is capital-intensive and remains at an early stage of deployment, limiting scalability. Conventional hydrogen delivery via trucks or trailers remains economically viable only for short distances and relatively small volumes. While blending hydrogen into existing natural gas pipelines can help reduce infrastructure costs, the allowable blending ratios are typically limited to approximately 10% due to material compatibility and safety constraints, which restricts the potential for large-scale hydrogen transmission through existing gas networks.

7.4.5 *Local Equilibrium*

This model is based on the local utilization of hydrogen produced from renewable energy sources on the power generation side. Its key advantages lie in its relatively low cost and higher overall efficiency, as the direct use of hydrogen eliminates the need for electricity–hydrogen–electricity reconversion and thereby significantly reduces energy losses associated with multiple conversion steps. By simplifying the energy pathway, this approach enhances system efficiency and improves the economic viability of hydrogen deployment. Nevertheless, the effectiveness of this model is strongly dependent on the availability of sufficient local hydrogen demand. In regions where hydrogen consumption is limited, surplus hydrogen produced from renewable sources must be transported to other areas, which can introduce additional costs and logistical challenges. As a result, hydrogen-intensive industries such as chemical processing, steel production, and fuel cell–based transportation are encouraged to strategically locate their operations in areas with abundant renewable energy potential. Such co-location can maximize the benefits of locally produced green hydrogen, facilitate industrial decarbonization, and support the efficient development of regional hydrogen ecosystems (Caldera et al. 2017).

7.5 Conclusion

Countries worldwide are experiencing a profound transformation in the global energy landscape, driven by the urgent need to mitigate climate change, reduce greenhouse gas emissions, and transition toward more sustainable and resilient energy systems. In this evolving context, hydrogen is increasingly recognized as a versatile energy carrier with high energy potential and favorable environmental attributes. Its integration into existing energy infrastructures offers significant opportunities, particularly for stabilizing intermittent renewable energy sources and providing long-term and seasonal energy storage solutions. At the same time, this transition entails substantial challenges related to infrastructure development, economic viability, and technological deployment. Hydrogen can be produced from a wide array of feedstocks—including natural gas, coal, biomass, and water through processes such as steam methane reforming (SMR), gasification, and electrolysis. Among these pathways, renewable hydrogen produced from wind, solar, and hydropower stands out as the most environmentally sustainable option, offering near-zero-carbon energy conversion. The growing utilization of hydrogen across multiple sectors transportation, industry, and residential heating—underscores its pivotal role in global decarbonization strategies, supporting both carbon emission reductions and enhanced energy security. The large-scale integration of hydrogen into power grids and energy systems requires substantial infrastructure development. This includes the construction of production facilities, as well as robust storage and transportation networks. Hydrogen storage technologies such as compressed gas, liquid hydrogen, and chemical carriers (e.g., ammonia, metal hydrides) involve distinct technical, safety, and economic considerations. Similarly, hydrogen transport often necessitates dedicated pipelines, which represent significant capital expenditures and can influence deployment timelines and overall costs. The widespread adoption of hydrogen is strongly influenced by its economic competitiveness. Currently, production costs, particularly for green hydrogen generated via electrolysis, remain higher than those of conventional fossil fuels. Ongoing research and development aim to reduce costs through economies of scale, enhanced system efficiencies, and technological innovation. Market adoption is also shaped by policy frameworks and industrial engagement, including financial incentives, regulatory support, and the development of hydrogen hubs, all of which are crucial for stimulating demand and attracting private investment. Integration of carbon capture, utilization, and storage (CCUS) with conventional SMR represents a promising pathway to transform traditional hydrogen production into a low-carbon technology. CCUS can potentially qualify hydrogen for carbon credits and other policy incentives, thereby improving economic viability. However, this approach introduces additional capital and energy costs, which must be balanced through supportive policies such as carbon pricing, direct subsidies, or tax incentives. In conclusion, hydrogen holds substantial potential as a key enabler of a sustainable, low-carbon energy future. Realizing this potential requires overcoming multiple economic, technological, and infrastructural barriers. Coordinated efforts among academia, industry, and government are essential to accelerate technological development, establish supportive policy frameworks, and deploy the necessary infrastructure. By prioritizing collaborative research, investment, and deployment strategies, hydrogen can emerge as a cornerstone of the global energy system, contributing decisively to climate change mitigation, energy security, and long-term sustainable development.

References

Abdin Z, Webb CJ, Gray EM (2015) Solar hydrogen hybrid energy systems for off-grid electricity supply: a critical review. Renew Sustain Energy Rev 52:1791–1808

Acar I, Dincer I (2018) 1.13 hydrogen energy. In: Comprehensive energy systems, vol 1. Elsevier Doyma, pp 568–605

Albatayneh A, Juaidi A, Jaradat M, Manzano-Agugliaro F (2023) Future of electric and hydrogen cars and trucks: an overview. Energies 16:3230. [CrossRef]

Alshehri F, Suárez VG, Rueda Torres JL, Perilla A, van der Meijden MAMM (2019) Modelling and evaluation of PEM hydrogen technologies for frequency ancillary services in future multi-energy sustainable power systems. Heliyon 5(4):e01396. https://doi.org/10.1016/j.heliyon.2019.e01396

Aziz M (2021) Liquid hydrogen: a review on liquefaction, storage, transportation, and safety. Energies 14(18):5917

Boretti A (2024) A high-efficiency internal combustion engine using oxygen and hydrogen. Int J Hydrog Energy 50:847–856

Caldera U, Bogdanov D, Afanasyeva S, Breyer C (2017) Role of seawater desalination in the management of an integrated water and 100% renewable energy based power sector in Saudi Arabia. Water 10:3

Cozzolino R, Bella G (2024) A review of electrolyzer-based systems providing grid ancillary services: current status, market, challenges and future directions. Front Energy Res 12:1358333. https://doi.org/10.3389/fenrg.2024.1358333

Da Silva DC, Kefsi L, Sciarretta A (2024) Closed-form expression to estimate the hydrogen consumption of a fuel cell hybrid electric vehicle. IEEE Trans Veh Technol 73:4717–4728

Dai S, Shen P, Deng W, Yu Q (2024) Hydrogen energy in electrical power systems: a review and future outlook. Electronics 13(17):3370. https://doi.org/10.3390/electronics13173370

Dall'Armi C, Pivetta D, Taccani R (2023) Hybrid PEM fuel cell power plants fuelled by hydrogen for improving sustainability in shipping: state of the art and review on active projects. Electronics 16:2022

Deng W, Zhang Y, Tang Y, Li Q, Yi Y (2023) A neural network-based adaptive power-sharing strategy for hybrid frame inverters in a microgrid. Front Energy Res 10:1082948

Endo N, Shimoda E, Goshome K, Yamane T, Nozu T, Maeda T (2019) Construction and operation of hydrogen energy utilization system for a zero emission building. Int J Hydrog Energy 44:14596–14604

Fu Z, Lu L, Zhang C, Xu Q, Zhang X, Gao Z, Li J (2023) Fuel cell and hydrogen in maritime application: a review on aspects of technology, cost, and regulations. Sustain Energy Technol Assess 57:103181

Ghorbani Y, Zhang SE, Nwaila GT, Bourdeau JE, Rose DH (2024) Embracing a diverse approach to a globally inclusive green energy transition: moving beyond decarbonization and recognizing realistic carbon reduction strategies. J Clean Prod 434:140–414

Gonzalez-Romero IC, Wogrin S, Gómez T (2020) Review on generation and transmission expansion co-planning models under a market environment. IET Gener Transm Distrib 14:931–944. [CrossRef]

Gu Z, Pan G, Gu W, Qiu H, Lu S (2023) Robust optimization of scale and revenue for integrated power-to-hydrogen systems within energy, ancillary services, and hydrogen markets. IEEE Trans Power Syst 39:5008–5023

He Y, Hu Y, Zhu Z, Li J, Huang Y, Zhang S, Balogun MS, Tong Y (2024) High-performance multidimensional-structured N-doped nickel modulated Mo2N/FeOxNy bifunctional electrocatalysts for efficient alkaline seawater splitting. Chem Eng J 489:151348

Hirscher M (2010) Handbook of hydrogen storage. WILEY-VCH

IEA (2023) Share of fuel cell electric vehicle (FCEV) and hydrogen refuelling station (HRS) stock by region, 2022. IEA, Paris. https://www.iea.org/data-and-statistics/charts/share-of-fuel-cell-electric-vehicle-fcev-and-hydrogen-refuelling-station-hrs-stock-by-region-2022. Licence: CC BY 4.0

IEA (2025) Global renewable power generation and change by technology, 2024 and 2030. IEA, Paris. https://www.iea.org/data-and-statistics/charts/global-renewable-power-generation-and-change-by-technology-2024-and-2030. Licence: CC BY 4.0

Ishaq H, Dincer I, Crawford C (2022) A review on hydrogen production and utilization: challenges and opportunities. Int J Hydrog Energy 47(62):26238–26264

Jain R, Nagasawa K, Veda S, Sprik S (2023) Grid ancillary services using electrolyzer-based power-to-gas systems with increasing renewable penetration. e-Prime Adv Electr Eng Electron Energy 6:100308

Johnsen AG, Mitridati L, Zarrilli D, Kazempour J (2025) The value of ancillary services for electrolyzers. Comput Chem Eng:109360

Khaleel MM, Adzman MR, Zali SM (2021) An integrated hydrogen fuel cell to distribution network system: challenges and opportunities for D-STATCOM. Energies 14:7073. [CrossRef]

Kovač A, Paranos M, Marciuš D (2021) Hydrogen in energy transition: a review. Int J Hydrog Energy 46(16):10016–10035

Maestre V, Ortiz A, Ortiz I (2021) Challenges and prospects of renewable hydrogenbased strategies for full decarbonization of stationary power applications. Renew Sustain Energy Rev 152:111628
Marquez RA, Espinosa M, Kalokowski E, Son YJ, Kawashima K, Le TV, Chukwuneke CE, Mullins CB (2024) A guide to electrocatalyst stability using lab-scale alkaline water electrolyzers. ACS Energy Lett 9:547–555. [CrossRef]
Mir MY (2025) Sustainable fuels and decarbonization: challenges and opportunities. In: Sustainable fuels production. Springer, pp 1–19
Mohammed MA, Ahmed MA, Hacimahmud AV (2023) Data-driven sustainability: leveraging big data and machine learning to build a greener future. Babylonian J Artif Intell 2023:17–23
Moore J, Shabani B (2016) A critical study of stationary energy storage polices in Australia in an international context: the role of hydrogen and battery technologies. Energies 9:674
Park YK, Kim BS (2023) Catalytic removal of nitrogen oxides (NO, NO2, N2O) from ammonia-fueled combustion exhaust: a review of applicable technologies. Chem Eng J 461:141958. [CrossRef]
Qiu Y, Li Q, Ai Y, Chen W, Benbouzid M, Liu S, Gao F (2023) Two-stage distributionally robust optimization-based coordinated scheduling of integrated energy system with electricity-hydrogen hybrid energy storage. Prot Control Mod Power Syst 8:542–555. [CrossRef]
Ragab A, Marei MI, Mokhtar M (2023) Comprehensive study of fuel cell hybrid electric vehicles: classification, topologies, and control system comparisons. Appl Sci 13:13057
Rocha C, Knöri T, Ribeirinha P, Gazdzicki P (2024) A review on flow field design for proton exchange membrane fuel cells: challenges to increase the active area for MW applications. Renew Sustain Energy Rev 192:114198. [CrossRef]
SaberiKamarposhti M, Kamyab H, Krishnan S, Yusuf M, Rezania S, Chelliapan S, Khorami M (2024) A comprehensive review of AI-enhanced smart grid integration for hydrogen energy: advances, challenges, and future prospects. Int J Hydrog Energy 67:1009–1025
Salama HS, Magdy G, Bakeer A, Vokony I (2022) Adaptive coordination control strategy of renewable energy sources, hydrogen production unit, and fuel cell for frequency regulation of a hybrid distributed power system. Prot Control Mod Power Syst 7:472–489. [CrossRef]
Sharma S, Agarwal S, Jain A (2021) Significance of hydrogen as economic and environmentally friendly fuel. Energies 14(21):7389
Sikiru S, Oladosu TL, Amosa TI, Olutoki JO, Ansari MNM, Abioye KJ, Rehman ZU, Soleimani H (2024a) Hydrogen-powered horizons: transformative technologies in clean energy generation, distribution, and storage for sustainable innovation. Int J Hydrog Energy 56:1152–1182
Sikiru S, Al-Yaseri A, Yekeen N, Soleimani HNBN, Hamza MF, Ghotbi MY (2024b) H2-quartz and cushion gas-quartz intermolecular interactions: implications for hydrogen geo-storage in sandstone reservoirs. Adsorption 30(6):631–650
Teoh YH, How HG, Le TD, Nguyen HT, Loo DL, Rashid T, Sher F (2023) A review on production and implementation of hydrogen as a green fuel in internal combustion engines. Fuel 333:126525
Ur Rehman Z, Rehman MA, Alomar SY, Rehman B, Awais M, Amjad M, Sikiru S, Ali EM, Hamad A (2024) Hydrogen storage capacity of lead-free perovskite NaMTH3 (MT= Sc, Ti, V): a DFT study. Int J Energy Res 2024(1):4009198
von Colbe JB, Ares J-R, Barale J, Baricco M, Buckley C, Capurso G, Gallandat N, Grant DM, Guzik MN, Jacob I, Jensen EH, Jensen T, Jepsen J, Klassen T, Lototskyy MV, Manickam K, Montone A, Puszkiel J, Sartori S, Sheppard DA, Stuart A, Walker G, Webb CJ, Yang H, Yartys V, Züttel A, Dornheim M (2019) Application of hydrides in hydrogen storage and compression: achievements, outlook and perspectives. Int J Hydrog Energy 44:7780–7808
Wang Y, Chen Y, Yun Y, Hong X, Huang Y, Ji H (2024) CoMoP hole transfer layer functionally enhances efficiency and stability of BiVO4 based photoanode for solar water splitting. Appl Catal B Environ Energy 358:124375
Yan Y (2021) Comparative analysis of the economics of hydrogen storage and transportation. Master's thesis. Huazhong University of Science Technology, Wuhan

Yan C, Zou Y, Wu Z, Maleki A (2024) Effect of various design configurations and operating conditions for optimization of a wind/solar/hydrogen/fuel cell hybrid microgrid system by a bio-inspired algorithm. Int J Hydrog Energy 60:378–391

Yang X, Sun J, Meng X, Sun S, Shao Z (2023) Cold start degradation of proton exchange membrane fuel cell: dynamic and mechanism. Chem Eng J 455:140823

Yu Z, Han J, Cao X (2011) Investigation on performance of an integrated solid oxide fuel cell and absorption chiller tri-generation system. Int J Hydrog Energy 36:12561–12573

Zainal BS, Ker PJ, Mohamed H, Ong HC, Fattah IMR, Rahman SA, Nghiem LD, Mahlia TI (2024) Recent advancement and assessment of green hydrogen production technologies. Renew Sustain Energy Rev 189:113941

Zhang S, Zhang N, Dai H, Liu L, Zhou Z, Shi Q, Lu J (2023) Comparison of different coupling modes between the power system and the hydrogen system based on a power–hydrogen coordinated planning optimization model. Energies 16:5374

Zhang SW, Sun BG, Lin SL, Li Q, Wu X, Hu T, Bao B-Z, Wang X, Luo Q-H (2024) Energy and exergy analysis for a turbocharged direct-injection hydrogen engine to achieve efficient and high-economy performances. Int J Hydrog Energy 54:601–612

Züttel A (2003) Materials for hydrogen storage. Mater Today:24–33

Chapter 8
Hydrogen-Powered Vehicles: A Transition to Sustainable Transportation Systems

Abstract Hydrogen is increasingly recognized as a cornerstone of sustainable energy systems, particularly for decarbonizing the transportation sector, which remains one of the largest global energy consumers. Owing to its high energy density, abundance, versatility in production pathways, and potential for zero or near-zero emissions when derived from renewable resources, hydrogen offers a compelling alternative to fossil fuels. This chapter provides a comprehensive assessment of hydrogen's viability as a primary transportation fuel, with a specific focus on hydrogen-fueled internal combustion engines (ICEs). While fuel cell vehicles exhibit lower tailpipe emissions, their higher costs, additional weight, and space requirements limit near term adoption, making hydrogen ICEs a practical transitional solution. The chapter reviews recent literature to examine performance, efficiency, emissions, and system redesign requirements associated with hydrogen combustion. A comparative evaluation of conventional, hybrid, electric, biofuel, fuel cell, and hydrogen ICE vehicles is conducted using indicators such as CO_2 and SO_2 emissions, social cost of carbon, energy and exergy efficiencies, fuel consumption and cost, and driving range. The results indicate that fuel cell vehicles achieve the highest overall performance ranking, followed closely by hydrogen-fueled ICEs and biofuel vehicles, whereas conventional vehicles perform poorest. These findings highlight the strategic role of hydrogen-based propulsion, particularly hydrogen ICEs as an effective pathway toward cleaner, more efficient, and sustainable transportation systems.

Keywords Hydrogen vehicles · Proton exchange membrane fuel cells · Refueling station · Sustainable energy

8.1 Introduction

The persistent growth of the global population, coupled with rising living standards, has led to a continuous and substantial increase in worldwide energy demand (Acar et al. 2016). At present, this demand is predominantly satisfied by fossil fuels,

D. M. Y. Mir et al., *Low-Carbon Hydrogen*, Clean Energy Production Technologies,
https://doi.org/10.1007/978-981-95-8142-9_8

despite their well-documented limitations and adverse impacts (Stougie et al. 2018). Key challenges associated with fossil fuel dependence include finite and unevenly distributed reserves, price volatility driven by geopolitical uncertainties and the rapid depletion of easily accessible resources, and significant greenhouse gas (GHG) emissions, which are widely recognized as the principal drivers of global climate change (Acar and Dincer 2017; Zhang 2017). These concerns have intensified the search for alternative, low-carbon energy carriers capable of ensuring long-term energy security and environmental sustainability. In this context, hydrogen has emerged as a promising candidate for replacing fossil fuels in future energy systems. By leveraging abundant renewable energy and material resources, hydrogen offers the potential for improved energy security, enhanced price stability, and reduced dependence on geopolitically sensitive fuel supplies (Dincer and Acar 2016). Moreover, hydrogen-based energy systems have the potential to mitigate major environmental and societal challenges associated with conventional energy use, particularly climate change and urban air pollution (Cipriani et al. 2014). Consequently, there is a growing demand for research, development, and demonstration activities focused on clean and sustainable energy technologies (Mir 2025). Such systems must adhere to the 3S framework—Source, System, and Service—which integrates sustainable energy and material resources, efficient energy conversion and distribution infrastructures, and suitable energy carriers for end-use applications (Dincer and Acar 2017). The transportation sector, as one of the largest consumers of global energy, is a critical area requiring a transition from conventional fossil-based systems to intelligent and sustainable energy solutions. Hydrogen is increasingly viewed as a strong candidate to support this transition and to enable a more sustainable future for transportation systems (Sherif et al. 2014). Recent studies have highlighted hydrogen's significant potential as an alternative energy carrier capable of partially or fully displacing fossil fuels. However, comprehensive and detailed analyses are still required to fully understand its advantages, limitations, and system-level implications prior to large-scale commercialization (Salvi and Subramanian 2015). Hydrogen is the cleanest known fuel, with a gravimetric heating value approximately three times higher than that of petroleum-based fuels. Nevertheless, hydrogen is not a primary energy source and must be produced using other energy and material inputs, which presents technical and economic challenges (Ehret and Bonhoff 2015). Currently, hydrogen production costs remain relatively high—often nearly three times those of conventional fossil fuels posing a major barrier to widespread adoption (Acar et al. 2015). Hydrogen generation from water, particularly through renewable energy-driven pathways, has been extensively investigated over recent decades due to its potential to provide a sustainable and renewable hydrogen supply. Despite this progress, significant challenges related to cost, efficiency, and scalability persist, especially when integrating hydrogen production with intermittent renewable energy sources (AlRafea et al. 2016). For hydrogen-powered vehicles to achieve true sustainability, the entire hydrogen value chain—including production, storage, distribution, and end-use—must be developed in an environmentally and economically sustainable manner. This requirement represents one of the central challenges of establishing a viable hydrogen economy. On the other hand, hydrogen offers several advantageous characteristics compared to conventional transportation fuels. It can be utilized in both internal combustion engines (ICEs) and fuel cells, providing considerable flexibility in vehicle design and deployment. Furthermore, hydrogen-fueled vehicles emit only water vapor at the point of use, making them particularly attractive from an environmental perspective. Owing to these advantages, hydrogen has become a focal point of extensive research and a rapidly expanding topic in the scientific literature (Larsson et al. 2015). The feasibility of hydrogen as a sustainable transportation fuel has been explored from multiple perspectives. For example, Cammack et al. (2001) developed an early roadmap for the comprehensive integration of hydrogen into energy systems, emphasizing the importance of learning from natural processes. Dutta (2014) examined critical safety measures required during hydrogen production, storage, and utilization, including safety considerations for refueling infrastructure and automotive applications. Dincer and Acar (2016) evaluated hydrogen's potential as an aviation fuel by comparing its energy consumption, emissions, costs, production pathways, and efficiencies with those of conventional and alternative aviation fuels. Cipriani et al. (2014) provided a broad assessment of hydrogen as an energy carrier for automotive applications, addressing technological, economic, and safety aspects. Salvi and Subramanian (2015) outlined pathways for the sustainable development of the road transportation sector through hydrogen-based energy systems, while Ehret and Bonhoff (2015) emphasized the role of coherent policies and strategic initiatives in facilitating the transition to hydrogen-powered air, land, and maritime transport. Regional assessments, such as the work by Larsson et al. (2015) for Sweden and Amoo and Fagbenle (2014) for Nigeria, further highlight the techno-economic and environmental implications of hydrogen integration across diverse national contexts. Based on these considerations, the primary objective of this study is to critically review and synthesize recent analyses of hydrogen-fueled vehicles and to comparatively evaluate emerging technological alternatives. The aim is to assess hydrogen's potential as a clean, reliable, safe, affordable, and efficient primary fuel for future transportation systems. Both hydrogen internal combustion engines (H_2ICE) and hydrogen fuel cell (H_2FC) technologies are examined in detail. In addition, this study systematically compares the social, environmental, economic, and thermodynamic performance of hydrogen-based vehicles with conventional, hybrid, battery electric (BEV), and biofuel-powered vehicles, with the goal of identifying key advantages, limitations, opportunities, and challenges associated with hydrogen-fueled transportation.

8.2 Proton Exchange Membrane Fuel Cells

Proton exchange membrane fuel cells (PEMFCs) represent one of the most thoroughly researched and widely deployed fuel cell (FC) technologies for vehicular applications. They are particularly valued for their high-power density, fast start-up capability, and dependable performance under low-temperature operating conditions. Typically operating within a temperature range of 60–80 °C, PEMFCs enable rapid start-up, simplified thermal management, and relatively high conversion efficiencies, which are critical requirements for automotive applications (Aminudin et al. 2023). Owing to these attributes, PEMFCs have garnered substantial interest for use not only in transportation but also in portable power devices and stationary energy systems. Recent developments have demonstrated electrical efficiencies approaching 65% and power outputs of up to 300 kW, underscoring their suitability for medium- and high-power vehicle platforms (Wang et al. 2022). The process initiates with the supply of hydrogen gas (H_2) to the anode, where a catalyst layer—commonly platinum-based—facilitates the dissociation of hydrogen molecules into protons (H^+) and electrons (e^-). The proton exchange membrane (PEM) selectively allows the passage of protons while blocking electrons, thereby forcing the electrons to flow through an external circuit and generate electrical power. At the cathode, oxygen gas (O_2) combines with the incoming protons and electrons in the presence of a catalyst to form water, which is released as the primary byproduct. The overall electrochemical reactions governing this process are summarized in Eqs. (8.1), (8.2), and (8.3).

Anode reaction

$$H_2 \rightarrow 2H^+ + 2e^- \tag{8.1}$$

Cathode reaction

$$O_2 + 4H^+ + 4e^- \rightarrow 2H_2O \tag{8.2}$$

Overall cell reaction

$$2H_2 + O_2 \rightarrow 2H_2O + \text{Electricity} + \text{Heat} \tag{8.3}$$

A typical PEMFC consists of several integrated components, including the proton exchange membrane, catalyst layers (CLs), gas diffusion layers (GDLs), bipolar plates (BPPs), current collectors, compression plates, and sealing gaskets. The PEM is sandwiched between two catalyst-coated electrodes, forming the membrane electrode assembly (MEA), which is the core functional unit of the fuel cell (Mancino et al. 2023). The membrane plays a pivotal role by enabling proton transport between the anode and cathode while simultaneously preventing electron conduction and gas crossover. Efficient PEMFC operation strongly depends on effective ion transport and precise water management within the membrane. Maintaining high proton conductivity under adequately hydrated conditions is essential, while excessive water accumulation or dehydration can lead to flooding or membrane drying, respectively, both of which negatively impact efficiency and durability (Mohsen and Basem 2024). In addition to high proton conductivity, the PEM must exhibit strong chemical, mechanical, and thermal stability to withstand harsh electrochemical and operational environments. Perfluorinated sulfonic acid (PFSA) polymers, such as Nafion, are the most commonly used membrane materials due to their excellent proton conductivity and acceptable thermal stability (Baroutaji et al. 2024). To address performance limitations under low-humidity or elevated-temperature conditions, advanced membrane concepts—including composite membranes and ionic liquid-infused membranes—have been developed to enhance durability and operational flexibility. Catalyst layers are responsible for driving the fundamental electrochemical reactions in PEMFCs, namely, hydrogen oxidation at the anode and oxygen reduction at the cathode. In addition to catalyzing these reactions, CLs facilitate proton transport from reaction sites to the PEM and enable electron conduction toward the current collectors. The structural design, composition, and distribution of catalyst materials most commonly platinum or platinum-based alloys are critical factors influencing fuel cell efficiency, durability, and cost (Fan et al. 2024). To reduce platinum loading while maintaining high catalytic activity, Pt is frequently alloyed with transition metals such as ruthenium or cobalt. Gas diffusion layers play a vital role in ensuring the uniform distribution of reactant gases across the catalyst layers, thereby enabling effective utilization of active catalytic sites. Proper gas transport through the GDL is essential for sustaining high power output and preventing local reactant starvation. Additionally, GDLs assist in managing water produced during cell operation by facilitating its removal from the catalyst layer and promoting efficient mass transport. Common GDL materials include carbon paper and carbon cloth, which provide an optimal balance between electrical conductivity, porosity, and mechanical durability (Kahraman and Akin 2024). Carbon paper is often treated with hydrophobic coatings to mitigate flooding while preserving gas permeability, whereas carbon cloth offers enhanced flexibility and a higher surface area, contributing to improved gas distribution. In some advanced designs, composite GDLs incorporating carbon with polymers or other reinforcing materials are employed to further enhance mechanical strength and long-term durability.

Bipolar plates (BPPs) play a central role in PEMFC operation by enabling the transport and distribution of reactant gases, facilitating effective heat management, establishing electrical connectivity between adjacent cells, and enabling the removal of reaction byproducts such as water (Fan et al. 2021). In addition to serving as structural support for the membrane electrode assembly (MEA), BPPs provide mechanical integrity to the entire fuel cell stack. Owing to their multifunctional role, BPPs account for a substantial fraction of the total weight, volume, and overall cost of PEMFC stacks. A variety of materials are employed in the fabrication of BPPs, with the choice largely dictated by the intended application and operating conditions. Graphite-based BPPs are widely used due to their excellent electrical conductivity and high resistance to corrosion in acidic fuel cell environments. However, graphite is inherently brittle and difficult to machine, which can pose challenges for large-scale manufacturing; these limitations can be partially mitigated through the use of graphite composite materials that offer improved machinability and mechanical robustness. Metallic BPPs, particularly those made from stainless steel, are also commonly adopted because of their superior mechanical strength, thin profile, and suitability for mass production. Nevertheless, metallic plates are susceptible to corrosion and interfacial contact resistance, necessitating the application of protective coatings or surface treatments to ensure long-term durability under PEMFC operating conditions. The design of flow field patterns within BPPs is a critical determinant of PEMFC performance. Properly engineered flow fields ensure uniform distribution of hydrogen and oxygen across the active area, promote effective water management by preventing both flooding and membrane dehydration, and maintain favorable pressure drop and flow dynamics throughout the cell. Optimized flow field geometries are therefore essential for achieving high efficiency, stable operation, and prolonged stack lifetime. In both single-cell and multi-cell stack configurations, PEMFC assembly is completed through the integration of current collectors and compression plates. Current collectors are responsible for efficiently conducting the electrical current generated by electrochemical reactions to the external circuit, thereby minimizing resistive losses and ensuring effective power extraction. Compression plates apply uniform clamping pressure across the fuel cell stack, which enhances interfacial electrical contact, prevents reactant gas leakage, and protects the MEA from mechanical damage. As a result, proper compression significantly contributes to improved performance, reliability, and durability of PEMFC systems. Additionally, gaskets are used to seal individual layers within the PEMFC stack, preventing cross-leakage of reactant gases, guiding reactants toward the active regions, and ensuring even pressure distribution throughout the assembly.

8.2.1 Recent Developments in PEMFCs

PEMFCs have emerged as a highly attractive clean energy conversion technology due to their high efficiency, low operating temperatures, and zero-emission characteristics, making them suitable for a wide range of applications including transportation, portable devices, and distributed power generation systems. Despite these advantages, the widespread commercialization of PEMFCs continues to be hindered by several persistent challenges, most notably elevated system costs, limited thermal operating windows, and mechanical as well as chemical degradation of critical components. These limitations have driven intensive research efforts aimed at enhancing membrane performance, durability, and economic feasibility, with particular focus on developing alternatives to conventional Nafion-based proton exchange membranes (PEMs) (Segale et al. 2024). Although Nafion membranes exhibit excellent proton conductivity under well-humidified conditions, their performance is highly sensitive to humidity, their operational temperature range is relatively narrow, and their manufacturing costs remain prohibitively high. Consequently, significant attention has been directed toward alternative polymeric membrane materials. Among these, sulfonated poly(ether ether ketone) (SPEEK) has gained prominence as a promising candidate owing to its lower cost, favorable chemical stability, and tunable proton conductivity. When engineered into composite membrane systems incorporating fluorinated polymers, carbon-based fillers, or metal oxide additives, SPEEK-based membranes exhibit marked improvements in proton transport, mechanical robustness, and long-term durability (Parekh 2022). Further progress in membrane technology has been achieved through the development of hybrid ionogel systems, which combine ionic liquids with solid polymer frameworks. These materials form interconnected nano-scale channels and extensive hydrogen-bonding networks that significantly enhance proton mobility while maintaining structural stability under demanding operating conditions, including elevated temperatures and low-humidity environments (Zhou et al. 2024). In parallel, the incorporation of functionalized inorganic nanoparticles—such as silicon dioxide (SiO_2), titanium dioxide (TiO_2), and cerium oxide (CeO_2)—has proven effective in improving water retention, mitigating membrane degradation, and reinforcing mechanical strength, thereby enhancing electrochemical performance and operational stability. Advanced nanomaterials, particularly graphene and MXenes, have also attracted growing interest for next-generation PEMs due to their exceptional thermal stability, high specific surface area, and superior electrical and proton transport properties. These attributes position them as highly promising components for durable, high-performance membranes suitable for large-scale industrial deployment (Solangi et al. 2024). For high-temperature PEMFC applications, polymer systems such as crosslinked polybenzimidazole (PBI) and sulfonated poly-based membranes have demonstrated excellent thermal resilience and long-term durability. Nevertheless, challenges associated with scalable fabrication, material consistency, and sustained performance uniformity remain unresolved and require further investigation (Salam et al. 2024). Beyond membrane development, substantial advancements have been made in catalyst and electrode engineering to enhance PEMFC efficiency and reduce reliance on precious metals. In particular, doped graphene/carbon black hybrid catalysts have shown significant improvements in oxygen reduction reaction (ORR) activity, corrosion resistance, and long-term durability. By leveraging the synergistic interaction between doped graphene and carbon black, these hybrid catalysts substantially lower platinum loading while maintaining high electrochemical performance. Experimental studies have reported peak power densities as high as 0.934 W cm^{-2} at 0.60 V approximately 1.55 times greater than those achieved using commercial Pt/C catalysts underscoring their strong potential for cost-effective PEMFC systems (Ji et al. 2022). Improvements in bipolar plate (BPP) materials have also played a critical role in enhancing PEMFC stack performance, durability, and cost efficiency. Graphite-based BPPs, valued for their corrosion resistance, electrical conductivity, and chemical stability, continue to face limitations related to high manufacturing costs, inherent porosity, and insufficient mechanical strength. To overcome these challenges, advanced processing techniques such as silicon compound impregnation, expandable graphite treatments, layered structural designs, and polymer impregnation strategies have been developed, resulting in improved flexibility, mechanical integrity, and gas impermeability (Gao et al. 2024). Metallic bipolar plates, including stainless steel and titanium alloys, offer superior mechanical strength and manufacturability, particularly for high-volume production. However, their susceptibility to corrosion in the acidic

PEMFC environment has necessitated the adoption of advanced surface modification strategies. Techniques such as noble metal coatings, nitride and carbide layers, and deposition methods including physical vapor deposition (PVD) and plasma-enhanced chemical vapor deposition (PECVD) have been widely employed to enhance corrosion resistance while maintaining low interfacial contact resistance (Liu et al. 2024). More recently, composite bipolar plates composed of polymer matrices reinforced with conductive fillers such as graphene, graphite, and carbon nanotubes have emerged as a promising alternative. While challenges related to achieving high electrical conductivity and uniform filler dispersion persist, innovative fabrication approaches—including hierarchical structuring, pre-impregnation methods, and surface functionalization—have demonstrated encouraging progress (Chen et al. 2024). Looking ahead, future research is expected to focus on the development of novel conductive fillers, optimized fabrication routes, and integrated material architectures to achieve an optimal balance between electrical performance, mechanical durability, and cost efficiency. Collectively, the continued advancement of membrane, catalyst, and bipolar plate technologies reflects a comprehensive, multi-material strategy aimed at realizing durable, efficient, and economically competitive PEMFC systems suitable for large-scale commercialization. The fabrication of the membrane electrode assembly (MEA) is another decisive factor influencing PEMFC performance and longevity, with each fabrication technique offering distinct advantages and limitations depending on the intended production scale. At the laboratory level, brush painting remains widely used due to its simplicity and low cost, although it often suffers from poor reproducibility and limited control over catalyst layer uniformity. Inkjet printing provides enhanced precision and design flexibility, enabling controlled catalyst placement and patterning; however, its broader application is constrained by strict ink rheological requirements and relatively low throughput (Da Costa et al. 2024). Spraying techniques offer versatility and can produce thin, relatively uniform catalyst layers, though multiple coating passes are typically required to achieve the desired thickness, increasing processing time and material usage. For industrial-scale manufacturing, roll-to-roll slot-die coating has emerged as the most widely adopted and scalable MEA fabrication technique. This continuous process enables the production of large-area MEAs with high coating efficiency, excellent reproducibility, and consistent catalyst layer quality—attributes essential for commercial deployment. Slot-die coating allows precise control over coating speed, ink delivery, and layer thickness, facilitating the fabrication of multi-meter-long catalyst layers with uniform properties (Miao et al. 2024). Moreover, the high level of automation associated with this technique significantly reduces human error and enhances process reliability, making it particularly suitable for early-stage commercialization of PEMFC technologies. Nonetheless, the approach requires substantial capital investment and careful optimization of key parameters, including ink rheology, vacuum pressure, and substrate characteristics. Future developments in MEA fabrication are therefore expected to emphasize advanced process control, innovative materials, and sustainable manufacturing strategies to further improve PEMFC performance while minimizing production costs and environmental impacts.

8.3 Hydrogen-Based Mobility

The promotion of hydrogen-powered vehicles as a future transportation platform is primarily driven by their zero tailpipe pollutant emissions and potential contribution to deep decarbonization of the mobility sector. Hydrogen fuel cell vehicle (HFCV) technology demonstrates broad applicability across multiple transportation segments, including light-duty vehicles, heavy goods vehicles, high-capacity passenger buses, rail transport, and unmanned or autonomous mobility systems. In particular, the adoption of hydrogen as a fuel for heavy- and very heavy-duty vehicles has gained significant momentum, with regions such as New Zealand and Paris identifying hydrogen mobility as a key pathway toward achieving their long-term climate neutrality commitments. According to the New Zealand Ministry of Business, Innovation and Employment (MBIE 2019), comprehensive feasibility assessments have highlighted the suitability of clean hydrogen for heavy vehicles exceeding 30 tons. Beyond advancements in hydrogen fuel production, a systemic transition toward clean hydrogen utilization is required across heavy-duty freight vehicles, public transport fleets, and passenger buses to realize meaningful emissions reductions (Topler 2017). In support of this transition, Air Liquide has announced the commissioning of Europe's first high-pressure hydrogen refueling station designed to serve long-haul hydrogen vehicle fleets. This strategic investment aligns with the company's broader objective of accelerating large-scale hydrogen adoption, particularly within the heavy transport segment. The evolving hydrogen mobility landscape, as discussed in vulnerabilities of hydrogen energy in emerging markets, reflects coordinated efforts by governments, research institutions, industry stakeholders, and end users at varying stages of technological and market development (Topler 2017). Against this backdrop, Germany has emerged as a global leader in the commercialization of hydrogen vehicles, supported by strong collaborations with Japan, South Korea, and the United States (Trencher and Edianto 2021). More broadly, hydrogen and fuel cell technologies hold significant promise for enabling an environmentally sustainable, low-emission transportation and energy system (Galich and Marz 2012). Consequently, policymakers and automotive manufacturers worldwide are actively pursuing strategies to accelerate the deployment of hydrogen-based electrification in road transport. Comparative analyses of these efforts have examined the key factors influencing hydrogen production pathways and the market penetration of privately owned fuel cell electric vehicles (FCEVs) as well as fuel cell electric buses (FCEBs) in public transportation systems (Trencher and Edianto 2021). Table 8.1 presents a comparative assessment of internal combustion engine vehicles, hybrid electric vehicles, battery electric vehicles, and hydrogen-powered fuel cell vehicles across key technological and sustainability indicators. While conventional and battery-based propulsion systems have contributed significantly to the decarbonization of road transport, hydrogen fuel cell vehicles demonstrate distinct advantages in terms of long driving range, rapid refueling, and zero tailpipe emissions. In particular, hydrogen-powered vehicles offer superior suitability for heavy-duty, long-haul, and high-utilization applications where battery weight

Table 8.1 Comparative analysis of internal combustion, hybrid, battery electric, and fuel cell vehicle technologies

Category	Vehicle class	Technology	Typical engine efficiency	Environmental impact	Challenges	References
Conventional	Internal Combustion Engine Vehicle (ICEV)	Petrol or diesel combustion engine	20–40%	High pollutant emissions (CO_2, NO_x); 130–180 g/km CO_2; 0.25–0.40 g/km NO_x	Mature and well-established technology; environmentally unsustainable	Johnson and Joshi (2018)
Hybrid systems	Hybrid Electric Vehicle (HEV)	ICE combined with battery-electric assistance (engine)	30–45% (engine)	Reduced emissions relative to ICE; ~50–100 g/km CO_2; 0.1–0.2 g/km NO_x	Improved fuel efficiency; increased system complexity	Boretti (2020)
Hydrogen internal combustion engine	Hydrogen Internal Combustion Engine Vehicle (H_2-ICEV)	Hydrogen-fueled ICE (CO_2-free, potential NO_x)	40–45%	Zero CO_2; possible NO_x formation (0.05–0.10 g/km NO_x)	Needs hydrogen production and refueling infrastructure	Rouleau et al. (2021)
Battery-based	Battery Electric Vehicle (BEV)	Lithium-ion battery with electric drivetrain	90–95%	Zero tailpipe emissions (grid dependent); 0 g/km CO_2; 0 g/km NO_x	Charging time, battery degradation, and raw material concerns	Zhang et al. (2022)
Fuel cell	Fuel Cell Electric Vehicle (FCEV)	Proton Exchange Membrane (PEM) fuel cell	50–60%	Zero tailpipe emissions; 0 g/km CO_2; 0 g/km NO_x	Hydrogen storage and refueling network still under development	Muthukumar et al. (2021); Pramuanjaroenkij and Kakaç (2023)
Hybrid fuel cell	Fuel Cell Hybrid Electric Vehicle (FCHEV)	Fuel cell combined with battery or supercapacitors	60–70% (system-level)	Zero emissions; optimized power utilization; 0 g/km CO_2; 0 g/km NO_x	Higher system complexity; improved efficiency and durability	Waseem et al. (2023)

and charging time remain limiting factors. The comparison highlights hydrogen's strategic role as a complementary energy carrier to electrification, enabling deep decarbonization of the transport sector. Consequently, hydrogen fuel cell technologies emerge as a promising pathway for achieving scalable and resilient sustainable mobility.

8.3.1 Hydrogen Vehicles

At present, railways represent one of the most economical modes of transportation for both passengers and freight. However, conventional railway systems rely heavily on fossil fuels, resulting in significant greenhouse gas (GHG) emissions and noise pollution. In response to the growing demand for zero-emission and low-noise transportation, Alstom introduced the world's first hydrogen-powered passenger train, Coradia iLint, at InnoTrans 2016 in Berlin. This train was commercialized and launched in 2020, operating over a 100 km route connecting Cuxhaven, Bremerhaven, Bremervoerde, and Buxtehude in northern Germany (Low et al. 2020). Recognizing the need to transition toward clean hydrogen-based transportation—particularly given that India possesses the second-largest railway network globally—Indian Railways has also initiated the development and testing of hydrogen fuel cell–based passenger trainsets (Jhunjhunwala et al. 2018). Similarly, the maritime sector faces considerable challenges due to emissions from conventional combustion-based marine engines, which contribute to air pollution and global warming. To regulate these harmful emissions, the International Maritime Organization (IMO) has introduced stringent provisions aimed at preventing pollution from ships. Advances in marine fuel cell efficiency for onboard applications have encouraged researchers to explore hydrogen-based propulsion and power generation systems for maritime use. Onboard ships require power for electricity generation, emergency backup, and propulsion, all of which can be sustainably supplied using hydrogen fuel cells as replacements for conventional fuels (Sattler 2000). Fuel cells offer substantial potential for both merchant and naval vessels. Their applications include (1) emergency power generation, (2) electrical power supply in environmentally regulated waters and harbors, (3) low-power propulsion during special operating conditions, such as silent or low-noise operation, and (4) power generation for shipboard electrical and propulsion networks, particularly in all-electric ship (AES) configurations used in naval applications (Sattler 2000). A comprehensive discussion on the requirements, advantages, and feasibility of replacing combustion engines with hydrogen fuel cell systems is provided by Tronstad et al. (2017). Among marine applications, submarines currently represent the most mature use case for fuel cell technology, where hydrogen/oxygen polymer electrolyte membrane (PEM) fuel cells are widely employed (Leo et al. 2010). To achieve emission reduction targets and mitigate climate change, the global aviation industry has also identified clean hydrogen as a promising energy carrier for both short- and long-range aviation, as well as space transportation (de-Troya et al. 2016). Continuous research and development efforts have been reported across the aviation and space sectors, particularly in the United States (Cecere et al. 2014; van Biert et al. 2016), as well as in Europe, India, China, and Russia, demonstrating significant progress toward hydrogen fuel adoption (Rao et al. 2019). Furthermore, advancements in transportation technologies have led to the emergence of unmanned mobility platforms, notably unmanned ground vehicles and unmanned aerial vehicles (UAVs) (Tan 2013). In UAV applications, several countries have focused on enhancing operational range and performance by replacing conventional fuels with hydrogen-based energy systems (Wang et al. 2013). Recent studies indicate a growing commercialization of hydrogen fuel cell vehicles, particularly in the United States, where approximately 5899 hydrogen-powered vehicles have been developed (Bayrak et al. 2020). To further promote hydrogen mobility, major automotive manufacturers including Toyota (Mirai), Honda (Clarity), Renault–Nissan, and General Motors have formed strategic alliances for joint production and technological development of hydrogen fuel cell vehicles (Dudek et al. 2021). Figure 8.1 summarizes the main-case trajectory of hydrogen and e-fuel consumption in the transport sector from 2016 to 2030.

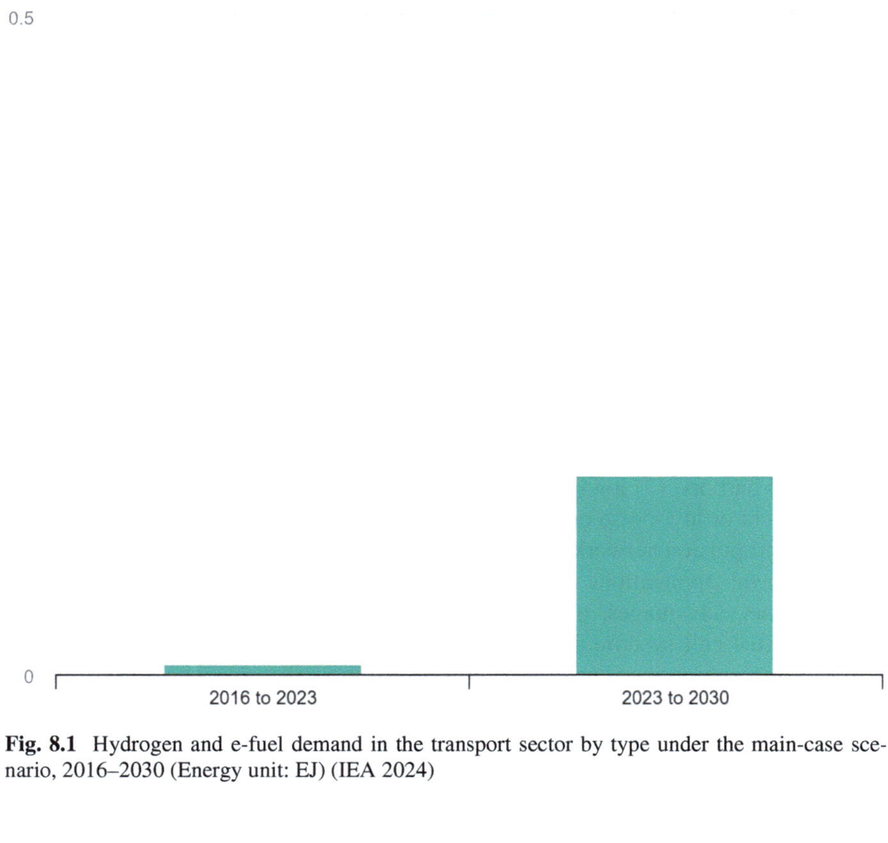

Fig. 8.1 Hydrogen and e-fuel demand in the transport sector by type under the main-case scenario, 2016–2030 (Energy unit: EJ) (IEA 2024)

8.3.2 Hydrogen Refueling into Vehicles

A hydrogen refueling station supplied with externally generated hydrogen typically consists of a compressor unit, high-pressure storage vessels, dispensing equipment, associated electrical and piping control systems, and integrated safety and monitoring infrastructure. The compressed hydrogen refueling process is conceptually similar to that used for compressed natural gas (CNG), in which the vehicle storage tank is filled through pressure equalization from a high-pressure stationary reservoir. For safe and effective operation, the capacity and pressure ratings of the stationary storage system are generally designed to be at least ten times greater than the maximum onboard tank volume and approximately 1.2 times higher than the nominal vehicle tank pressure, respectively (Maus et al. 2008). Such refueling stations operate in accordance with the standardized procedures defined under the SAE J2601 protocol (Maus et al. 2008). The refueling sequence involves several well-defined steps. Initially, the vehicle hydrogen tank is connected to the high-pressure stationary storage system via the dispenser. A brief initial gas pulse is then introduced to equalize the pressure within the refueling line, allowing estimation of the onboard tank volume. Based on this information, the target filling pressure is determined, and hydrogen is supplied until the specified pressure is achieved. To avoid overfilling and ensure safe operation, the final state of charge is typically maintained below the nominal maximum tank capacity. In order to remain competitive with refueling durations of internal combustion engine vehicles (ICEVs), automotive manufacturers target hydrogen refueling times of less than 3 min for passenger vehicles (Schneider et al. 2005), highlighting the importance of rapid refueling for the widespread adoption of hydrogen fuel cell electric vehicles (HFCEVs). To address this requirement, the Society of Automotive Engineers (SAE) developed the SAE J2601 standard, which enables the delivery of approximately 5–7 kg of hydrogen to light-duty HFCEVs within 3–5 min. In addition, the SAE J2601-2 guideline specifies refueling of roughly 20–30 kg of hydrogen for heavy-duty vehicles within 5–10 min. During high-rate refueling, the rapid inflow of high-pressure hydrogen causes a pronounced temperature increase inside the onboard storage cylinder. This temperature rise is primarily attributed to the isenthalpic expansion of hydrogen across throttling components in the dispenser and tank inlet, commonly referred to as the Joule–Thomson effect. Changes in refueling temperature directly affect the pressure rise rate within the cylinder and, consequently, the overall refueling performance (Li et al. 2023). Excessive thermal loading during refueling can lead to several adverse effects, including material degradation or structural failure of the storage tank, as well as a reduction in usable hydrogen capacity—reported to be up to 20% lower than that of CNG systems—thereby limiting vehicle driving range (Suryan et al. 2012). The temperature evolution following refueling can be predicted using simplified thermodynamic models (Liu et al. 2010). Furthermore, both experimental studies and computational fluid dynamics (CFD) analyses have been employed to investigate the influence of key operating parameters, such as initial tank temperature, initial pressure, mass flow rate, and inlet gas temperature, on thermal behavior in Type III and

Type IV hydrogen storage cylinders. Type III cylinders comprise a metallic inner liner overwrapped with carbon fiber, whereas Type IV cylinders utilize a non-metallic liner—typically high-density polyethylene or composite materials—reinforced with an external carbon fiber shell (Zheng et al. 2013). Owing to their high strength-to-weight ratios and improved gravimetric storage efficiency, both Type III and Type IV cylinders offer substantial advantages over fully metallic Type I and Type II designs and are therefore widely regarded as the most suitable options for automotive hydrogen storage applications (Stetson et al. 2016). In particular, Type IV cylinders can achieve weight reductions of up to 70% while providing hydrogen storage capacities more than four times those of conventional Type I cylinders, making them especially attractive for next-generation HFCEVs.

8.3.3 Effects of Initial Pressure

An increase in the initial pressure of the onboard fuel tank or storage cylinder has been consistently shown to mitigate the temperature rise experienced during hydrogen refueling. Both experimental measurements and computational fluid dynamics (CFD) simulations reported by Liu et al. (2010), Kim et al. (2010), Zheng et al. (2012), and Li et al. (2022) systematically investigated the effect of initial cylinder pressure on thermal behavior and uniformly observed a linear reduction in peak temperature rise with increasing initial pressure. These findings exhibited strong statistical consistency, with coefficients of determination (R^2) exceeding 0.923. The attenuation of temperature rise at elevated initial pressures is primarily attributed to the restricted ingress of hydrogen, which limits the effective mass flow rate and thereby suppresses excessive thermal accumulation within the storage cylinder. Complementary insights were provided by De Miguel et al. (2016), who performed CFD simulations of hydrogen refueling in both Type III and Type IV cylinders. Their results indicated that, when hydrogen is delivered at a precooled inlet temperature of −40 °C, the initial cylinder pressure has a negligible effect on the resulting temperature rise. This observation underscores the dominant influence of inlet gas temperature under precooled refueling conditions, where thermal effects associated with pressure variation are significantly diminished. Beyond the role of initial pressure, the temperature evolution during refueling has also been shown to depend on the instantaneous pressure within the cylinder. Multiple studies have demonstrated that the temperature increases nonlinearly as the cylinder pressure rises, with the rate of temperature increase progressively decreasing at higher pressures. As a result, this behavior is more accurately described by a logarithmic correlation. Although most investigations reported strong agreement between experimental or numerical data and the fitted correlations, Sadi and Deymi-Dashtebayaz (2019) observed comparatively weaker fits, reflected by lower R^2 values. In their study, Sadi et al. developed a thermodynamic model to simulate hydrogen refueling using the same system configuration as the experimental setup of Liu et al. (2010), while assuming adiabatic conditions. The modest discrepancy between the predicted and measured temperature rise can be attributed to this assumption, as adiabatic refueling inherently leads to higher temperature elevations than non-adiabatic scenarios in which heat transfer to the surroundings occurs (Sadi and Deymi-Dashtebayaz 2019).

8.3.4 Effects of Initial Cylinder, Inlet Gas, and Ambient Temperature

The maximum temperature rise within a hydrogen storage cylinder during refueling is strongly influenced by the initial cylinder temperature, inlet hydrogen temperature, and ambient temperature. De Miguel et al. (2015) examined the thermal response of Type III and Type IV hydrogen tanks during refueling and assessed the role of initial cylinder temperature. Their results demonstrated a reduction in temperature rise with increasing initial temperature. Based on the reported data, the relationship between temperature rise and initial cylinder temperature can be described by a linear correlation with a coefficient of determination (R^2) exceeding 0.925. Similar linear trends were also reported by Li et al. (2022) and Melideo and Baraldi (2015), confirming the mitigating effect of higher initial cylinder temperatures on thermal accumulation during refueling. In contrast, inlet hydrogen temperature and ambient temperature exert an opposite influence on the thermal behavior of the cylinder. An increase in either parameter leads to a higher temperature rise during refueling, with several studies reporting a linear relationship between these variables and the resulting temperature increase (Zhao et al. 2010). Experimental investigations by Zhao et al. (2010) and Zheng et al. (2012) revealed that a 1 °C rise in ambient temperature results in an approximate 0.3 °C increase in the refueling-induced temperature rise within the cylinder. However, Liu et al. (2010) reported a comparatively minor influence of ambient temperature on cylinder temperature evolution, suggesting that its effect may be secondary under certain refueling conditions. The inlet hydrogen temperature has been identified as a particularly influential parameter. Observed that a 1 °C increase in inlet gas temperature led to an approximately 1 °C increase in the cylinder gas temperature during refueling. Refueling with hydrogen at elevated inlet temperatures reduces gas density, thereby decreasing the mass of hydrogen admitted into the cylinder for a given pressure level. This reduction in mass inflow contributes to enhanced temperature rise within the cylinder due to increased compressive and thermal effects during refueling (2014).

8.3.5 Effects of Mass Flow Rate

Hydrogen refueling time can be significantly reduced by increasing the mass flow rate; however, this acceleration limits the duration available for heat transfer between the hydrogen gas and the storage cylinder, resulting in a pronounced increase in cylinder temperature that may exceed the recommended upper safety limit of 85 °C (Li et al. 2012). Experimental and numerical investigations by Zhao et al. (2010), Liu et al. (2010), De Miguel et al. (2015), and Li et al. (2022) consistently reported an increase in temperature rise with increasing hydrogen mass flow rate. The relationship between mass flow rate and temperature rise is inherently nonlinear. While the maximum temperature rise increases with higher flow rates, the rate of temperature increase progressively diminishes, indicating a saturation-like behavior at elevated flow conditions. Accordingly, the combined data from these studies are best described by a logarithmic correlation between hydrogen mass flow rate and maximum temperature rise. Supporting this observation, Zhao et al. (2010) conducted numerical simulations across a wide range of mass flow rates (12, 18, 25, 40, 65, 130, and 200 g s^{-1}) and found that the difference in temperature rise between flow rates of 130 and 200 g s^{-1} was negligible, suggesting a threshold beyond which further increases in flow rate provide limited thermal benefit while potentially exacerbating safety concerns.

8.3.6 Effects of Other Parameters

In addition to operating conditions, structural factors such as cylinder type and geometric dimensions significantly influence heat transfer characteristics and, consequently, the temperature rise during hydrogen refueling. De Miguel et al. (2015) experimentally examined refueling behavior in Type III and Type IV hydrogen storage tanks under varying initial temperatures and hydrogen mass flow rates and reported consistently higher gas temperatures in Type IV tanks. This behavior is primarily attributed to differences in liner material properties. Type III tanks incorporate an aluminum inner liner with relatively high thermal conductivity, which enhances heat dissipation from the compressed hydrogen to the tank walls. In contrast, Type IV tanks employ a polymer-based liner with substantially lower thermal conductivity, resulting in reduced heat transfer and higher gas temperatures during refueling. Cylinder dimensions also play a critical role in thermal behavior. Zheng et al. (2012) investigated the influence of cylinder size on temperature rise and observed that larger cylinders exhibited more pronounced non-uniform temperature distributions compared to smaller cylinders. This spatial temperature variation arises from differences in surface-to-volume ratio and internal flow dynamics, which affect both heat transfer efficiency and thermal stratification during the refueling process.

8.4 Recent Technological Advancements-Mobile/Portable Refueling Station

The availability of hydrogen refueling infrastructure is a fundamental requirement for the large-scale deployment and commercialization of hydrogen-powered vehicles. Nevertheless, the substantial capital investment and installation costs associated with conventional stationary refueling stations continue to represent a significant obstacle to the widespread adoption of hydrogen fuel cell electric vehicles (HFCEVs). In response to this challenge, China has achieved notable progress in advancing hydrogen refueling infrastructure development under the guidance of the national technical standard GB 50516–2010. A key early milestone was the development of the world's first mobile hydrogen refueling station in 2004 by Tongji University in collaboration with Shanghai Sunwise Energy System Company, designed to support China's initial HFCEV demonstration programs (Li et al. 2014). Around the same period, Niedwiecki et al. (2004) proposed and patented an alternative concept for a portable hydrogen refueling station. Mobile hydrogen refueling stations constitute an integrated and flexible infrastructure solution that offers several advantages over conventional stationary installations, including compact system design, reduced capital and installation costs, rapid deployment capability, lower maintenance requirements, and enhanced operational reliability. Although specific system configurations differ among developers, the mobile refueling station developed by Tongji University and Shanghai Sunwise Energy System Company typically comprises interconnected high-pressure tube vessels for hydrogen storage, pneumatically driven boosters replacing traditional compressors for pressure amplification, a nitrogen system to actuate the boosters, and a dispensing unit equipped with a control panel and chiller. Comprehensive safety provisions—including pressure and temperature regulators, hydrogen sensors, safety valves, rupture discs, and emergency shutdown systems—are integral components of the design (Sun et al. 2014). Several alternative mobile refueling concepts have since been introduced by industry. For example, Hylium Industries developed a mobile system incorporating a liquid hydrogen storage tank, pump, gasifier, buffer tank, and dispenser (Hylium Industries 2022), while the FUELINGAS system integrates a compressor, high-pressure storage vessels, and a dispensing unit (FUELINGAS 2022). Building on earlier designs, Tongji University and Shanghai Sunwise Energy System Company introduced a third-generation mobile refueling station in 2007, featuring enhanced refueling efficiency, upgraded safety systems, and improved operational protocols. This system was deployed during the Expo Shanghai 2010, where two portable refueling stations successfully serviced 196 HFCEVs—including buses, passenger cars, and sightseeing vehicles—delivering a total of 15,204 kg of hydrogen across 20,302 refueling events without any major safety incidents (Pan et al. 2010). Subsequent safety assessments indicated that the associated risks were well below the acceptable limits established for stationary refueling stations (Pan et al. 2013). In 2014, China further strengthened regulatory support by introducing a dedicated standard for mobile hydrogen refueling stations (GBT 31139–2014), covering operating pressures from 15 to 70 MPa (Li et al. 2014). According to this standard, the working pressure of onboard hydrogen storage systems must be maintained at 1.25–1.5 times the nominal refueling pressure. In recent years, several companies have commercialized mobile hydrogen refueling technologies and now provide associated operational and maintenance services. Notably, Wystrach's mobile refueling system, “WyRefueler,” has been deployed in multiple European countries—including Belgium, Germany, France, and the Netherlands—under the H2Share project to support hydrogen refueling for heavy-duty trucks (Wystrach 2022). The same system has also been utilized within the HyBayern project for refueling passenger vehicles and buses. Air Products has implemented self-contained portable hydrogen refueling stations with storage capacities of up to 150 kg at various locations worldwide. In Australia, Queensland University of Technology initiated research activities in 2021 to develop a portable hydrogen refueling system in collaboration with H2H Energy Pty Ltd. and the Queensland Government (CSIRO 2022). Additionally, MAXIMATOR has developed a portable refueling station capable of operating at pressures up to 105 MPa, while OneH2 has introduced a mobile refueler capable of dispensing hydrogen directly from high-pressure trailers at pressures between 35 and 70 MPa (OneH2 2023). NanoSUN has developed a portable hydrogen refueling system known as the “Pioneer 145,” which can refuel approximately 60 delivery vans equipped with 5 kg onboard storage at 35 MPa within 10–13 min, or up to seven heavy-duty trucks with 40 kg storage capacity within 20–60 min (NanoSUN 2022). At 70 MPa, the system is capable of refueling up to 95 passenger vehicles in approximately 10 min. Its modular architecture enables scalable expansion of the refueling network by deploying additional units in line with fleet growth. Moreover, the system employs a flexible cascade refueling strategy rather than conventional decant refueling, eliminating the need for a compressor and increasing the number of complete refueling cycles. Hylium Industries, based in South Korea, has further advanced mobile refueling technology by developing the world’s first mobile liquid hydrogen refueling station capable of directly servicing up to 25 vehicles per day (Hylium Industries 2022). By replacing conventional compressors and chillers used in gaseous hydrogen stations with a liquid hydrogen pump and gasifier, the system achieves approximately 70% lower energy consumption. Collectively, these developments demonstrate the technical feasibility, safety, and practicality of mobile hydrogen refueling stations, particularly for early-stage infrastructure deployment and small- to medium-scale fleet operations. Figure 8.2 presents a comparative overview of fuel cell vehicle deployment and hydrogen refueling station distribution by region in 2020, highlighting both the challenges and opportunities associated with the expansion of hydrogen mobility infrastructure.

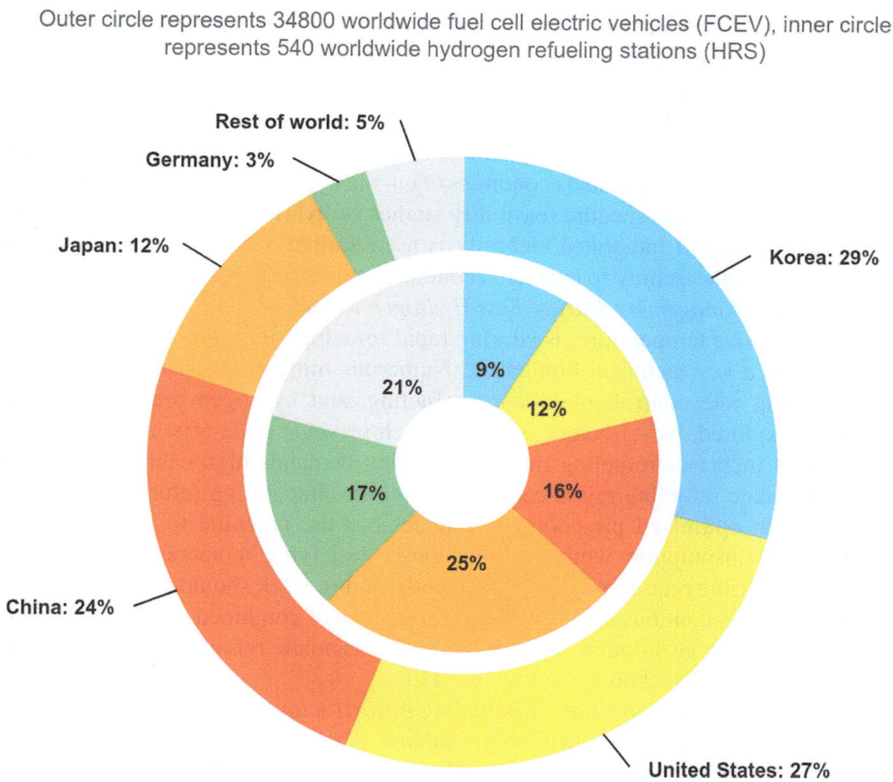

Fig. 8.2 Regional stock of fuel cell vehicles and hydrogen refueling Infrastructure, 2020 (IEA 2021)

8.5 Critical Analysis and Future Research Areas

Research spanning hydrogen production through to its utilization in the transportation sector has attracted substantial attention in recent years, leading to notable technological progress. Despite these advancements, several critical challenges remain unresolved and must be addressed to enable large-scale commercial deployment. Accordingly, this study highlights key directions for future research, as summarized below.

1. *Hydrogen Production and Delivery Infrastructure*
   The transportation of hydrogen from centralized production facilities to refueling stations remains a major challenge due to the high infrastructure investment required, which significantly increases the levelized cost of hydrogen refueling. On-site, small- to medium-scale hydrogen production using mature technologies such as water electrolysis powered by renewable electricity offers a viable alternative by eliminating hydrogen transport and reducing end-user costs. Several commercial systems with production capacities ranging from approximately

20 kg day^{-1} to more than 1000 kg day^{-1} have already been developed, with pilot-scale and early commercial deployments underway in Europe, the USA, and Canada (McPhy 2023; Hydrogen Central 2023). However, renewable energy resources are inherently variable and location dependent, with fluctuations driven by geography and seasonal conditions. This variability can significantly influence the reliability and economics of on-site hydrogen generation. Therefore, country- or region-specific feasibility studies of hybrid energy systems such as solar–wind–grid integrated electrolysis are required, explicitly accounting for renewable uncertainty to evaluate economic returns and investment viability.

2. *Thermal Management During Fast Hydrogen Refueling*

   Excessive temperature rise during rapid refueling of high-pressure hydrogen remains a key technical bottleneck. Numerous mitigation strategies including refueling rate control, multi-stage refueling, and hydrogen pre-cooling have been explored. Nevertheless, each approach presents trade-offs: mass flow rate control increases refueling time, pre-cooling demands high energy input, and multi-stage refueling raises capital expenditure. Integrating refueling rate control with optimized pre-cooling strategies has the potential to reduce overall energy consumption while simultaneously satisfying temperature limits and refueling time requirements. Consequently, future work should focus on experimental investigations assessing the performance of combined strategies, including energy consumption, compliance with standard refueling protocols, and comprehensive techno-economic feasibility.

3. *Planning and Optimization of Refueling Infrastructure*

   The lack of adequate hydrogen refueling infrastructure remains a significant barrier to the widespread adoption of hydrogen fuel cell vehicles, particularly for long-haul and heavy-duty transport applications. Strategic planning and spatial assessment are essential prerequisites for effective infrastructure deployment. Although some studies have addressed national-level refueling infrastructure planning (Yaïci and Longo 2022), further geographically diverse investigations are required. The development of advanced GIS-based methodologies that incorporate existing vehicle distributions, projected fleet growth, hydrogen delivery pathways, and customer demand would enable identification of optimal refueling station locations and support informed policy development for phased infrastructure expansion.

4. *Role of Mobile Hydrogen Refueling Stations*

   Mobile hydrogen refueling stations represent a promising alternative to stationary stations for small and emerging fleets, where the high capital cost of permanent infrastructure is economically prohibitive. While current deployments demonstrate their feasibility for limited-scale operations, future research should focus on the supply chain design and network optimization of mobile refueling systems for larger fleets and heavy-duty vehicles. Comparative techno-economic assessments between mobile and stationary refueling solutions considering scalability, operating costs, safety, and system flexibility are essential to determine their long-term role within evolving hydrogen refueling networks.

8.6 Conclusion

The transportation sector represents one of the largest sources of global CO_2 emissions, thereby substantially exacerbating climate change and global warming. Within this context, hydrogen fuel cell vehicles (HFCVs) have emerged as a strategically important pathway toward deep decarbonization and sustainable mobility, offering a low-emission alternative to conventional diesel- and gasoline-powered internal combustion engine (ICE) vehicles. This study presents a comparative evaluation of passenger and commercial HFCVs against ICE vehicles, plug-in hybrid electric vehicles (PHEVs), and battery electric vehicles (BEVs), using key performance metrics including well-to-wheel (WTW) efficiency, driving range, fuel or energy consumption, refueling or recharging time, tailpipe and life-cycle emissions, and overall vehicle cost. The comparative results indicate that BEVs currently dominate the passenger vehicle segment, largely driven by lower initial purchase costs, improving battery performance, and the rapid expansion of electric charging infrastructure. In contrast, HFCVs exhibit clear advantages in commercial and heavy-duty transport applications, where rapid refueling and long driving range are critical requirements for high-utilization, long-haul, and fleet-based operations. These attributes position hydrogen-based propulsion as a particularly attractive solution for sectors in which battery mass, charging duration, and operational downtime remain limiting factors. Despite these advantages, large-scale deployment of HFCVs is constrained by the limited availability of hydrogen refueling infrastructure compared with the well-established gasoline and electric charging networks, as well as by the relatively high cost of hydrogen production. However, the increasing penetration of renewable energy sources presents a significant opportunity to reduce the cost of electricity used in water electrolysis, thereby lowering hydrogen production costs and accelerating the transition toward green hydrogen. Concurrently, continued technological progress in hydrogen production, storage, transport, and distribution systems is expected to further enhance the economic competitiveness of hydrogen-based mobility solutions. In recent years, a marked global expansion of hydrogen refueling infrastructure and electrolysis capacity has been observed, with particularly strong growth in the Asia–Pacific region. Government policies, regulatory frameworks, and targeted financial incentives play a decisive role in driving this momentum. Direct subsidies for consumers and manufacturers, coupled with sustained investment in research, development, and demonstration activities, are essential to advancing hydrogen production technologies, scaling refueling infrastructure, and reducing system-level costs. Looking forward, future research priorities should include large-scale renewable hydrogen production, the development of cost-effective fuel cell materials and manufacturing routes, improvements in hydrogen storage and distribution technologies, and advances in materials science aimed at reducing fuel cell stack costs. In addition, comprehensive life-cycle assessments of HFCVs are required to fully quantify their long-term environmental performance, with particular emphasis on recycling strategies and end-of-life management of proton exchange membrane (PEM) components, as well as broader environmental trade-offs. Stronger collaboration among governments, industry stakeholders, and academic institutions will be critical to overcoming remaining technical, economic, and infrastructural challenges. Through coordinated action, HFCVs can play a transformative role in the transition to a decarbonized transportation sector, contributing substantially to global climate mitigation goals and long-term energy sustainability.

References

Acar C, Dincer I (2017) Testing and performance evaluation of a hybrid photoelectrochemical hydrogen production system. Int J Hydrog Energy 42(6):3605–3613

Acar C, Ghosh S, Dincer I, Zamfirescu C (2015) Evaluation of a new continuous type hybrid photo-electrochemical system. Int J Hydrog Energy 40(34):11112–11124

Acar C, Dincer I, Naterer GF (2016) Clean hydrogen and power from impure water. J Power Sources 331:189–197

AlRafea K, Fowler M, Elkamel A, Hajimiragha A (2016) Integration of renewable energy sources into combined cycle power plants through electrolysis generated hydrogen in a new designed energy hub. Int J Hydrog Energy 41(38):16718–16728

Aminudin MA, Kamarudin SK, Lim BH et al (2023) An overview: current progress on hydrogen fuel cell vehicles. Int J Hydrog Energy 48(11):4371–4388. https://doi.org/10.1016/j.ijhydene.2022.10.156

Amoo LM, Fagbenle RL (2014) An integrated impact assessment of hydrogen as a future energy carrier in Nigeria's transportation, energy and power sectors. Int J Hydrog Energy 39(24):12409–12433

Baroutaji A, Arjunan A, Robinson J et al (2024) Additive manufacturing for Proton Exchange Membrane (PEM) hydrogen technologies: merits, challenges, and prospects. Int J Hydrog Energy 52:561–584. https://doi.org/10.1016/j.ijhydene.2023.07.033

Bayrak ZU, Kaya U, Oksuztepe E (2020) Investigation of PEMFC performance for cruising hybrid powered fixed-wing electric UAV in different temperatures. Int J Hydrog Energy 45(11):7036–7045. https://doi.org/10.1016/j.ijhydene.2019.12.214

Boretti A (2020) Series BEV with a small battery pack and high-efficiency ICE onboard electricity production: B-class, high-roof hatchback and Le Mans Hypercar applications (No. 2020-01-2250). SAE Technical Paper

Cammack R, Frey M, Robson R (eds) (2001) Hydrogen as a fuel: learning from nature, 1st edn. CRC Press. https://doi.org/10.1201/9780203471043

Cecere D, Giacomazzi E, Ingenito A (2014) A review on hydrogen industrial aerospace applications. Int J Hydrog Energy 39(20):10731–10747. https://doi.org/10.1016/j.ijhydene.2014.04.126

Chen J, Fan R, Qin N et al (2024) Recent advances on materials and technologies of composite bipolar plate for proton exchange membrane fuel cell. J Power Sources 594:234009. https://doi.org/10.1016/j.jpowsour.2023.234009

Cipriani G, Di Dio V, Genduso F, La Cascia D, Liga R, Miceli R et al (2014) Perspective on hydrogen energy carrier and its automotive applications. Int J Hydrog Energy 39(16):8482–8494

CSIRO (2022) Mobile hydrogen refueller. https://research.csiro.au/hyresearch/mobile-hydrogen-refueller/. Accessed 2 Dec 2022

Da Costa GP, Garcia DMG, Nguyen THV et al (2024) Advancements in printed components for proton exchange membrane fuel cells: a comprehensive review. Int J Hydrog Energy 69:710–728. https://doi.org/10.1016/j.ijhydene.2024.05.072

De Miguel N, Ortiz Cebolla R, Acosta B, Moretto P, Harskamp F, Bonato C (2015) Compressed hydrogen tanks for onboard application: thermal behaviour during cycling. Int J Hydrog Energy 40:6449–6458. https://doi.org/10.1016/J.IJHYDENE.2015.03.035

De Miguel N, Acosta B, Baraldi D, Melideo R, Ortiz Cebolla R, Moretto P (2016) The role of initial tank temperature on refuelling of on-board hydrogen tanks. Int J Hydrog Energy 41:8606–8615. https://doi.org/10.1016/J.IJHYDENE.2016.03.158

de-Troya JJ, Álvarez C, Fernández-Garrido C, Carral L (2016) Analysing the possibilities of using fuel cells in ships. Int J Hydrog Energy 41(4):2853–2866. https://doi.org/10.1016/j.ijhydene.2015.11.145

Dincer I, Acar C (2016) A review on potential use of hydrogen in aviation applications. Int J Sustain Aviat 2(1):74–100

Dincer I, Acar C (2017) Smart energy systems for a sustainable future. Appl Energy 194:225–235

Dudek M, Lis B, Raźniak A, Krauz M, Kawalec M (2021) Selected aspects of designing modular PEMFC stacks as power sources for unmanned aerial vehicles. Appl Sci 11(2):675. https://doi.org/10.3390/app11020675

Dutta S (2014) A review on production, storage of hydrogen and its utilization as an energy resource. J Ind Eng Chem 20(4):1148–1156

Ehret O, Bonhoff K (2015) Hydrogen as a fuel and energy storage: success factors for the German Energiewende. Int J Hydrog Energy 40(15):5526–5533

Fan L, Tu Z, Chan SH (2021) Recent development of hydrogen and fuel cell technologies: a review. Energy Rep 7:8421–8446. https://doi.org/10.1016/j.egyr.2021.08.00

Fan R, Chang G, Xu Y, Zhang Y, Wei P (2024) Enhancing PEM fuel cell dynamic performance: co-optimization of cathode catalyst layer composition and operating conditions using a novel surrogate model. Renew Energy 231:120993. https://doi.org/10.1016/j.renene.2024.120993

FUELINGAS (2022) Hydrogen refuelling station (portable/skidmounted). https://www.fuelingas.com/product/hydrogen_refueling_station_c_series_general_type. Accessed 1 Dec 2022

Galich A, Marz L (2012) Alternative energy technologies as a cultural endeavor: a case study of hydrogen and fuel cell development in Germany. Energy Sustain Soc 2(1):2. https://doi.org/10.1186/2192-0567-2-2

Gao X, Chen J, Xu R et al (2024) Research progress and prospect of the materials of bipolar plates for proton exchange membrane fuel cells (PEMFCs). Int J Hydrog Energy 50:711–743. https://doi.org/10.1016/j.ijhydene.2023.09.005

Hydrogen Central (2023) Hydra energy breaks ground on world's largest hydrogen refuelling station for heavy-duty trucks. https://hydrogen-central.com/category/refuellingstations/. Accessed 6 Feb 2023

Hylium Industries (2022) Mobile hydrogen refueling station. http://hylium-industries.com/hrs. Accessed 1 Dec 2022

IEA (2021) Fuel cell vehicles and hydrogen refueling station stock by region, 2020. IEA, Paris. https://www.iea.org/data-and-statistics/charts/fuel-cell-vehicles-and-hydrogen-refueling-station-stock-by-region-2020. Licence: CC BY 4.0

IEA (2024) Transport sector renewable fuel growth by type, main case, 2016–2030. IEA, Paris. https://www.iea.org/data-and-statistics/charts/transport-sector-renewable-fuel-growth-by-type-main-case-2016-2030. Licence: CC BY 4.0

Jhunjhunwala A, Kaur P, Mutagekar S (2018) Electric vehicles in India: a novel approach to scale electrification. IEEE Electrific Mag 6(4):40–47. https://doi.org/10.1109/mele.2018.2871278

Ji Z, Chen J, Pérez-Page M et al (2022) Doped graphene/carbon black hybrid catalyst giving enhanced oxygen reduction reaction activity with high resistance to corrosion in proton exchange membrane fuel cells. J Energy Chem 68:143–153. https://doi.org/10.1016/j.jechem.2021.09.031

Johnson T, Joshi A (2018) Review of vehicle engine efficiency and emissions. SAE Int J Engines 11:1307–1330

Kahraman H, Akin Y (2024) Recent studies on proton exchange membrane fuel cell components, review of the literature. Energy Convers Manag 304:118244. https://doi.org/10.1016/j.enconman.2024.118244

Kim SC, Lee SH, Yoon KB (2010) Thermal characteristics during hydrogen fueling process of type IV cylinder. Int J Hydrog Energy 35:6830–6835. https://doi.org/10.1016/J.IJHYDENE.2010.03.130

Larsson M, Mohseni F, Wallmark C, Gronkvist S, Alvfors P (2015) Energy system analysis of the implications of hydrogen fuel cell vehicles in the Swedish road transport system. Int J Hydrog Energy 40(35):11722–11729

Leo TJ, Durango JA, Navarro E (2010) Exergy analysis of PEM fuel cells for marine applications. Energy 35(2):1164–1171. https://doi.org/10.1016/j.energy.2009.06.010

Li Q, Zhou J, Chang Q, Xing W (2012) Effects of geometry and inconstant mass flow rate on temperatures within a pressurized hydrogen cylinder during refueling. Int J Hydrog Energy 37:6043–6052. https://doi.org/10.1016/J.IJHYDENE.2011.12.020

Li Z, Pan X, Sun K, Zhou W, Gao D, Liu S et al (2014) Development of safety standard for mobile hydrogen refueling facilities in China. Int J Hydrog Energy 39:13935–13939. https://doi.org/10.1016/J.IJHYDENE.2014.02.017

Li JQ, Li JCL, Park K, Kwon JT (2022) Investigation on the changes of pressure and temperature in high pressure filling of hydrogen storage tank. Case Stud Therm Eng 37:102143. https://doi.org/10.1016/J.CSITE.2022.102143

Li JQ, Chen Y, Ma YB, Kwon JT, Xu H, Li JC (2023) A study on the Joule-Thomson effect of during filling hydrogen in high pressure tank. Case Stud Therm Eng 41:102678. https://doi.org/10.1016/J.CSITE.2022.102678

Liu YL, Zhao YZ, Zhao L, Li X, Chen H, gang, Zhang LF et al (2010) Experimental studies on temperature rise within a hydrogen cylinder during refueling. Int J Hydrog Energy 35:2627–2632. https://doi.org/10.1016/J.IJHYDENE.2009.04.042

Liu J, Zhang L, Yuan B et al (2024) Design and development of coating for metallic bipolar plates in proton exchange membrane fuel cell (PEMFC): a review. Mater Des 246:113338. https://doi.org/10.1016/j.matdes.2024.113338

Low J, Haszeldine RS, Mouli-Castillo J (2020) Comparative evaluation of battery electric and hydrogen fuel cell electric vehicles for zero carbon emissions road vehicle fuel in Scotland. engrXiv

Mancino AN, Menale C, Vellucci F et al (2023) PEM fuel cell applications in road transport. Energies 16(17):6129. https://doi.org/10.3390/en16176129

Maus S, Hapke J, Ranong CN, Wüchner E, Friedlmeier G, Wenger D (2008) Filling procedure for vehicles with compressed hydrogen tanks. Int J Hydrog Energy 33:4612e21. https://doi.org/10.1016/J.IJHYDENE.2008.06.052

MBIE (2019) Vision for hydrogen in New Zealand. Energy strategies for New Zealand. Available at: https://www.mbie.govt.nz/building-and-energy/energy-and-natural-resources/energy-strategies-for-new-zealand/. Accessed 19 Aug 2021

McPhy (2023) Fill up on clean energy in just a few minutes. https://mcphy.com/en/equipment-services/hydrogenstations/. Accessed 6 Feb 2023

Melideo D, Baraldi D (2015) CFD analysis of fast filling strategies for hydrogen tanks and their effects on key-parameters. Int J Hydrog Energy 40:735–745. https://doi.org/10.1016/J.IJHYDENE.2014.10.138

Miao M, Li-Xiao S, Jing L et al (2024) Recent advances in Pt catalysts and membrane electrode assemblies fabrication for proton exchange membrane fuel cells. Rare Met 43(9):4198–4221. https://doi.org/10.1007/s12598-024-02698-6

Mir MY (2025) Comparative perspectives of biofuel generations: new insights into challenges and opportunities. In: Sustainable fuels production. Springer, pp 61–78

Mohsen AM, Basem A (2024) Enhancing heat removal and H2O retention in passive aircooled polymer electrolyte membrane fuel cells by altering flow field geometry. Sustain Times 16:4666. https://doi.org/10.3390/su16114666

Muthukumar M, Rengarajan N, Velliyangiri B, Omprakas M, Rohit C, Raja UK (2021) The development of fuel cell electric vehicles–a review. Mater Today Proc 45:1181–1187

NanoSUN (2022) Looking to fuel a hydrogen powered vehicle fleet? https://www.nanosun.co.uk/pioneer/pioneerhydrogen-refuelling-station. Accessed 1 Dec 2022

Niedwiecki A, Sirosh N, Abele A (2004) Transportable hydrogen refueling station. US6755225B1, Washington, DC

OneH2 (2023) Dispensers. https://oneh2.com/solutions/dispensers/. Accessed 5 Feb 2023

Pan X, Lv H, Zhou W, Ma J, Gao D (2010) Design and operation of the hydrogen supply chain for fuel-cell vehicles in Expo Shanghai 2010. World Electr Veh J 4:671–676. https://doi.org/10.3390/WEVJ4040671

Pan X, Ma J, Li Z (2013) Risk analysis on mobile hydrogen refueling stations in the world Expo Shanghai. Int Conf Hydrog Saf Hydrog Knowl Cent:1–12

Parekh A (2022) Recent developments of proton exchange membranes for PEMFC. A review. Front Energy Res 10. https://doi.org/10.3389/enrg.2022.95613

Pramuanjaroenkij A, Kakaç S (2023) The fuel cell electric vehicles: the highlight review. Int J Hydrog Energy 48:9401–9425

Rao Y, Shao Z, Ahangarnejad AH, Gholamalizadeh E, Sobhani B (2019) Shark smell optimizer applied to identify the optimal parameters of the proton exchange membrane fuel cell model. Energy Convers Manag 182:1–8. https://doi.org/10.1016/j.enconman.2018.12.057

Rouleau L, Duffour F, Walter B, Kumar R, Nowak L (2021) Experimental and numerical investigation on hydrogen internal combustion engine (No. 2021-24-0060). SAE Technical Paper

Sadi M, Deymi-Dashtebayaz M (2019) Hydrogen refueling process from the buffer and the cascade storage banks to HV cylinder. Int J Hydrog Energy 44:18496–18504. https://doi.org/10.1016/J.IJHYDENE.2019.05.023

Salam A, Zholobko O, Wu X (2024) Roles of functionalized nanoparticles in the performance improvement of proton-exchange membranes used in low- and intermediate-temperature hydrogen fuel cells: a review. Prog Nat Sci Mater Int 34(3):437–453. https://doi.org/10.1016/j.pnsc.2024.04.004

Salvi BL, Subramanian KA (2015) Sustainable development of road transportation sector using hydrogen energy system. Renew Sustain Energy Rev 51:1132–1155

Sattler G (2000) Fuel cells going onboard. J Power Sources 86(1–2):61–67. https://doi.org/10.1016/s0378-7753(99)00414-0

Schneider J, Suckow T, Lynch F, Ward J, Caldwell M, Tillman J, Mathison S, Stephanian G, Richards M, Liss B, Quong S, Durán A, Friedlmeier G, Maus S, King J, Canteen Walla Z, Moorhead B, Adler R, Chernicoff W, Sloan C, Steele M, Cherry J (2005) Optimizing hydrogen vehicle refueling. In: Proceedings of the National Hydrogen Association Annual Hydrogen Conference (NHA 2005), Washington, DC, USA

Segale M, Seadira T, Sigwadi R et al (2024) A new frontier towards the development of efficient SPEEK polymer membranes for PEM fuel cell applications: a review. Mater Adv 5(20):7979–8006. https://doi.org/10.1039/D4MA00628C

Sherif SA, Goswami DY, Stefanakos EL, Steinfeld A (eds) (2014) Handbook of hydrogen energy. CRC Press

Solangi NH, Mubarak NM, Karri RR et al (2024) Recent development of graphene and MXene-based nanomaterials for proton exchange membrane fuel cells. Int J Hydrog Energy 73:905–931

Stetson NT, McWhorter S, Ahn CC (2016) Introduction to hydrogen storage. Compend Hydrog Energy:3–25. https://doi.org/10.1016/B978-1-78242-362-1.00001-8

Stougie L, Giustozzi N, van der Kooi H, Stoppato A (2018) Environmental, economic and exergetic sustainability assessment of power generation from fossil and renewable energy sources. Int J Energy Res 42:2916–2926

Sun K, Pan X, Li Z, Ma J (2014) Risk analysis on mobile hydrogen refueling stations in Shanghai. Int J Hydrog Energy 39:20411–20419. https://doi.org/10.1016/J.IJHYDENE.2014.07.098

Suryan A, Kim HD, Setoguchi T (2012) Three dimensional numerical computations on the fast filling of a hydrogen tank under different conditions. Int J Hydrog Energy 37:7600–7611. https://doi.org/10.1016/J.IJHYDENE.2012.02.019

Tan YH (2013) Research on large thrust liquid rocket engine. Yuhang Xuebao/J Astronaut 34:1303–1308

Topler J (2017) Hydrogen technology and economy in Germany-history and present state. In: Lordache I (ed) Hydrogen in an international context: vulnerabilities of hydrogen energy in emerging markets. River Publishers, Gistrup; Delft, pp 3–48

Trencher G, Edianto A (2021) Drivers and barriers to the adoption of fuel cell passenger vehicles and buses in Germany. Energies 14(4):833. https://doi.org/10.3390/en14040833

Tronstad T, Åstrand HH, Haugom GP, Langfeldt L (2017) Study on the use of fuel cells in shipping. Report to European Maritime Safety Agency by DNV GL; DNV GL, Oslo

van Biert L, Godjevac M, Visser K, Aravind PV (2016) A review of fuel cell systems for maritime applications. J Power Sources 327:345–364. https://doi.org/10.1016/j.jpowsour.2016.07.007

Wang H, Wang Z, Sun M, Wu H (2013) Combustion modes of hydrogen jet combustion in a cavity-based supersonic combustor. Int J Hydrog Energy 38(27):12078–12089

Wang Y, Pang YH, Xu H, Martinez A, Chen KS (2022) PEM fuel cell and electrolysis cell technologies and hydrogen infrastructure development – a review. Energy Environ Sci 15:2288–2328. https://doi.org/10.1039/d2ee00790h

Waseem M, Amir M, Lakshmi GS, Harivardhagini S, Ahmad M (2023) Fuel cell-based hybrid electric vehicles: an integrated review of current status, key challenges, recommended policies, and future prospects. Green Energy Intell Transp 2:100121

Wystrach (2022) WyRefueler- high pressure hydrogen refuelling. https://www.wystrach.gmbh/en/products/wyrefueler/. Accessed 1 Dec 2022

Yaïci W, Longo M (2022) Feasibility investigation of hydrogen refuelling infrastructure for heavy-duty vehicles in Canada. Energies 15:2848. https://doi.org/10.3390/EN15082848

Zhang XR (2017) Natural future of energy utilization. Int J Energy Res 41(6):757–760

Zhang X, Wang Y, Yuan X, Shen Y, Lu Z (2022) Adaptive dynamic surface control with disturbance observers for battery/supercapacitor-based hybrid energy sources in electric vehicles. IEEE Trans Transp Electrif 9:5165–5181

Zhao L, Liu Y, Yang J, Zhao Y, Zheng J, Bie H et al (2010) Numerical simulation of temperature rise within hydrogen vehicle cylinder during refueling. Int J Hydrog Energy 35:8092–8100. https://doi.org/10.1016/J.IJHYDENE.2010.01.027

Zheng J, Liu X, Xu P, Liu P, Zhao Y, Yang J (2012) Development of high pressure gaseous hydrogen storage technologies. Int J Hydrog Energy 37:1048–1057. https://doi.org/10.1016/J.IJHYDENE.2011.02.125

Zheng J, Guo J, Yang J, Zhao Y, Zhao L, Pan X et al (2013) Experimental and numerical study on temperature rise within a 70 MPa type III cylinder during fast refueling. Int J Hydrog Energy 38:10956–10962. https://doi.org/10.1016/J.IJHYDENE.2013.02.053

Zhou Y, Wang B, Ling Z et al (2024) Advances in ionogels for proton-exchange membranes. Sci Total Environ 921:171099. https://doi.org/10.1016/j.scitotenv.2024.171099

Chapter 9
Hydrogen for Industrial Decarbonization: Technological Advances and Sociotechnical Dynamics

Abstract Industrial decarbonization is critical for achieving global climate targets, yet it remains particularly challenging for energy-intensive and hard-to-abate sectors such as steel, cement, and chemical production. Hydrogen has emerged as a key energy carrier with the potential to support deep emissions reductions by providing high-temperature process heat, acting as a feedstock for industrial processes, and enabling long-duration energy storage. This chapter examines the role of hydrogen in industrial decarbonization, focusing on recent technological advances in low-carbon and zero-carbon hydrogen production pathways, including renewable-based electrolysis and hydrogen systems integrated with carbon capture technologies. Beyond technical considerations, this chapter adopts a sociotechnical perspective to assess economic feasibility, infrastructure requirements, policy frameworks, and social acceptance issues that shape hydrogen deployment across industrial sectors. By analyzing both technological readiness and broader systemic dynamics, this chapter highlights the opportunities and constraints associated with hydrogen adoption and outlines pathways for accelerating its integration into industrial decarbonization strategies.

Keywords Hydrogen · Industrial decarbonization · Sociotechnical systems · Energy impact · Climate impacts

9.1 Introduction

The deployment of hydrogen within the industrial sector has the potential to play a transformative role in the global energy transition, particularly in the decarbonization of hard-to-abate industries. Hydrogen generated from renewable energy sources is especially attractive, as it can function both as a feedstock integrated into industrial processes and as a fuel, either partially or fully substituting natural gas. However, the diffusion of new energy technologies typically follows an incremental rather than disruptive pathway, with innovations often building upon existing industrial systems and infrastructures. In industrial applications requiring process

D. M. Y. Mir et al., *Low-Carbon Hydrogen*, Clean Energy Production Technologies, https://doi.org/10.1007/978-981-95-8142-9_9

temperatures above 1500 °C, such as steelmaking and glass production, full electrification of heat generation remains technically and economically challenging. As a result, combustion-based technologies are expected to continue playing a central role in industrial heating. Achieving CO_2-neutral heat in these contexts, therefore, necessitates the development and deployment of low-emission combustion solutions, among which hydrogen combustion emerges as a particularly promising option. Green hydrogen, produced via renewable-powered electrolysis, is increasingly recognized as a viable carbon-free fuel for high-temperature industrial processes (Mayrhofer et al. 2021; Sorgulu et al. 2023). One pragmatic pathway for introducing hydrogen into industrial energy systems is through blending with conventional fuels. Such blending strategies facilitate the gradual integration of renewable energy into existing energy mixes, thereby enhancing both energy security and system sustainability (Franco and Giovannini 2023). Importantly, hydrogen blending allows industries to leverage existing infrastructure while progressively reducing dependence on fossil fuels, positioning it as an effective transitional strategy toward a hydrogen-based energy system. In this context, the use of hydrogen in blended combustion for burners, boilers, and furnaces has attracted growing attention (von Schéele 2022). Nevertheless, the use of hydrogen as a co-fuel with methane introduces significant technical challenges due to its distinct physical and chemical properties. Hydrogen's higher flame speed and broader flammability range can substantially alter combustion behavior, resulting in more compact and turbulent flames and increasing the risks of backfire and flashback (Amaduzzi et al. 2021). These effects necessitate modifications in combustion control strategies, such as optimized air staging, to stabilize flames and manage ignition characteristics. While hydrogen blending can significantly reduce CO_2 emissions, the elevated flame temperatures may increase NO_x formation, highlighting the need for advanced burner designs. Consequently, future combustion systems are increasingly oriented toward Dry Low NO_x and flameless combustion concepts to mitigate emissions while maintaining thermal efficiency. The development of such advanced combustion technologies for natural gas–hydrogen co-fired furnaces and boilers is therefore a key research and engineering priority (Xu et al. 2024). Beyond combustion dynamics, hydrogen's small molecular size and its propensity to interact with steel alloys pose additional challenges for pipelines, valves, and storage systems, requiring careful material selection and infrastructure adaptation. Moreover, replacing or blending fuels affects the operating range and stability of combustion systems. Differences in gas density between hydrogen and natural gas, which can vary by an order of magnitude under atmospheric conditions, significantly influence volumetric flow rates through fixed nozzles, heat release rates, and overall heat transfer performance. Since heat release rate is a critical parameter for boilers and furnaces, comparable volumetric flow rates of different gases can yield markedly different thermal outputs, complicating system design and control. Recent studies have examined the feasibility of hydrogen deployment in hard-to-abate sectors such as steel, glass, and paper manufacturing, emphasizing its multifaceted role as a combustion fuel, a reducing agent, and an energy storage medium coupled with process electrification and renewable energy integration (Franco and Giovannini 2023).

Collectively, these findings underscore both the significant potential of hydrogen in industrial decarbonization and the complex technical, infrastructural, and operational challenges that must be addressed to enable its large-scale adoption. Several studies have provided strong support for hydrogen deployment in industrial combustion systems through advanced computational approaches, including CFD and multiphysics modeling of hydrogen-rich fuel combustion in industrial boilers, as well as the potential integration of catalytic units in hard-to-abate technologies (Taamallah et al. 2015). Recent literature further suggests that low-energy catalytic processes could be leveraged to optimize hydrogen–methane combustion, enhancing reaction stability and efficiency while mitigating emissions (Osman et al. 2024). For instance, catalysts that improve hydrogen production efficiency or stabilize hydrogen–methane fuel blends may substantially enhance the technical feasibility and overall effectiveness of industrial decarbonization pathways. In parallel, several contributions have examined hydrogen adoption in specific industrial sectors—such as steel, glass, and related high-temperature manufacturing processes—focusing on its role as an alternative fuel in burners and identifying practical upper limits for hydrogen blending ratios (Zhan et al. 2022). Experimental investigations by Luzzo et al. evaluated hydrogen–natural gas mixtures containing 30% and 50% hydrogen by volume in existing industrial burners, demonstrating that both blends are technically viable, although combustion stability was notably higher at a 30% hydrogen concentration (Luzzo et al. 2022). Complementary work by Glanville et al. emphasized the potential benefits and relevance of higher hydrogen blending ratios in industrial combustion systems (Glanville et al. 2022). Moreover, Leicher et al. systematically assessed the impact of hydrogen utilization on burner performance metrics, including heat transfer rates, combustion efficiency, flame morphology, and NO_x emissions, while also evaluating the role of advanced measurement and control technologies in managing ignition behavior and excess air ratios (Leicher et al. 2018). Alongside technological advancements, policy support has emerged as a critical driver of hydrogen adoption in industry, particularly in the United States. The US government has made substantial investments to foster a hydrogen economy, with a strategic emphasis on low-carbon (clean) hydrogen as a means to reduce industrial greenhouse gas (GHG) emissions. In 2021, the U.S. Department of Energy (DOE) launched the Hydrogen Earth shot Challenge, aiming to reduce the unsubsidized cost of clean hydrogen by 80% within a decade—from approximately USD 5 to USD 1 per kilogram—while simultaneously enabling new market opportunities (DOE 2023). This initiative has been reinforced by the Bipartisan Infrastructure Law of 2023, which allocates USD 7 billion to the Regional Clean Hydrogen Hubs program to establish interconnected networks for clean hydrogen production, distribution, and end use (DOE 2023). Additionally, the Inflation Reduction Act of 2023 provides production tax credits for clean hydrogen based on lifecycle GHG emissions, further strengthening the economic case for deployment (DOE 2023). Under the Infrastructure Investment and Jobs Act of 2023, the DOE has also introduced targeted grant programs, including USD 1 billion for clean hydrogen electrolysis to reduce production costs and USD 500 million for clean hydrogen manufacturing to accelerate the development of hydrogen-related technologies (DOE 2023). Given the growing policy momentum and financial incentives supporting hydrogen in US industry, it is essential to critically assess hydrogen's role within a broader and more holistic industrial decarbonization strategy. Such an assessment must identify near-term investment opportunities while also addressing persistent technical, economic, and infrastructural challenges that require further research and development. The United States has committed to achieving net-zero GHG emissions by 2050, in alignment with the objective of limiting the increase in global average temperature to below 1.5 °C above preindustrial levels (IPCC 2023). Since industrial activities account for approximately 30% of total U.S. GHG emissions (EPA 2023), emissions reductions in this sector must closely track economy-wide decarbonization trajectories if national climate targets are to remain achievable. In many of these sectors, energy expenditures represent a substantial share of production costs for example, accounting for roughly one-third of total costs in basic steel production (Bettinger 2011). This strong economic incentive has already driven significant improvements in energy efficiency, with several core industrial processes, such as blast furnace pig iron production, approaching their practical efficiency limits (Gutowski et al. 2013). Although additional efficiency gains remain possible, they are widely acknowledged to be insufficient on their own to deliver net-zero emissions (Allwood et al. 2010). Consequently, a range of complementary decarbonization strategies has been proposed for hard-to-abate industries. These include direct electrification of industrial processes (DOE 2022; IEA 2021a, b), carbon capture, utilization, and storage (CCUS), material efficiency improvements, and the substitution of conventional fuels, energy sources, and feedstocks with low-carbon alternatives, including hydrogen (IEA 2021a, b).

9.2 Research Design and Conceptual Approach

This chapter adopts a critical and systematic study design grounded in a sociotechnical perspective. Specifically, a sociotechnical lens is employed to structure and guide the assessment of hydrogen-based pathways for industrial decarbonization, while a critical review approach supported by a systematic search protocol is used to address the research questions outlined in the Introduction. This combined framework enables an integrated evaluation of technological developments alongside institutional, policy, economic, and social dimensions that shape hydrogen deployment in industrial contexts. Notwithstanding efforts to ensure comprehensive and systematic coverage, the review is subject to several acknowledged limitations. The most prominent constraint relates to the length of the manuscript. Given the breadth and complexity of the evidence base, a substantial volume of literature was required to adequately capture the state of the art. Interpreting and synthesizing this evidence necessitated an extended review, for which special editorial approval was obtained. A second limitation is the restriction of the literature search to English-language publications, which may result in the omission of relevant contributions from non-English-speaking regions, particularly from Asia and South America, where hydrogen-related research and deployment activities are rapidly evolving. Third, systematic evidence collection was concluded in January 2021, and consequently, more recent studies may not be fully reflected in the analysis. Finally, as demonstrated throughout the review, applying a sociotechnical lens to industrial decarbonization via hydrogen occasionally requires consideration of related applications and developments in adjacent sectors. Such cross-sectoral discussions are intentionally included, as they are essential for capturing the interconnected and dynamic nature of hydrogen systems and for constructing a robust and holistic sociotechnical assessment (Kim et al. 2021; Khoshgoftar et al. 2022).

9.2.1 Searching Protocol and Analytical Parameters

To structure a rigorous and transparent critical review, this chapter adopted a systematic search strategy built around three complementary sets of keywords that collectively reflect the core themes of hydrogen-enabled industrial decarbonization. Combining these keyword groups yielded 48 distinct search permutations. Each permutation was applied across 12 carefully selected data repositories, encompassing both peer-reviewed academic sources and relevant gray literature, resulting in a total of 576 individual search strings. The intent was to capture the breadth and depth of state-of-the-art research and policy discourse in this rapidly evolving field. Initial generic searches returned more than 9.7 million potentially relevant documents. Through iterative sorting and screening procedures, this corpus was progressively refined to a subset of 4674 studies considered most relevant to the scope of this review. Further filtering was then applied based on recency, relevance, timeliness, and uniqueness. Specifically, temporal constraints were imposed in 11 databases or search engines by restricting publication dates to "after 01/01/2000" or the "2000–2021" period, while results from the National Academies Publications database required manual temporal filtering. Studies were retained only if they directly addressed the mitigation of climate impacts from industrial processes, with preference given to more recent publications that update or extend earlier work. Duplicate records were systematically removed. Following this multistage screening, the core dataset was reduced to 2134 studies, from which additional short-listing was undertaken as the review progressed. Ultimately, the final reference list comprises more than 700 cited sources. Notably, approximately 25.4% of the studies included in the meta-review were published from 2020 onward, underscoring the recency and dynamism of the field. In terms of publication types, original research articles and scholarly books constitute the majority of sources, while gray literature is predominantly represented by reports, including institutional studies, position papers, and technical communications. From a thematic perspective, engineering and natural sciences dominate the subject areas of the selected literature. Although social science perspectives are present across a range of journals, they account for a relatively small share of the total publications compared with technical disciplines. This imbalance highlights the strong technical orientation of existing research on hydrogen in industry and suggests significant opportunities for deeper investigation of social, economic, and broader sociotechnical dimensions in future studies.

9.2.2 Analytical Frame of Sociotechnical Systems

To systematically organize and interpret the insights drawn from the final corpus of literature on hydrogen-enabled industrial decarbonization, this review adopts the analytical lens of sociotechnical systems (Sovacool and Hess 2017; Hess and Sovacool 2020). Within this framework, hydrogen is understood not merely as an isolated fuel or energy carrier, but as a component embedded in a broader constellation of social, technical, institutional, and economic systems that collectively influence how it is produced, distributed, and utilized (McDowall 2014; Upham et al. 2018). Conceptually, this sociotechnical perspective encompasses not only the material and energy inputs required for hydrogen production, but also the supporting technology platforms, intermediate carriers, conversion pathways, and the wide range of industrial and end-use applications that hydrogen enables. Crucially, these technical elements are examined alongside institutional and user-driven factors, including policy and regulatory frameworks, financial mechanisms, and broader economic enablers, which together shape the trajectories of hydrogen adoption and deployment. In this review, the hydrogen sociotechnical system is conceptualized as an interconnected network comprising five core elements: hydrogen production, the hydrogen supply chain, industrial hydrogen use, institutional drivers, and end-use or societal drivers. The interactions within this network are represented by a system of identifiers, where numerical labels denote the originating element and lowercase letters indicate the element affected by a given interaction. This representation emphasizes the *complexity and interdependence* of hydrogen sociotechnical dynamics. Hydrogen production itself can be further disaggregated into multiple sub-elements, including material feedstocks such as fossil fuels, biomass, and water; the energy sources used both for production and downstream applications; processing routes converting hydrogen into intermediates like ammonia, methanol, or liquid organic hydrogen carriers; energy generation processes; and the underlying technology platforms. These production and energy components form a tightly coupled nexus with bidirectional flows of materials and energy across industrial sectors. Some feedstocks, including biomass and fossil fuels, may simultaneously function as both material and energy sources, adding layers of complexity to system boundaries (Gordon et al. 2023). While certain industrial sectors, including steelmaking, chemical manufacturing, and oil refining, rely on in situ hydrogen production for internal operations, large-scale deployment necessitates a robust supply chain infrastructure. The hydrogen supply chain is therefore examined as a distinct sociotechnical element, encompassing transmission, distribution, storage, and conversion processes. The diversity of hydrogen feedstocks, production pathways, technology platforms, and end-use applications makes the supply chain a highly interconnected and nonlinear network. Hydrogen can be converted into and from intermediate carriers, act as a storage medium to balance variable renewable energy supply and demand, or circulate multiple times across sub-elements, reflecting the cyclical nature of hydrogen flows. Industrial hydrogen applications constitute another critical element of the system. In some sectors, hydrogen serves simultaneously as a feedstock and fuel in the production of materials such as steel, cement, and chemicals, which themselves have downstream applications. In other contexts, hydrogen functions primarily as an energy vector, for instance in transport, power generation, or heating and cooling systems. In these cases, hydrogen adoption often occurs alongside other energy vectors, enabling renewable energy integration and the gradual displacement of fossil-based systems (Akrami et al. 2018; Shahbaz et al. 2022). Collectively, hydrogen production, supply chains, and industrial use capture the tangible flows of materials, energy, and technology from source to end use. Beyond the technological dimensions, institutional drivers play a crucial role in shaping hydrogen adoption and deployment. Governance elements, including policy, regulatory, and fiscal frameworks, as well as stakeholder and decision-making structures determine the pace and direction of hydrogen system development. At finer scales, governmental institutions influence the emergence of hydrogen economies and manage the opportunities and challenges of implementation across industrial systems, operating at supranational, national, regional, and local levels. Finally, end-use and societal drivers interface directly with industrial systems, reflecting cultural, symbolic, and market influences on hydrogen uptake. Social norms, environmental awareness, consumer preferences, and behavioral inertia all shape hydrogen adoption trajectories, highlighting the importance of nontechnical factors in the broader sociotechnical system. Incorporating these dimensions enables a more comprehensive understanding of the industrial ecology that could emerge from a transition toward a hydrogen-based economy (Massaro et al. 2023).

9.3 Energy and Climate Impacts of Industrial Hydrogen Use

Current industrial consumption of high-purity hydrogen defined as hydrogen containing minimal contaminants or impurities is largely concentrated in oil refining and ammonia synthesis, with the latter being predominantly used for fertilizer production (IEA 2019). In 2019, global demand for pure hydrogen was estimated at approximately 75 million metric tons (Mt). In addition, around 45 Mt. per year of hydrogen was consumed in the form of gas mixtures used either as fuels or as feedstocks, commonly referred to as non-pure hydrogen (IEA 2020). The primary industrial applications of this non-pure hydrogen include methanol synthesis and steel manufacturing. The overwhelming majority of pure hydrogen production is currently fossil-based. Natural gas reforming accounts for the largest share, producing so-called gray hydrogen, followed by coal-based production routes, often referred to as black or brown hydrogen. Together, these pathways consume roughly 275 million tonnes of oil equivalent (Mtoe) of fossil energy, corresponding to about 2% of total global primary energy demand. This reliance on fossil fuels results in substantial carbon dioxide emissions, with approximately 10 metric tons of CO_2 emitted per metric ton of hydrogen produced from natural gas, and up to 19 metric tons of CO_2 per metric ton of hydrogen produced from coal. Overall, global hydrogen production is associated with CO_2 emissions of approximately 830 Mt. CO_2 per year (IEA 2019). While the majority of these emissions are released directly into the atmosphere, a portion of the CO_2 generated during hydrogen production particularly in ammonia and urea synthesis is captured and utilized as an input for fertilizer production. This practice partially offsets emissions at the process level but does not eliminate the broader climate impacts associated with fossil-based hydrogen pathways. The integration of hydrogen across industrial sectors offers a range of complementary technological opportunities that can accelerate decarbonization. Various processes, from hydrogen-based direct reduction in steelmaking to hydrogen-powered heat and power systems, exhibit synergistic effects when combined with electrification, carbon capture, and renewable energy integration. Understanding these interactions is essential for optimizing system-level performance, reducing emissions, and enhancing energy efficiency. Table 9.1 summarizes the key technological synergies that enable hydrogen-driven industrial decarbonization, highlighting their mechanisms, application domains, and potential impact on carbon reduction pathways.

9.3.1 Energy Impacts

Although hydrogen currently plays a relatively limited role in the global energy system, future deployment scenarios point toward a markedly expanded scope of applications. The Hydrogen Council—an industry-led consortium comprising established fossil-based hydrogen producers, sectoral investors, and major industrial hydrogen users—has identified 35 use cases in which hydrogen could exert a substantial impact both in the near term and over longer time horizons (Hydrogen Council 2020). These applications span both emerging and existing markets across transport, buildings, industrial heat, and industrial feedstock use, and collectively account for approximately 60% of global energy- and process-related greenhouse gas emissions. Projections under the International Energy Agency's Sustainable Development Scenario (SDS) suggest that hydrogen demand and end-use patterns could diverge significantly from those observed today by 2070. In particular, the direct use of hydrogen, alongside hydrogen-derived synthetic fuels and ammonia, could play a central role in meeting transportation energy needs. Under this scenario, transport-related applications alone could account for nearly 70% of total hydrogen demand, with overall hydrogen consumption potentially exceeding 500 Mt. per year (IEA 2020). If these SDS trajectories materialize, low-carbon hydrogen produced through a balanced portfolio of fossil-based pathways equipped with carbon capture, utilization, and storage (CCUS) and renewable-powered water electrolysis could supply more than 20% of final energy demand in the chemical sector

Table 9.1 Hydrogen-enabled technologies and their synergistic role in industrial decarbonization

Industry sector	Hydrogen technology	Synergistic integration	Decarbonization mechanism	Estimated CO_2 reduction potential	References
Steel manufacturing	Hydrogen-based direct reduction (H-DRI)	H_2 replaces coke in direct reduction ironmaking; link with renewable electrolysis systems	Reduces process CO_2 emissions by substituting fossil reductants	Significant CO_2 reduction relative to traditional blast furnace—basic oxygen furnace	Ji et al. (2024)
Cement industry	Hydrogen fuel combustion in kilns	Use of H_2 as alternative fuel in high-temperature calcination and heat	Reduces fossil fuel combustion emissions in clinker production	20–30% + CO_2 reduction with hydrogen substitution	Williams et al. (2024)
Chemical & refining	Hydrogen feedstock + CCUS	Green H_2 replaces gray H_2 in synthesis processes (e.g., methanol, ammonia)	Avoids fossil-derived hydrogen emissions and integrates CCUS	High CO_2 saving vs. SMR baseline	Ganter et al. (2024)
Glass & ceramics	Hydrogen–natural gas blending/ pure H_2 combustion	Hydrogen blended with natural gas in industrial furnaces; CFD and combustion assessments	Lowers process CO_2 in high-temperature furnaces	Up to ~25–77% CO_2 reduction depending on blend	Daurer et al. (2025)
Aluminum & metals recycling	Green hydrogen in melting/ recycling	Hydrogen combustion or H_2-NG blends for metal melting and reheat	Substitutes natural gas for high-temperature heat	Up to ~90% CO_2 reduction when replacing natural gas	Novelis Inc. (2025)

and approximately 15% of final energy demand in iron and steel production. Among the various end-use opportunities, the most significant potential may lie in maritime transport, where hydrogen and hydrogen-derived fuels offer one of the few scalable options for deep decarbonization in the absence of viable direct electrification pathways.

9.3.2 *Climate Impacts*

Industrial greenhouse gas emissions can generally be grouped into three main categories, though not every industrial sector contributes to all three. The first category, direct energy-related emissions, arises from the on-site combustion of fuels to generate power and heat for industrial operations. The second category, indirect emissions, is associated with the consumption of electricity and heat produced externally. The third category, direct process emissions, results from inherent chemical reactions within industrial processes, such as the metal oxide reduction reactions in cement and steel manufacturing. Mitigating these emissions often requires fundamental transformations in industrial production systems, including the substitution of carbon-intensive feedstocks, the development of alternative production pathways, and the design of novel product formulations (Amin et al. 2022). In heavy industry, direct energy-related emissions and process emissions are widely recognized as particularly challenging to abate and together account for the majority of residual industrial emissions projected under the IEA Sustainable Development Scenario (SDS). For sectors that use hydrogen as a feedstock, such as oil refining and chemical production, decarbonizing hydrogen itself represents one of the most effective strategies for reducing overall process carbon intensity. Likewise, in industries that require high-temperature heat—such as cement, metals, glass, and ceramics—substituting fossil fuels with low-carbon hydrogen provides a practical pathway for emissions reduction. The potential CO_2 mitigation achieved per unit of hydrogen varies depending on the application. In industrial heat and transport fuel applications, emissions reductions are primarily realized by avoiding CO_2 released from conventional fossil fuel combustion. In contrast, for process applications where hydrogen functions as a feedstock, the decarbonization effect stems largely from replacing fossil-based hydrogen with low-carbon alternatives. When decarbonized hydrogen is incorporated into new products, such as methanol or urea in the chemical sector, the relative emissions reduction per unit of hydrogen may be lower because avoided emissions are limited primarily to upstream hydrogen production (Tiedje et al. 2022). These distinctions highlight the need to simultaneously address both hydrogen feedstock demand and broader industrial energy requirements to fully leverage the decarbonization potential of low-carbon hydrogen across industrial systems.

9.4 Technological Synergies Within Industrial Applications for Decarbonization via Hydrogen

Industrial sectors are among the most challenging domains for deep decarbonization due to their reliance on high-temperature heat, carbon-based chemical feedstocks, and energy-intensive processes that are difficult to electrify directly. In this context, hydrogen emerges as a uniquely versatile decarbonization vector, serving both as a low-carbon energy carrier and as a critical chemical feedstock. According to International Energy Agency (IEA) assessments, industry accounts for roughly 35% of global final energy consumption and nearly 40% of energy-related CO_2 emissions, with iron and steel, chemicals, cement, and oil refining representing the most emissions-intensive sectors. Hydrogen enables deep emissions reductions across these industries through multiple technological pathways, including the direct reduction of iron ore in steelmaking, the substitution of fossil hydrocarbons in chemical synthesis, and the provision of high-temperature process heat via combustion or fuel cell systems. When deployed within broader industrial ecosystems, hydrogen can generate substantial system-level synergies that enhance energy efficiency, operational flexibility, and economic performance (Sun et al. 2022). Industrial clusters—geographically concentrated networks of industries with interconnected energy and material flows—provide particularly favorable conditions for large-scale hydrogen deployment. Shared infrastructure, coordinated investment strategies, and collective risk management within such clusters allow for economies of scale, while co-locating hydrogen production with industrial consumers reduces transport requirements and facilitates the valorization of industrial by-products, such as waste heat and oxygen. Through these integrated approaches, hydrogen evolves from a standalone decarbonization option into a central pillar of circular industrial ecosystems, maximizing resource efficiency while minimizing environmental impacts (Swennenhuis et al. 2022). Hydrogen deployment in industrial sectors is advancing through pilot plants and early commercial projects that demonstrate process integration, operational feasibility, and measurable emissions reductions. In the iron and steel sector, hydrogen-based direct reduced iron (H_2-DRI) has reached technology readiness levels (TRLs) of 7–9, supported by pilot-scale sponge iron production and announced commercial capacities exceeding 1 Mt. per year by the mid-2020s. Concurrently, hydrogen injection into existing blast furnaces is being explored as a transitional measure to reduce coal consumption while dedicated DRI infrastructure scales up. These initiatives have helped define operational parameters for shaft furnaces, gas flow management strategies, and integration with downstream electric arc furnaces. In the chemical sector, ammonia production is the closest to full commercial deployment, with several projects replacing conventional steam methane reforming with electrolysis-based hydrogen at TRLs of 8–9. Methanol synthesis and refinery processes are progressing through partial hydrogen substitution in existing units, supported by multiple retrofit projects under construction. By contrast, cement and glass remain at earlier stages of technological maturity. Cement applications are typically assessed at TRLs of 6–7, involving kiln fuel substitution and hybrid oxy-fuel combustion trials, while glass and ceramics applications are at TRLs of 5–6, requiring significant redesign of burners and furnaces. Notably, steel and chemical projects provide quantitative evidence of achievable emissions reductions and integration with existing process units, whereas cement and glass demonstrations primarily highlight engineering requirements for stable hydrogen firing in kilns and furnaces. Together, these projects establish a clear baseline for subsequent analyses of industrial hydrogen pathways (Tackie-Otoo and Haq 2024). Among industrial sectors, steel, cement, and glass manufacturing are particularly significant contributors to global CO_2 emissions, collectively accounting for over 20% of the total (International Energy Agency 2023). These industries rely on continuous high-temperature heat and complex chemical transformations that are difficult to decarbonize via electrification alone. Global steel production exceeds 1.8 Gt per year, with specific emissions ranging from 1.8 to 2.2 tCO_2 per tonne of steel, making it one of the largest single industrial sources of CO_2. Although glass production is smaller in absolute volume, it exhibits relatively high emission intensities (0.5–0.8 tCO_2 per tonne) due to sustained thermal demands. Cement production contributes approximately 2.5 Gt of CO_2 annually, with roughly half arising from limestone calcination rather than energy use (International Energy Agency 2020, 2022). Green hydrogen offers a viable pathway to mitigate emissions in these industries by providing carbon-free thermal energy and, in steelmaking, serving as a chemical reductant. In H_2-DRI processes, hydrogen substitutes coking coal, whereas in cement kilns and glass furnaces it can replace natural gas through oxy-fuel and hybrid combustion configurations compatible with carbon capture and renewable energy integration. Under the Net Zero Scenario, projections for 2030 indicate substantial expansion of the low-carbon hydrogen project pipeline, closely aligned with rising industrial demand. Accelerated deployment of green and blue hydrogen production facilities is expected to support key sectors such as steel, chemicals, refining, and cement, where direct electrification remains challenging. This growing pipeline reflects increasing policy support, investment commitments, and technological maturity, positioning hydrogen as a strategic enabler of deep industrial decarbonization. Overall, the projected balance between supply capacity and industrial demand underscores low-carbon hydrogen as a cornerstone of net-zero-compatible industrial transformation by 2030 (Fig. 9.1).

9.4.1 Decarbonization Pathways in Heavy Industry

Building on the sector-specific challenges and opportunities associated with hydrogen deployment in heating, smart grids, and policy frameworks, the industrial sector stands out as both the most complex and the most impactful domain for large-scale hydrogen integration. Adoption in heavy industry faces significant technical and operational barriers that necessitate sustained research and development efforts (Seboldt et al. 2021). Across multiple sectors, processes must be adapted to accommodate hydrogen's distinct physical and chemical properties compared with conventional fuels and feedstocks. In steelmaking, hydrogen-based direct reduction exhibits slower reaction kinetics than carbon-based routes, requiring modified operating conditions, novel reactor designs, and optimized gas flow management. Cement production is constrained by process-related emissions from limestone calcination, which must be addressed through combinations of oxy-fuel combustion and carbon capture technologies. In glass manufacturing, hydrogen's high flame speed and elevated combustion temperatures necessitate redesigned burners,

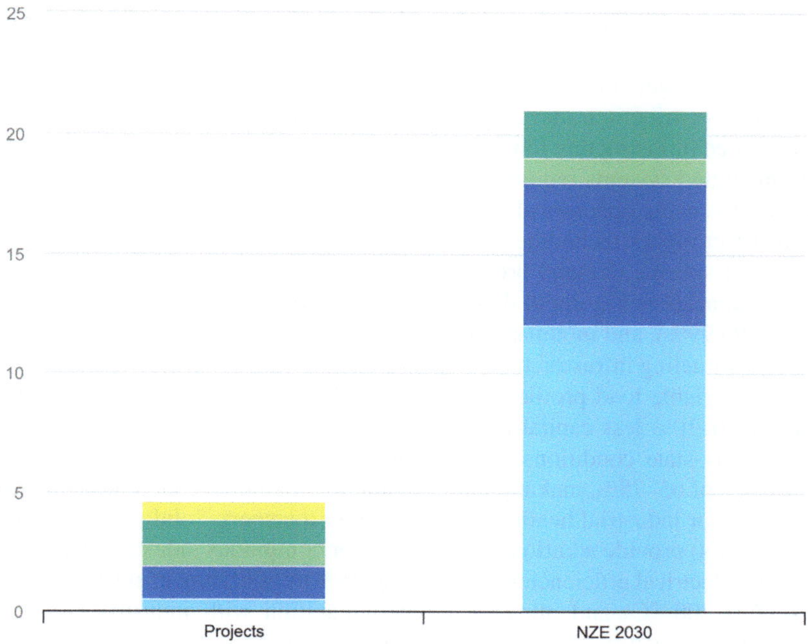

Fig. 9.1 Projected low-carbon hydrogen project pipeline and industrial demand in 2030 within the Net Zero Scenario (IEA 2021a, b)

enhanced control strategies, and improved refractory materials to ensure stable and durable furnace operation. These challenges are increasingly being addressed through coordinated research programs that bring together industrial operators, equipment manufacturers, and research institutions to validate process modifications and assess long-term component performance. The deep decarbonization of heavy industry is best understood as a sequential, multiphase transition aligned with international roadmaps and net-zero targets. In the initial phase, industrial production remains largely carbon-intensive, with emission intensities ranging from approximately 0.5 tCO_2 per tonne of glass to more than 2.2 tCO_2 per tonne of steel (Mendecka and Lombardi 2019). Transitional phases introduce partial electrification of thermal processes, expanded use of electric arc furnaces in steelmaking, supplementary cementitious materials in cement, and electric melting technologies in glass production, achieving emission reductions of roughly 50–75% relative to conventional processes (Perpiñán et al. 2023). The final phase requires transformative process changes, including hydrogen-based direct reduction in steel, oxy-fuel combustion with carbon capture in cement, and renewable-powered hybrid furnaces in glass, ultimately reducing residual emissions to below 0.1 tCO_2 per tonne across sectors (Sasiain et al. 2021). Hydrogen integration further enables cross-sectoral synergies through industrial ecology approaches. Waste heat generated during electrolysis can be utilized in district heating networks or adjacent industrial processes, while oxygen co-produced during electrolysis can be valorized in wastewater treatment, chemical processing, and metallurgical applications. The concept of "hydrogen valleys"—geographically concentrated clusters encompassing hydrogen production, distribution, and end use—is particularly relevant in industrial hubs, where shared infrastructure lowers unit costs and spreads investment risk. Realizing such integrated systems requires standardized measurement, metering, and safety protocols to ensure operational reliability, interoperability, and equitable value allocation. Electrolyzer technology selection is critical to aligning hydrogen supply with sector-specific demand profiles. Proton exchange membrane (PEM) electrolyzers, characterized by rapid dynamic response (below 1 second), high hydrogen purity (>99.99%), and modular design, are well suited for distributed applications, including refueling infrastructure, decentralized heating systems, and transport sectors with variable load profiles. In contrast, alkaline water electrolyzers (AWEs), typically 20–30% less capital-intensive than PEM systems, offer stable operation under steady-state conditions, low degradation rates (<2% per 1000 hours), and efficiencies of 65–75%, making them suitable for centralized, large-scale hydrogen production for industrial heating and heavy-duty transport. Solid oxide electrolysis cells (SOECs) provide additional advantages for continuous industrial applications, achieving electrical efficiencies approaching 85% by operating at elevated temperatures (700–800 °C) and allowing direct integration with industrial waste heat streams. Consequently, SOECs are particularly promising for energy-intensive sectors such as steel, ammonia, and cement manufacturing. Selecting the optimal electrolyzer technology thus requires a multi-objective assessment that considers efficiency, dynamic performance, degradation behavior, hydrogen quality, and levelized cost of hydrogen (LCOH). Application-specific deployment strategies based on such frameworks are essential to maximize emissions reductions while ensuring economic competitiveness and scalability. The trajectory of industrial hydrogen adoption will be shaped by harmonized standards, certification schemes, and sector-specific process adaptations. International standards organizations, such as ISO, are developing requirements for hydrogen quality, handling equipment, and safety compliance to support large-scale deployment and interoperability. Certification frameworks for green steel, cement, and chemicals establish verifiable carbon-intensity benchmarks that create market differentiation and provide investment signals. Beyond retrofitting existing facilities, hydrogen deployment enables the development of novel reactor designs and manufacturing pathways. Policy frameworks, exemplified by the European Green Deal, demonstrate how coordinated public funding, infrastructure development, and market-based mechanisms can accelerate industrial transformation. Hydrogen-based processes are increasingly transitioning from pilot-scale demonstrations to early commercial deployment in emissions-intensive sectors, including steel, cement, glass, and chemicals. While cost and infrastructure challenges remain, advances in reactor engineering, burner design, electrolyzer scaling, process integration, and supportive regulatory environments are steadily improving commercial viability. Industrial actors that align hydrogen adoption with efficiency improvements, digital monitoring, and circular economy principles are likely to gain a competitive advantage. As hydrogen technologies mature and capital costs decline, hydrogen is expected to assume a structural role in the deep decarbonization of heavy industry, supporting climate objectives while sustaining industrial competitiveness. In this way, industrial decarbonization complements hydrogen pathways in heating, smart grids, and urban systems, forming a cohesive foundation for comprehensive energy transition strategies.

9.5 Cross-Cutting Solutions for Decarbonization of Hydrogen Production and Use

This chapter has examined in depth the potential for hydrogen to decarbonize a broad spectrum of industrial activities. Achieving genuinely low- or zero-carbon hydrogen pathways, however, requires integrated, cross-cutting solutions that span multiple components of the hydrogen sociotechnical system. Among these, material and design improvements, coupled with enhanced resource efficiency, emerge as the most pervasive enablers. For hydrogen production technologies at low technology readiness levels (TRLs), progress toward commercial maturity is closely linked to advances in materials science, including the development of novel catalysts, membranes, reaction media, and optimized reactor designs. In contrast, for mature, high-TRL processes, incremental material and design improvements primarily drive cost reductions, improved durability, and more economically viable deployment. Similarly, robust hydrogen distribution, storage, and dispensing infrastructure depend on continued innovation in physical infrastructure materials. From both safety and economic perspectives, improved materials are essential for hydrogen transmission under compressed or liquefied conditions, as well as for alternative carrier-based transport pathways. Advances in stationary hydrogen storage, initially motivated by supply chain requirements, can also be leveraged for downstream applications, including hydrogen refueling stations (Balasooriya et al. 2021). Resource efficiency constitutes a second major cross-cutting solution, manifested in two complementary trends. First, from a waste minimization and reuse perspective, repurposing existing fossil gas infrastructure—where technically and economically feasible—offers a pragmatic route to accelerate hydrogen deployment across the sociotechnical system. Second, for newly developed infrastructure, circular design approaches that emphasize material recycling, upcycling, and reduced resource demand are increasingly important for progressing toward zero-waste systems. Together, these strategies can significantly reduce the economic and environmental burdens of phasing out carbon-intensive energy vectors while scaling hydrogen technologies. Waste energy recovery and utilization represents another recurring cross-cutting solution. Although energy losses are inevitable throughout hydrogen production, conversion, storage, and use, technologies that capture and reuse waste energy streams can indirectly improve overall system efficiency. At the system level, recovering energy losses associated with hydrogen conversion and regeneration from intermediate carriers is particularly important for enhancing the viability of hydrogen-based storage and transmission options (Sharma et al. 2020). Energy efficiency improvements, combined with greater use of renewable and low-carbon energy sources, form a fourth cross-cutting solution that benefits all elements and sub-elements of the sociotechnical system. Enhanced energy performance reduces indirect energy-related emissions, while deeper integration of renewable and low-carbon electricity supports the decarbonization of hydrogen production and downstream applications. System-wide efficiency gains are especially critical for energy storage pathways such as power-to-hydrogen-to-power, where conversion losses can otherwise constrain effectiveness (Yilmaz et al. 2020). Carbon capture, utilization, and storage (CCUS) represent a further cross-cutting solution with relevance across multiple hydrogen and CO_2 value chains. CCUS-enabled hydrogen pathways provide substantial decarbonization potential for hard-to-abate sectors, particularly where fossil-based hydrogen production remains economically or technically necessary (Kreutz et al. 2020). From a sociotechnical perspective, coupling CCUS with hydrogen production and utilization offers direct emissions reductions while enabling blue hydrogen and blue hydrogen-derived intermediates (Berger et al. 2020). Finally, hydrogen itself acts as a cross-cutting enabler within its sociotechnical system. The tight coupling between hydrogen deployment and renewable or low-carbon energy sources creates reinforcing feedback loops: expanded hydrogen use stimulates demand for cleaner hydrogen production, which in turn supports broader decarbonization of power generation, heating, and fuel supply throughout the hydrogen value chain (Chandrasekar and Flynn 2021). In this sense, hydrogen functions as both an enabler and a beneficiary of system-wide decarbonization. Collectively, these cross-cutting solutions, including material and design improvements, resource efficiency, waste energy recovery, energy efficiency coupled with renewable and low-carbon energy integration, CCUS, and hydrogen deployment, span the technological elements of the hydrogen sociotechnical system. Complementing policy interventions, they highlight priority research areas and deployment pathways essential for advancing low-carbon hydrogen at scale. Hydrogen is considered here only where it constitutes a direct intervention, avoiding overlap or double counting with renewable and low-carbon energy sources.

9.6 Policies Related to Hydrogen

The decarbonization of industrial systems through hydrogen will depend critically on policy frameworks that simultaneously stimulate hydrogen supply, create sustained demand, and enable the development of dedicated supply chain infrastructure. Although existing policy instruments designed to promote renewable energy generation and deployment provide a useful foundation, they are insufficient on their own to support large-scale hydrogen integration. Instead, tailored hydrogen-specific policy mechanisms are required to address the unique technological, economic, and infrastructural challenges associated with hydrogen production, transport, storage, and end use (Bleischwitz and Bader 2010). Hydrogen policy frameworks must therefore encompass a broad and coordinated set of interventions spanning the entire value chain. These include support for early-stage and applied research to accelerate technology maturation; regulatory measures to establish standards, safety protocols, and market rules; and fiscal incentives to reduce cost gaps relative to incumbent fossil-based alternatives. In addition, public financing instruments such as grants, loan guarantees, and risk-sharing mechanisms play a critical role in de-risking first-of-a-kind projects and enabling private sector participation during early deployment phases.

9.6.1 Policies on Hydrogen Technology R&D

Hydrogen-focused research and development (R&D) policies are essential for advancing emerging technologies toward commercial readiness and overcoming scale-up barriers that currently limit widespread industrial adoption (Rissman et al. 2020). The central objective of hydrogen R&D policy is to accelerate innovation cycles while avoiding technological "lock-ins" that perpetuate carbon-intensive industrial processes. By supporting continuous learning, experimentation, and validation, these policies enable the transition from laboratory-scale concepts to pilot, demonstration, and ultimately commercial-scale deployment. National and regional governments have implemented targeted funding mechanisms to support hydrogen R&D across both academic and industrial domains. Representative examples include Australia's CSIRO Hydrogen Roadmap; the European Fuel Cells and Hydrogen Joint Undertaking (FCH 2 JU) and the Horizon Europe 2021–2027 framework; hydrogen programs supported by US federal agencies, such as the Department of Energy (DOE) and Department of Transportation (DOT); and Japan's hydrogen development initiatives coordinated by the Ministry of Economy, Trade and Industry (METI). Beyond individual research projects, public funding has also been directed toward the establishment of dedicated hydrogen research centers and long-term institutional programs. Notable examples include Hydrogen Europe research platforms within the European Union, Germany's Hydrogen Technology Center, hydrogen-focused programs embedded within Japanese national universities and research institutes under METI, and hydrogen research conducted across the U.S. DOE National Laboratory system. Public–private partnerships play a critical role in bridging the gap between R&D and commercialization by enabling large-scale demonstrations and first-of-a-kind projects. Prominent initiatives include Australia's Hydrogen Energy Supply Chain (HESC) program; Germany's Seventh Energy Research Programme; the European Union's H2ME, H2FUTURE, and H2PORTS projects; the Japan Hydrogen Association; and Sweden's HYBRIT project for fossil-free steelmaking. Financial support for these demonstration and scale-up activities is provided through a range of European and national funding instruments, including the European Clean Hydrogen Alliance (ECHA), the InvestEU Programme, the European Regional Development Fund, the Cohesion Fund, the Just Transition Mechanism for carbon-intensive regions, and infrastructure-oriented schemes such as the Connecting Europe Facility (CEF) and the Connecting Europe Facility Transport (CEF-T) (Barnes 2020; IEA 2020).

9.6.2 Regulation and Certification of Hydrogen

Within the hydrogen sociotechnical system, regulatory, standardization, and certification frameworks span the full value chain, encompassing hydrogen production, supply and distribution infrastructure, and industrial end use. The relevant standards development organizations and the categories of standards developed by each body in relation to the hydrogen sociotechnical system are described in this section. At the international level, hydrogen-related standardization is primarily coordinated by the International Organization for Standardization (ISO). In 1990, ISO established a dedicated technical committee for hydrogen technologies, ISO/TC 197, initially focused on fuel cells and subsequently expanded to address broader hydrogen systems. The current mandate of ISO/TC 197 covers standardization of systems and equipment for the production, storage, transportation, measurement, and utilization of hydrogen (Chelhaoui and Serre-Combe 2006). To date, this committee has published 17 international standards, with an additional 15 under active development, collectively addressing most of the technical pathways associated with hydrogen deployment. These standards include foundational safety principles for hydrogen systems; performance, safety, and quality requirements for hydrogen production via electrolysis; safety standards for hydrogen separation and purification; specifications for storage systems; and safety and performance standards for hydrogen distribution. A significant proportion of existing standards focuses on end-use applications in the transport sector, particularly fuel cell electric vehicles and hydrogen refueling infrastructure, where safety, interoperability, and fuel quality are critical (ISO n.d.). However, notable gaps remain within the international standardization landscape. Harmonized ISO standards for hydrogen transmission infrastructure, including pipelines and marine transport, as well as for several categories of industrial hydrogen use, are still lacking. In practice, globally recognized guidance for these elements is largely provided by industry-led organizations such as the Compressed Gas Association (CGA) and the European Industrial Gases Association (EIGA). While these standards are widely applied in industry, their voluntary nature and regional specificity limit their role in creating fully harmonized international regulatory regimes. The translation of hydrogen policy objectives into enforceable regulations has progressed unevenly across regions. Many developing economies remain at early stages of hydrogen policy formulation and implementation, as observed in Mexico and other parts of Latin America (Ávalos Rodríguez et al. 2019). Nevertheless, existing regulatory, certification, and standardization frameworks have often been leveraged to support the introduction of hydrogen technologies in emerging markets, even in the absence of comprehensive national hydrogen strategies, roadmaps, or action plans (ESMA 2019). As a result, both top-down approaches driven by national policy and bottom-up approaches shaped by industrial demand can be observed in standards development and regulatory adoption. Regardless of the pathway taken, alignment with internationally harmonized standards is increasingly critical. National and subnational regulatory authorities that fail to adopt compatible certification schemes and technical regulations risk limiting domestic industry participation in global hydrogen value chains and international markets. Consequently, regulatory coherence and standard harmonization emerge as enabling conditions for scalable hydrogen deployment and cross-border trade within the evolving hydrogen economy.

9.6.3 Energy and Environmental Regulation

Because hydrogen functions as a foundational platform for industrial decarbonization, policy instruments that stimulate hydrogen production are intrinsically relevant to industrial transformation, even when they do not explicitly target industrial end uses. Accordingly, this section considers policy measures that support the development of the broader hydrogen ecosystem, recognizing that such an ecosystem is a prerequisite for achieving cost reductions in hydrogen production and for deploying the enabling infrastructure required to make hydrogen a viable option for industrial applications at scale. Hydrogen policy adoption has frequently been linked to renewable energy policy frameworks, particularly through hydrogen's role as a long-duration energy storage medium capable of mitigating the variability and intermittency of solar and wind power generation (Anderson and Leach 2004). In this context, certification systems for renewable electricity provide an important institutional foundation for low-carbon hydrogen production. Within the European Union, robust certification schemes have already been established through the European Energy Certificate System (EECS). This harmonized framework governs the issuance, transfer, and cancellation of guarantees of origin (GO) for renewable electricity, thereby enabling cross-border trade and traceability of green power among EU Member States. Such mechanisms are increasingly relevant for certifying renewable electricity inputs to electrolysis-based hydrogen production. In the transport sector, near-term technological constraints and infrastructure lock-ins continue to favor liquid fuels, implying that hydrogen-derived energy carriers such as ammonia and synthetic fuels are likely to play a complementary role alongside other low-carbon alternatives over longer time horizons. Supporting the deployment of these hydrogen-based fuels requires not only regulatory and certification frameworks but also substantial fiscal and financial policy interventions. These include public funding for research, development, and demonstration of synthetic fuel production pathways, as well as investment in enabling infrastructure, such as the adaptation and upgrading of existing pipelines and distribution networks to accommodate hydrogen and hydrogen-derived fuels (Carnevale and Sachs 2019). While the interaction between energy regulation and the transition from fossil fuels to renewable energy systems has been extensively examined in the hydrogen policy literature, the environmental implications of hydrogen deployment are often addressed indirectly through policy instruments in adjacent sectors. This is particularly evident in transport-focused regulatory frameworks, including the European Union's Directive 2014/94/EU on the deployment of alternative fuels infrastructure and the Low Carbon Fuel Standard (LCFS) administered by the California Air Resources Board (CARB) in the United States (Lowenthal 2006). These policies incorporate life-cycle emissions considerations and market-based incentives that can significantly influence hydrogen demand and investment signals, even when hydrogen is not the primary policy target.

9.7 Conclusion

Despite the rapid growth in policy attention and strategic ambitions surrounding hydrogen in recent years, its deployment at the global scale remains modest and continues to rely predominantly on fossil-based production pathways. Within the industrial sector, decarbonization creates two closely linked avenues for hydrogen adoption. The first involves the substitution of existing fossil-derived hydrogen with low-carbon or renewable hydrogen in established applications such as refining and chemical production. The second concerns the expansion of hydrogen into new roles as a decarbonization vector in industrial processes that currently depend on carbon-intensive fuels or feedstocks. Beyond industrial applications, the prospective use of hydrogen in transportation, power generation, and heating further underscores the need to evaluate hydrogen deployment within an integrated value-chain perspective. These cross-sectoral linkages highlight how decarbonization opportunities and constraints emerge across multiple, interconnected sociotechnical systems rather than within isolated end-use domains. Consequently, innovation within the hydrogen sociotechnical system must address challenges spanning all system elements, including production, infrastructure, end use, and institutional frameworks. Creating sustained market demand for hydrogen will require robust and coordinated regulatory and standards frameworks that align with national and international net-zero transition objectives. Among the most powerful enabling mechanisms would be an effective and internationally harmonized carbon pricing regime that minimizes emissions leakage across borders. In parallel, targeted financial incentives and public support measures remain essential to accelerate hydrogen uptake, reduce investment risk, and foster the emergence of hydrogen-based industrial clusters that can exploit economies of scale and shared infrastructure. From a geopolitical standpoint, the emergence of a global hydrogen economy has the potential to fundamentally reshape international energy relations. Unlike the current fossil fuel system characterized by concentrated geopolitical power linked to natural resource endowments, the hydrogen sociotechnical system expands opportunities for renewable energy deployment across a wider range of regions, albeit with uneven geographical distribution. At the same time, geopolitical competition is likely to intensify around electricity system expansion and the development and manufacturing of key enabling technologies, particularly electrolyzers. Nevertheless, the parallel evolution of circular carbon economy concepts may sustain or even expand the role of low-carbon, nonrenewable hydrogen pathways, notably hydrogen produced from fossil fuels coupled with carbon capture technologies. A growing contest may emerge in the coming years between nations strongly committed to green hydrogen such as countries within the European Union and those aiming to capitalize on exporting hydrogen or its derivatives produced through other pathways, particularly blue hydrogen, including countries like Saudi Arabia and Russia.

References

Akrami E, Khazaee I, Gholami A (2018) Comprehensive analysis of a multi-generation energy system by using an energy-exergy methodology for hot water, cooling, power and hydrogen production. Appl Therm Eng 129:995–1001. https://doi.org/10.1016/j.applthermaleng.2017.10.095

Allwood JM, Cullen JM, Milford RL (2010) Options for achieving a 50% cut in industrial carbon emissions by 2050. ES&T 44(6):1888–1894

Amaduzzi R, Ferrarotti M, Parente A (2021) Strategies for hydrogen-enriched methane flameless combustion in a quasi-industrial furnace. Front Energy Res 8:590300

Amin M, Shah HH, Fareed AG, Khan WU, Chung E, Zia A et al (2022) Hydrogen production through renewable and non-renewable energy processes and their impact on climate change. Int J Hydrogen Energy 47(77):33112–33134. https://doi.org/10.1016/j.ijhydene.2022.07.172

Anderson D, Leach M (2004) Harvesting and redistributing renewable energy: on the role of gas and electricity grids to overcome intermittency through the generation and storage of hydrogen. Energy Policy 32:1603–1614. https://doi.org/10.1016/S0301-4215(03)0013

M.L. Ávalos Rodríguez, J.J. Alvarado Flores, J.V. Alcaraz Vera, J.G. Rutiaga Quinones, ˜ J.E. Valencia, The legal regulation of the H2 as a strategy for public policy in Mexico from the consolidation of the National Council of the hydrogen, Int J Hydrogen Energy 44 (2019) 12303–12308, https://doi.org/10.1016/j.ijhydene.2018.0

Balasooriya W, Clute C, Schrittesser B, Pinter G (2021) A review on applicability, limitations, and improvements of polymeric materials in high-pressure hydrogen gas atmospheres. Polym Rev:1–36. https://doi.org/10.1080/15583724.2021.18

Barnes KY (2020) EU hydrogen vision: regulatory opportunities and challenges. Oxford Institute for Energy Studies

Berger M, Radu D, Fonteneau R, Deschuyteneer T, Detienne G, Ernst D (2020) The role of power-to-gas and carbon capture technologies in cross-sector decarbonisation strategies. Electr Power Syst Res 180:106039. https://doi.org/10.1016/j.epsr.2019.106039

Bettinger D (2011) Increasing the energy efficiency of iron and steelmaking. In: Presentation to the steel Business Briefing green steel Conference. [Brussels]

Bleischwitz R, Bader N (2010) Policies for the transition towards a hydrogen economy: the EU case. Energy Policy 38:5388–5398. https://doi.org/10.1016/j.enpo

Carnevale P, Sachs JD (2019) Roadmap to 2050: a manual for nations to decarbonize by mid-century, Fondazione Eni Enrico Mattei (FEEM). Sustainable Development Solutions Network (SDSN)

Chandrasekar D, Flynn E (2021) Syron, operational challenges for low and high temperature electrolyzers exploiting curtailed wind energy for hydrogen production. Int J Hydrog Energy. https://doi.org/10.1016/j.ijhydene.2020.1

Chelhaoui S, Serre-Combe P (2006) Overview of European and international regulation and standardisation activities, Lyon, France

Daurer G, Schwarz S, Demuth M, Gaber C, Hochenauer C (2025) On the use of hydrogen in oxy-fuel glass melting furnaces: an extensive numerical study of the fuel switching effects based on coupled CFD simulations. Fuel 380:133576

DOE (2022) Industrial decarbonization roadmap. Retrieved from, https://www.energy.gov/industrial-technologies/doe-industrial-decarbonization-roadmap

DOE (2023) Pathways to commercial liftoff: clean hydrogen. Retrieved from, https://liftoff.energy.gov/clean-hydrogen/

Emirates Authority for Standardization & Metrology (ESMA) (2019) Technical regulation on hydrogen and fuel cell vehicles https://members.wto.org/crnattachments/2019

EPA (2023) Inventory of U.S. Greenhouse gas emissions and Sinks: 1990–2021. U.S. Environmental Protection Agency. EPA 430-R-23-002. https://www.epa.gov/ghgemissions/inventory-us-greenhouse-gas-emissions-andsinks-1990-2021

Franco A, Giovannini C (2023) Routes for hydrogen introduction in the industrial hard-to-abate sectors for promoting energy transition. Energies 16:6098

Ganter A, Gabrielli P, Goericke H, Sansavini G (2024) Minimum-regret hydrogen supply chain strategies to foster the energy transition of European hard-to-abate industries. arXiv preprint arXiv:2407.05988

Glanville P, Fridlyand A, Sutherland B, Liszka M, Zhao Y, Bingham L, Jorgensen K (2022) Impact of hydrogen/natural gas blends on partially premixed combustion equipment: Nox emission and operational performance. Energies 15:1706

Gordon JA, Balta-Ozkan N, Nabavi SA (2023) Socio-technical barriers to domestic hydrogen futures: repurposing pipelines, policies, and public perceptions. Appl Energy 336:120850. https://doi.org/10.1016/j.apenergy.2023.120850

Gutowski TG, Sahni S, Allwood JM, Ashby MF, Worrell E (2013) The energy required to produce materials: constraints on energy-intensity improvements, parameters of demand. Phil Trans Math Phys Eng Sci 371(1986):20120003

Hess DJ, Sovacool BK (2020) Sociotechnical matters: reviewing and integrating science and technology studies with energy social science. Energy Res Soc Sci 65:101462

Hydrogen Council (2020) Path to hydrogen competitiveness: a cost perspective. Hydrogen Council, Belgium

IEA (2019) The future of hydrogen: seizing today's opportunities. International Energy Agency, Paris

IEA (2020) Energy technology perspectives 2020. International Energy Agency, Paris

IEA (2021a) Low-carbon hydrogen project pipeline and demand in industry by 2030 in the net zero scenario, IEA, Paris. https://www.iea.org/data-and-statistics/charts/low-carbon-hydrogen-project-pipeline-and-demand-in-industry-by-2030-in-the-net-zero-scenario, Licence: CC BY 4.0

IEA (2021b) Net zero by 2050. A roadmap for the global energy sector: IEA special report. Retrieved from https://www.iea.org/reports/net-zero-by-2050

International Energy Agency (2020) Iron and steel technology roadmap. Tech. rep., IEA. reports annual steel production exceeding 1.8 Gt with emissions intensities of 1.8–2.2 tCO2/t

International Energy Agency (2022) Tracking report: cement 2022. Tech. rep., IEA. cement sector accounts for 2.5 Gt CO2 annually, with 50% from limestone calcination

International Energy Agency (2023) Tracking industrial emissions 2023. Tech. rep., IEA. reports that industry accounts for nearly 30% of global CO2 emissions, with steel, cement, and glass among the major contributors

IPCC Summary for policymakers (2023) Climate change 2023: synthesis report. A report of the intergovernmental panel on climate change. In: Lee H, Romero J (eds) Contribution of working Groups I, II and III to the Sixth Assessment report of the intergovernmental panel on climate change. Core Writing team. IPCC, p 36

ISO, Standards by ISO/TC 197 – Hydrogen technologies, ISO (n.d.). https://www.iso.org/committee/54560/x/catalogue/p/1/u/1/w/0/d/0. Accessed 19 Apr 2021

Ji Y, Chi Z, Yuan S, Chen Y, Li Y, Jiang T, Liu X, Zhang W (2024) Development and application of hydrogen-based direct reduction iron process. PRO 12(9):1829. https://doi.org/10.3390/pr12091829

Khoshgoftar MMH, Ghorbani S, Blanco-Marigorta AM (2022) Optimal design and analysis of a combined freshwater-power generation system based on integrated solid oxide fuel cell-gas turbine-organic Rankine cycle-multi effect distillation system. Appl Therm Eng 211:118438. https://doi.org/10.1016/j.applthermaleng.2022.118438

Kim J, Yu J, Lee S, Tahmasebi A, Jeon C-H, Lucas J (2021) Advances in catalytic hydrogen combustion research: catalysts, mechanism, kinetics, and reactor designs. Int J Hydrog Energy 46(80):40073–40104. https://doi.org/10.1016/j.ijhydene.2021.09.236

Kreutz TG, Larson ED, Elsido C, Martelli E, Greig C, Williams RH (2020) Technoeconomic prospects for producing Fischer-Tropsch jet fuel and electricity from lignite and woody biomass with CO2 capture for EOR. Appl Energy 279:115841. https://doi.org/10.1016/j.apenergy.2020.115841

Leicher J, Nowakowski T, Giese A, Gorner KK (2018) Hydrogen in natural gas: how does it impact industrial end users? In Proceedings of the world gas conference 2018, Washington, DC, 25–29 June 2018

Lowenthal (2006) SB-1505 Fuel: hydrogen alternative fuel. https://leginfo.legislature.ca.gov/faces/billTextClient.xhtml?bill_id=200520060SB1505. Accessed 19 Apr 2021

Luzzo I, Cirilli F, Jovhler G, Gambato A, Longhi J, Rampinini G (2022) Feasibility study for the utilization of natural gas and hydrogen blends on industrial furnaces. Mater Tech 109:306

Massaro MC, Biga R, Kolisnichenko A, Marocco P, Monteverde AHA, Santarelli M (2023) Potential and technical challenges of on-board hydrogen storage technologies coupled with fuel cell systems for aircraft electrification. J Power Sources 555:232397. https://doi.org/10.1016/j.jpowsour.2022.232397

Mayrhofer M, Koller M, Seemann P, Prieler R, Hochenauer C (2021) Assessment of natural gas/hydrogen blends as an alternative fuel for industrial heat treatment furnaces. Int J Hydrog Energy 46:21672–21686

McDowall W (2014) Exploring possible transition pathways for hydrogen energy: a hybrid approach using socio-technical scenarios and energy system modelling. Futures 63:1–14. https://doi.org/10.1016/j.futures.2014.07.004

Mendecka B, Lombardi L (2019) Life cycle environmental impacts of wind energy technologies: a review of simplified models and harmonization of the results. Renewable Sustain Energy Rev 111:462–480

Novelis Inc. (2025) Significant step toward carbon-neutral production: Novelis successfully tests industrial scale use of hydrogen in a recycling furnace. Press release, Feb 20, 2025. Novelis Inc., Zurich. Available at: https://novelis.com/novelis-successfully-tests-industrial-scale-use-of-hydrogen-in-a-recycling-furnace/. Accessed 23 Jan 2026

Osman AI, Ayati A, Krivoshapkin P, Tanhaei B, Farghali M, Yap PS, Abdelhaleem A (2024) Coordination-driven innovations in low-energy catalytic processes: advancing sustainability in chemical production. Coord Chem Rev 514:215900

Perpiñán J, Peña B, Bailera M, Eveloy V, Kannan P, Raj A et al (2023) Integration of carbon capture technologies in blast furnace based steel making: a comprehensive and systematic review. Fuel 336:127074

Rissman J, Bataille C, Masanet E, Aden N, Morrow WR, Zhou N, Elliott N, Dell R, Heeren N, Huckestein B, Cresko J, Miller SA, Roy J, Fennell P, Cremmins B, Koch Blank T, Hone D, Williams ED, de la Rue du Can S, Sisson B, Williams M, Katzenberger J, Burtraw D, Sethi G, Ping H, Danielson D, Lu H, Lorber T, Dinkel J, Helseth J (2020) Technologies and policies to decarbonize global industry: Review and assessment of mitigation drivers through 2070. Appl Energy 266:114848. https://doi.org/10.1016/j.apenergy.2020.114848

Sasiain CA, Rechberger K, Spanlang A, Wolfmeir H, Harris C (2021) Decarbonization of the steel industry. A techno-economic analysis. Mater Tech 109(3–4):305

Seboldt D, Mansbart M, Grabner P, Eichlseder H (2021) Hydrogen engines for future passenger cars and light commercial vehicles. MTZ Worldwide 82(2):42–47

Shahbaz M, Al-Ansari T, Inayat A, Inayat M (2022) Technical readiness level of biohydrogen production process and its value chain. In: Value-chain of Biofuels. Elsevier, p 335e55. https://doi.org/10.1016/B978-0-12-824388-6.00017-8

Sharma S, Basu S, Shetti NP, Aminabhavi TM (2020) Waste-to-energy nexus for circular economy and environmental protection: recent trends in hydrogen energy. Sci Total Environ 713:136633. https://doi.org/10.1016/j.scitotenv.2020.136633

Sorgulu F, Ozturk M, Javani N, Dincer I (2023) Experimental investigation for combustion performance of hydrogen and natural gas fuel blends. Int J Hydrog Energy 48:34476–34485

Sovacool BK, Hess DJ (2017) Ordering theories: typologies and conceptual frameworks for sociotechnical change. Soc Stud Sci 47:703–750

Sun H, Xu X, Kim H, Jung W, Zhou W, Shao Z (2022) Electrochemical water splitting: bridging the gaps between fundamental research and industrial applications. Energy Environ Mater. https://doi.org/10.1002/eem2.12441

Swennenhuis F, de Gooyert V, de Coninck H (2022) Towards a CO industry: justice aspects of CO 2 2-neutral steel capture and storage, biomass- and green hydrogen- based emission reductions. Energy Res Soc Sci 88:102598. https://doi.org/10.1016/j.erss.2022.102598

Taamallah S, Vogiatzaki K, Alzahrani FM, Mokheimer EM, Habib MA, Ghoniem AF (2015) Fuel flexibility, stability and emissions in premixed hydrogen-rich gas turbine combustion: technology, fundamentals, and numerical simulations. Appl Energy 154:1020–1047

Tackie-Otoo BN, Haq MB (2024) A comprehensive review on geo-storage of H 2 in salt caverns: prospect and research advances. Fuel 15(356):129609. Available from: https://doi.org/10.1016/j.fuel.2023.129609

Tiedje JM, Bruns MA, Casadevall A, Criddle CS, Eloe-Fadrosh E, Karl DM, Nguyen NK, Zhou J (2022) Microbes and climate change: a research prospectus for the future. MBio 13(3):e0080022. https://doi.org/10.1128/mbio.00800-22

Upham P, Dütschke E, Schneider U, Oltra C, Sala R, Lores M, Klapper R, Bogel P (2018) Agency and structure in a sociotechnical transition: hydrogen fuel cells, conjunctural knowledge and structuration in Europe. Energy Res Soc Sci 37:163–174. https://doi.org/10.1016/j.erss.2017.09.040

von Schéele J (2022) Advancing use of hydrogen as fuel in steelmaking; Millenium steel service. LLC, Princeton, pp 20–22

Williams F, Yang A, Nhuchhen DR (2024) Decarbonisation pathways of the cement production process via hydrogen and oxy-combustion. Energy Convers Manag 300:117931

Xu S, Tian Z, Cjhen Y, Liang S, Tu Y, Liu H (2024) Effect of hydrogen-blending ratio and wall temperature on establishment, NO formation, and heat transfer of hydrogen-enriched methane MILD combustion. Fuel 369:131787

Yilmaz F, Ozturk M, Selbas R (2020) Design and thermodynamic modeling of a renewable energy based plant for hydrogen production and compression. Int J Hydrog Energy 45:26126–26137. https://doi.org/10.1016/j.ijhydene.201

Zhan X, Chen Z, Qin C (2022) Effect of hydrogen-blended natural gas on combustion stability and emission of water heater burner. Case Stud Therm Eng 37:102246

Chapter 10
Global Strategies for Hydrogen Energy: Policy Directions and Technology Pathways

Abstract Hydrogen energy is increasingly recognized as a cornerstone of the global transition toward low-carbon and resilient energy systems, particularly for decarbonizing hard-to-electrify sectors such as industry, transport, and long-duration energy storage. This chapter provides a comprehensive global assessment of hydrogen energy systems by examining supply–demand dynamics, policy frameworks, and emerging technological pathways. It analyzes current and projected hydrogen production routes, end-use applications, and regional demand patterns shaping the global hydrogen economy. A strategic evaluation of hydrogen energy policies is presented, highlighting national roadmaps, regulatory instruments, market incentives, and international cooperation initiatives driving hydrogen deployment. This chapter further reviews the global technological landscape, covering advancements in hydrogen production, storage, transportation, and utilization technologies. By integrating policy analysis with technological insights, this chapter identifies key challenges, opportunities, and strategic directions necessary to scale hydrogen energy sustainably. The findings contribute to a deeper understanding of global hydrogen strategies and provide actionable insights for policymakers, industry stakeholders, and researchers advancing the hydrogen transition.

Keywords Hydrogen energy · Demand · Supply · Energy policy · Strategic planning

10.1 Introduction

The global energy system is undergoing a fundamental transformation as countries intensify efforts to transition toward sustainable, low-carbon energy pathways. Within this context, green hydrogen (GH_2) produced via water electrolysis using renewable electricity has rapidly emerged as a pivotal element of the energy transition. Owing to its capacity to substantially reduce greenhouse gas emissions while enhancing energy security and reducing dependence on fossil fuel imports, GH_2 is increasingly recognized as a critical instrument in national and international climate

D. M. Y. Mir et al., *Low-Carbon Hydrogen*, Clean Energy Production Technologies,
https://doi.org/10.1007/978-981-95-8142-9_10

change mitigation strategies. This study comparatively examines hydrogen deployment pathways adopted by leading economies, including the United States, China, Australia, and Germany. Each of these countries has pursued a distinct hydrogen development trajectory shaped by its unique resource availability, economic structure, policy priorities, and long-term energy and climate objectives. These contextual differences strongly influence the design, pace, and implementation mechanisms of national hydrogen strategies. Collectively, the four countries have developed a broad spectrum of policy instruments and strategic roadmaps, ranging from early-stage vision-setting to advanced implementation frameworks, reflecting different levels of maturity in their respective hydrogen transitions. Germany has positioned green hydrogen as a cornerstone of its long-term energy transition strategy, with the dual objective of strengthening national energy security and enabling large-scale decarbonization in transportation and industrial mobility (Lux et al. 2022). However, constraints associated with limited domestic renewable energy potential mean that Germany cannot meet its projected hydrogen demand solely through domestic production. Consequently, long-term reliance on imported GH_2 has become a central component of its strategy. This import-oriented approach not only enhances supply security but is also expected to generate substantial green employment opportunities both domestically and in partner countries over the coming decade. By fostering strategic international partnerships and investing in regions with high GH_2 production potential, Germany is positioning itself as a global leader in the emerging hydrogen economy, promoting mutual economic benefits while contributing to a more balanced and resilient global energy system. In June 2020, Germany formalized this vision through the launch of its National Hydrogen Strategy, which aims to establish the country as a frontrunner in GH_2 production, technological innovation, and system integration (Quitzow et al. 2024). The strategy sets ambitious domestic electrolyzer capacity targets of 10 GW, with 5 GW to be installed by 2030 and a further 5 GW planned for deployment between 2035 and 2040 among the most ambitious hydrogen production goals in Europe. In parallel, the strategy emphasizes large-scale infrastructure development, including import terminals, shipping facilities, and pipeline networks, as well as bilateral and multilateral cooperation across the hydrogen value chain. Together, these measures reflect Germany's commitment to building a robust, interconnected, and resilient GH_2 ecosystem. Australia's hydrogen sector has similarly emerged as a major growth opportunity and a key pillar of the country's renewable energy transition. Although still at an early stage of development, Australia's hydrogen strategy aims to support global decarbonization by establishing a commercially competitive hydrogen industry that integrates both domestic consumption and export-oriented value chains by 2030 (Andeobu et al. 2024). In the Australian context, GH_2 is envisioned as a multifunctional energy carrier capable of serving as a feedstock for energy-intensive industries, reducing emissions from high-temperature industrial processes to support net-zero targets by 2050, enabling the production of chemicals and fertilizers, and providing flexible energy storage and grid-balancing services. Australia's vast renewable resource base further positions it as a leading potential exporter of GH_2, particularly in the form of liquefied hydrogen and green ammonia. At the same time, domestic value creation is emphasized through the production of green iron and alumina. This dual focus on domestic utilization and international exports reinforces Australia's strategic role in the global green hydrogen economy and aligns with projections under the Paris Climate Agreement, which suggest that hydrogen could account for approximately 15% of global final energy consumption by 2050. Assessments by the International Energy Agency (IEA) and Australia's Department of Climate Change, Energy, the Environment and Water (DCCEEW) indicate that robust climate policies could enable hydrogen deployment to displace nearly 14 billion cubic meters of natural gas, 20 million tonnes of coal, and 360 kilobarrels of oil per day by 2030, highlighting hydrogen's transformative potential.

In the United States, green hydrogen has been identified by the Department of Energy (US-DOE) as a key enabling technology for decarbonizing hard-to-abate sectors such as steelmaking, ammonia production, long-duration energy storage, and heavy-duty transportation (Bade and Tomomewo 2024). Clean hydrogen forms a central pillar of the DOE's strategy to achieve the Biden administration's targets of a carbon-free electricity sector by 2035 and net-zero economy-wide emissions by 2050. A defining element of this strategy is the "1–1–1" goal introduced under the Hydrogen Shot initiative in June 2021, which aims to reduce the cost of GH_2 to USD 1 per kilogram within a decade. To operationalize this vision, the DOE released a draft National Clean Hydrogen Strategy and Roadmap structured around three priorities: accelerating deployment in high-impact applications, reducing production and supply-chain costs, and rapidly developing regional clean hydrogen hubs. The passage of the Inflation Reduction Act in August 2022 further strengthened the US hydrogen ecosystem through substantial financial incentives, including production tax credits that significantly improve the economic viability of GH_2. Importantly, the US approach emphasizes inclusive and coordinated stakeholder engagement involving federal and state governments, industry, academia, national laboratories, local and Tribal communities, and environmental and environmental-justice organizations. This inclusive governance model is viewed as essential for ensuring equitable outcomes, rapid deployment, and sustained market growth, thereby reinforcing US leadership in clean hydrogen innovation and commercialization. China, as the world's largest hydrogen producer, also possesses considerable potential to leverage GH_2 to enhance energy security and meet its climate commitments by reducing emissions across major industrial and energy-consuming sectors (Brown and Grünberg 2022; Li et al. 2022a, b). The Chinese government is actively advancing the development of a large-scale and integrated hydrogen economy, underpinned by the country's extensive renewable energy capacity. GH_2 is positioned as a key enabler across transportation, grid-scale energy storage, and industrial applications, forming a central component of China's long-term sustainable energy strategy. In March 2022, China released a national hydrogen development plan for the period 2021–2035, signaling strong policy commitment and long-term strategic intent. The plan outlines five core objectives: establishing a centralized innovation platform for GH_2 technologies; systematically developing hydrogen production, storage, and transport infrastructure; scaling hydrogen applications in transportation, energy storage, and industry; optimizing policy frameworks, particularly electricity pricing mechanisms for GH_2 production; and developing comprehensive standards for hydrogen quality, safety, and infrastructure. Hydrogen has already been incorporated into development strategies across nearly all Chinese provinces, with more than 120 GH_2-related projects currently under development. China's phased approach prioritizes the use of industrial by-product hydrogen in the near term, complemented by localized renewable hydrogen production, enabling gradual integration into the energy system while supporting balanced regional development (Zou et al. 2022). More broadly, China is pursuing a comprehensive green transformation across its industrial, energy, transportation, and urban–rural systems, supported by large-scale deployment of wind and solar power, enhanced grid infrastructure, advanced energy storage technologies, and international cooperation on green energy standards. Beyond major economies, hydrogen also presents significant opportunities for emerging and developing countries. In Bangladesh, for example, hydrogen as a renewable energy carrier could play a critical role in reducing reliance on costly natural gas imports and environmentally damaging coal-based energy systems. Hydrogen demand in Bangladesh is rising across sectors such as transportation, power generation, petroleum refining, steel production, methanol synthesis, electronics, and food processing (Al-tabatabaie et al. 2022). The strategic deployment of GH_2 in these sectors could substantially enhance energy security, lower emissions, and support long-term sustainable economic development, demonstrating the broad global relevance of green hydrogen beyond advanced industrial economies.

10.2 Global Hydrogen Energy: Supply and Demand

10.2.1 Supply-Side Hydrogen Strategies

The multifunctionality of hydrogen as an energy carrier has led many policymakers to regard it as a promising solution for deep decarbonization; however, two fundamental challenges related to hydrogen supply necessitate well-defined and supportive governmental policy frameworks. First, hydrogen is not a primary energy source and is rarely available in a pure form in nature. Its production therefore depends heavily on chemical, physical, or biological conversion processes and associated technologies. Second, hydrogen's intrinsic chemical properties such as low volumetric energy density and high flammability require advanced and reliable technologies to ensure its safe storage, transportation, and distribution (Somenzi et al. 2021). By 2023, a total of 41 countries had adopted national hydrogen strategies, reflecting the growing global policy momentum behind hydrogen deployment (OECD 2023). These national strategies primarily aim to stimulate hydrogen production, with a strong emphasis on green and blue hydrogen pathways. In many cases, the announcement of hydrogen strategies has served as a key trigger for public funding and investment support. For instance, India's National Green Hydrogen

Mission proposes subsidies of approximately INR 2 billion to support the production of 5 Mt. of hydrogen for domestic consumption and 10 Mt. for export by 2030. Similarly, the United States has updated its hydrogen strategy to align with newly introduced tax credits and funding mechanisms for carbon capture, renewable energy, and sustainable aviation fuels (SAFs), culminating in the adoption of the U.S. National Clean Hydrogen Strategy and Roadmap. In parallel, countries such as Belgium, Germany, Japan, and South Korea have revised their existing hydrogen strategies, while several European nations are currently in the process of updating their policy frameworks. Hydrogen can be produced through a variety of pathways, including the reforming of fossil fuels (natural gas and coal), the conversion of biomass sources (solid biofuels, organic waste, and biogas), and electricity-driven water electrolysis (Megía et al. 2021). The selection of production pathways is strongly influenced by national resource availability and existing energy infrastructure. In Europe, where fossil fuel resources are relatively limited, hydrogen strategies predominantly focus on electrolyzer deployment powered by renewable electricity. As a result, electrolyzers are considered the most viable and attractive hydrogen production technology across the region. In terms of planned investments in electrolyzer capacity, the European Union appears to be in a leading position globally (Hesel et al. 2022). According to Table 10.1, Germany, France, Spain, and Italy have collectively committed nearly €30 billion by 2030 to expand electrolyzer capacity by approximately 20 GW (García-Herrero et al. 2021). In contrast, the United States has adopted a more diversified approach, supporting both fossil fuel–based and renewable-based hydrogen production pathways (European Commission 2023). Given its continued reliance on natural gas, the US strategy emphasizes the use of carbon capture and storage (CCS) technologies to mitigate emissions associated with hydrogen production. Under the Infrastructure Investment and Jobs Act (IIJA), the federal government plans to establish at least four regional clean hydrogen hubs aimed at producing and utilizing hydrogen across manufacturing, heating, and transportation sectors (Bade and Tomomewo 2024). The legislation specifies that at least two hubs will be located in regions with abundant natural gas resources, with one hub producing hydrogen from fossil fuels, another relying on renewable

Table 10.1 Leading national hydrogen strategies: goals and committed funding

Country/ region	Key 2030 targets	Committed funding	References
European Union	40 GW electrolyzer capacity; 10 Mt. renewable hydrogen production	€24–42 bn	European Commission (2020, 2022, 2023)
Germany	≥10 GW electrolyzer capacity	€9 bn	BMWK (2020, 2023)
France	~6.5 GW electrolysis capacity	€7 bn	Ministère de l'Économie (2020), Bouacida (2024)
Spain	4 GW electrolysis capacity	€9 bn	Burgess (2021)
Italy	~5 GW electrolysis; ~2% final energy share	€3.19 bn	Carboni (2023)
Netherlands	3–4 GW electrolysis capacity	~€9 bn	Ken Research (2025)

energy, and a third using nuclear power as its primary energy source. The potential use of coal as a feedstock is also acknowledged. Notably, the IIJA does not set explicit targets for scaling renewable-based hydrogen production, but it is expected to mobilize investments of up to USD 9 billion between 2022 and 2026. China currently ranks as the world's largest hydrogen producer; however, the majority of its hydrogen output is derived from coal rather than renewable sources (Liu et al. 2022). In China's 14th Five-Year Plan, hydrogen has been identified as a strategic emerging industry (Meng et al. 2022), with the objective of increasing the share of renewable-based hydrogen to 50% of total hydrogen production by 2030. Achieving this target remains challenging due to China's heavy reliance on coal-based hydrogen production. As in the United States, CCS technologies are expected to play a critical role in decarbonizing hydrogen production in China (Li et al. 2022a, b). A closer examination of national hydrogen strategies reveals that research and development (R&D) support and financial subsidization remain central priorities for governments worldwide (Somenzi et al. 2021). These efforts are widely viewed as essential for reducing costs, improving system efficiency, and enabling large-scale market adoption of hydrogen in the coming decades.

10.2.2 Demand-Side Hydrogen Strategies

In 2022, global hydrogen demand reached a record high; however, this growth was largely driven by traditional applications rather than emerging low-carbon uses (Hassan et al. 2024). Worldwide hydrogen consumption rose to approximately 95 Mt. in 2022, representing an increase of nearly 3% compared to the previous year. Demand expanded across most major consuming regions, with the notable exception of Europe, where industrial activity was constrained by a sharp escalation in natural gas prices (International Energy Agency 2023). Importantly, this overall increase in hydrogen use does not necessarily reflect the effectiveness of policy initiatives aimed at expanding hydrogen into new sectors; instead, it largely mirrors broader global energy and industrial trends. Hydrogen demand remains heavily concentrated in conventional sectors such as refining and industrial feedstock applications. New uses in heavy industry, transportation, and power generation currently account for less than 0.1% of total hydrogen consumption. Moreover, the penetration of low-emission hydrogen into existing applications has been slow, representing only about 0.7% of total hydrogen demand. As a result, hydrogen production and consumption in 2022 were associated with more than 900 million tonnes of CO_2 emissions, underscoring the continued dominance of fossil-based hydrogen pathways (International Energy Agency 2023). Policy attention toward stimulating demand for low-emission hydrogen has only recently emerged and remains insufficient to align hydrogen deployment with long-term climate targets (Evro et al. 2024). To date, government interventions have largely focused on supporting the supply side of the hydrogen value chain, while comparatively less emphasis has been placed on stimulating demand. Although aggregated government targets for low-emission hydrogen production range between 27 and 35 Mt., corresponding demand targets amount to only about 14 Mt., with less than half explicitly linked to the decarbonization of current hydrogen uses (International Energy Agency 2023). Even if these targets are fully realized, they would represent only around 20% of the low-emission hydrogen consumption required under the Net Zero Emissions (NZE) Scenario by 2050. At the same time, insufficient demand poses a significant barrier to investment, as producers of low-emission hydrogen are unlikely to commit substantial capital without clear market signals, thereby jeopardizing the economic viability of the emerging hydrogen sector (Azadnia et al. 2023). While supply-side incentives such as tax credits and public funding for research, development, and demonstration can effectively reduce hydrogen production costs, further investment in production capacity ultimately depends on addressing uncertainties on the demand side. Demand-related risks can deter project developers from securing financing needed to scale production and associated midstream infrastructure. For example, by the end of 2022, approximately 12 million metric tonnes per year of clean hydrogen production capacity had been announced in the United States; however, only around 10% of these projects had reached the final investment decision (FID) stage, where substantial financial commitments are made. The absence of guaranteed near-term demand or long-term offtake agreements significantly increases project risk, discouraging private capital investment in clean hydrogen facilities and related infrastructure (Craen 2023). This situation highlights the critical role of demand-side policy measures in accelerating the early growth of the hydrogen market. Consequently, the development of hydrogen hubs requires a balanced mix of supply-side and demand-side policy instruments. Economic literature suggests that market failures on the demand side can significantly impede the diffusion of innovative technologies, even when supply-side support is substantial (Edquist and Hommen 1999; Lee et al. 2022). While traditional innovation policies have predominantly emphasized supply-side mechanisms, demand-side instruments such as guaranteed offtake agreements, public procurement, and market-creation policies can serve as powerful complementary tools and may be particularly urgent in the case of clean hydrogen. Investors in hydrogen hubs must contend with considerable uncertainty regarding future demand levels and the cost trajectories of emerging hydrogen technologies (Shirizadeh et al. 2023). To mitigate these risks, project developers often rely on long-term contracts to secure debt financing, as such agreements provide predictable revenue streams and reduce investment uncertainty. The experience of solar and other clean energy technologies illustrates the effectiveness of this approach, where power purchase agreements (PPAs) played a crucial role in enabling large-scale deployment by guaranteeing demand and ensuring stable cash flows (Słotwinski 2022). However, potential industrial offtakers may be reluctant to enter long-term contracts with clean hydrogen producers if they anticipate rapid declines in technology costs (Nygrén et al. 2015). As the clean hydrogen industry expands, learning-by-doing and economies of scale are expected to significantly reduce costs, with estimates suggesting that the cost of renewable-based hydrogen production via electrolysis could decline by approximately 50% by 2030 (Zun and McLellan 2023). This expectation creates hesitation among end users, who may be unwilling to commit to long-term agreements at prices perceived as higher than future market levels, further complicating the establishment of stable demand in the early stages of market development. Consequently, clean hydrogen hubs are likely to face significant short-term challenges arising from insufficient demand. These demand-side constraints are particularly acute for emerging applications of clean hydrogen, especially in the industrial and transportation sectors, such as the production of low-carbon steel and the development of sustainable aviation fuels (SAFs) (Qazi 2022). In such conditions, there is a clear opportunity for governments to deploy demand-pull policy instruments to support nascent industries that hold strategic, environmental, or social importance, particularly where private-sector actors are hesitant to commit to long-term offtake agreements (Taylor 2008). A range of demand-pull mechanisms is available to policymakers, including advance market commitments (AMCs), innovation prizes and awards, and contracts for difference (CfD) schemes. Economic evidence demonstrates that these instruments can be highly effective in accelerating innovation and market formation, as illustrated by their successful application in sectors such as vaccine development, particularly in low- and middle-income countries (Postigo 2022). Although the market conditions for vaccines differ from those of clean energy technologies, the underlying economic principles that drove the effectiveness of demand-pull strategies—namely, reducing demand uncertainty and improving investment incentives—are equally applicable to emerging climate and clean energy solutions. Indeed, economic analyses suggest that instruments such as AMCs, which provide greater certainty regarding future demand, can play a pivotal role in accelerating investment and deployment in clean energy sectors (Brunner et al. 2012). Furthermore, several studies indicate that output-based or performance-oriented incentives may be more efficient than traditional supply-side subsidies alone (Aldy et al. 2023). Direct demand-side support can enhance market confidence by signaling that substantial and sustained demand will materialize as production costs decline. In this context, the adoption of clean hydrogen technologies could enable sectors such as natural gas–fired power generation and heavy-duty transportation to comply with proposed greenhouse gas emission standards. Although the implementation of such regulatory frameworks may take several years, their announcement and gradual enforcement can provide developers with stronger long-term market signals. Anticipation of future compliance requirements, combined with expected cost reductions in clean hydrogen technologies, can significantly improve investor confidence and stimulate early-stage market development. Figure 10.1 compares projected global hydrogen demand across sectors in the Net Zero Emissions Scenario for 2050 (IEA 2023).

10.3 Strategic Analysis of Hydrogen Energy Policy

At present, leading global economies including the United States, the European Union, Japan, and several other industrialized regions are proactively accelerating the development of robust hydrogen energy ecosystems. These economies have formally introduced comprehensive national and regional hydrogen strategies accompanied by detailed roadmaps that provide clear, long-term visions for hydrogen

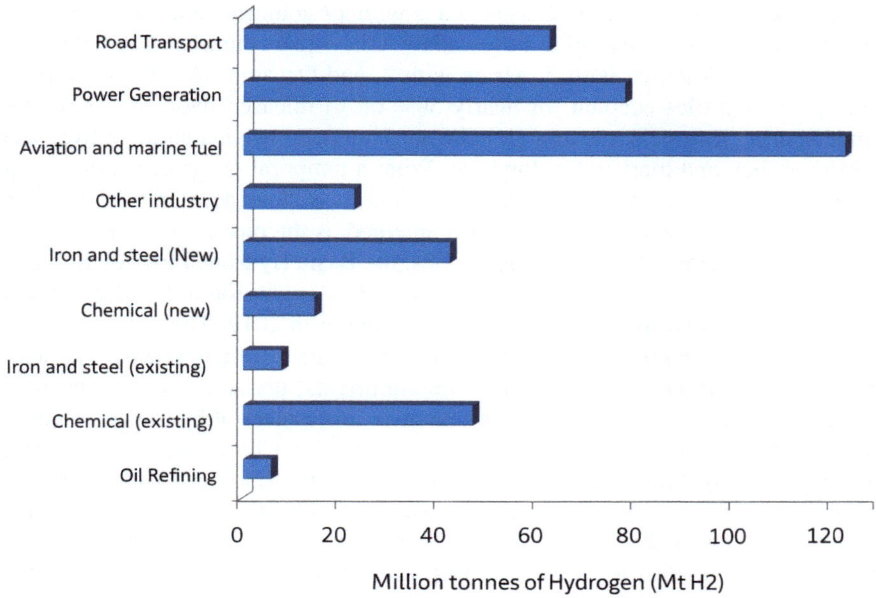

Fig. 10.1 Global hydrogen demand in the Net Zero Emissions Scenario for 2050

deployment. Such policy frameworks emphasize coordinated and systematic actions across the entire hydrogen value chain. This includes scaling up low-carbon and renewable hydrogen production pathways, establishing efficient and safe storage and transportation infrastructure, and promoting diverse end-use applications in sectors such as industry, transportation, power generation, and buildings. In addition, these strategies place strong emphasis on technological innovation, research and development, workforce development, and targeted investment mechanisms to enhance cost competitiveness and market readiness. Collectively, these integrated roadmaps aim to reduce greenhouse gas emissions, strengthen energy security, foster industrial competitiveness, and position hydrogen as a cornerstone of the global energy transition toward a sustainable and carbon neutral future.

10.3.1 Analysis of Global Hydrogen Energy Strategic Planning

With the rapid progress of sustainable energy transitions and the maturation of emerging energy technologies, hydrogen has gained widespread policy attention worldwide. By 2022, a total of 35 countries and regions had formally issued hydrogen-related strategies or policy frameworks, reflecting the growing recognition of hydrogen as a key pillar of future low-carbon energy systems. An analysis of the global spatial distribution of these hydrogen strategy documents shows a clear regional concentration. Europe leads with 17 countries publishing national or regional hydrogen strategies, followed by Asia with 6 countries, South America with 5, North America with 2, Africa with 2, and Oceania with 2. Collectively, European countries account for nearly 50% of all released hydrogen strategies, underscoring Europe's role as a global frontrunner and agenda-setter in hydrogen energy policy and market development. From a temporal perspective, the evolution of hydrogen strategy formulation highlights a clear acceleration in policy momentum over the past decade. Japan emerged as the pioneer by launching the world's first national-level hydrogen policy, the Basic Hydrogen Energy Strategy, in 2017. This milestone was followed by a gradual expansion of hydrogen policy initiatives globally, with a notable surge beginning in 2019. The period between 2020 and 2021 represents the peak phase of hydrogen strategy releases, coinciding with strengthened global climate commitments, net-zero targets, and post-pandemic green recovery plans. A comparative assessment of hydrogen strategies adopted by major economies and regions including China, the United States, Japan, Germany, France, the United Kingdom, and the European Union reveals a high degree of convergence in strategic objectives. Despite differences in resource endowments and industrial structures, these strategies are predominantly driven by shared goals such as achieving deep decarbonization, enhancing long-term energy security, fostering industrial competitiveness, and stimulating economic growth through the development of new hydrogen value chains. Technologically, these national and regional roadmaps place strong emphasis on a set of priority pathways and application domains. Key areas include hydrogen production via proton exchange membrane (PEM) electrolysis, advanced storage and transport solutions such as organic liquid hydrogen carriers and cryogenic liquid hydrogen, hydrogen-based power generation systems, and integrated renewable hydrogen configurations that couple hydrogen production with wind and solar energy resources. Together, these strategic priorities reflect a coordinated global effort to accelerate hydrogen technology innovation, scale up infrastructure, and enable widespread deployment across multiple sectors of the energy system (Al Humaidan et al. 2023).

10.3.2 Analysis of Hydrogen Energy Policies in Major Countries

10.3.2.1 United States

To sustain its competitive edge in next-generation energy technologies, the United States places considerable strategic emphasis on strengthening capabilities across both the upstream and downstream segments of the hydrogen energy value chain. In line with this objective, federal policies and programs consistently prioritize research, development, and demonstration (RD&D) activities in hydrogen production, storage, distribution, and utilization, alongside continued advancements in fuel cell technologies. Policy analyses indicate that hydrogen is increasingly regarded not only as an emerging clean energy carrier but also as a strategic energy reserve within the broader US energy system, capable of enhancing system flexibility and resilience. The national hydrogen strategy therefore adopts a comprehensive and application-oriented approach, identifying five key domains for hydrogen deployment. These include its use as a transportation fuel, its role in power generation and grid stabilization, applications in residential and commercial buildings, utilization as an industrial feedstock and for long-distance transport, and its direct use as an industrial fuel. Collectively, these focus areas reflect the United States' intent to integrate hydrogen across multiple sectors, accelerate technology commercialization, and reinforce energy security while supporting long-term decarbonization goals (Ullah et al. 2025).

10.3.2.2 European Union

Hydrogen energy development within the European Union is largely underpinned by the rapid expansion of renewable energy capacity, supported by the region's extensive and mature natural gas pipeline network. This dual advantage enables a highly integrated approach to the hydrogen economy, facilitating the large-scale production of green hydrogen while simultaneously supporting the establishment of efficient hydrogen storage and transport infrastructures. Together, these elements create favorable conditions for the decarbonization of energy-intensive sectors, including transportation, industry, and the built environment. These coordinated initiatives are designed to reinforce the EU's broader energy transition agenda and to accelerate the shift toward a green, low-carbon economic model. Policy analyses consistently indicate that the European Union regards hydrogen as a pivotal instrument for achieving deep decarbonization and as a central pillar of its long-term climate and energy strategy (Rehman et al. 2024). In line with this strategic positioning, the EU's hydrogen deployment roadmap prioritizes several key application areas: decarbonization of industrial processes; large-scale adoption of hydrogen fuel cell vehicles, particularly in heavy-duty transport such as trucks and coaches; the integration of hydrogen in aviation and maritime sectors; hydrogen-based power generation to support system flexibility; and the utilization of hydrogen in residential and commercial buildings. Collectively, these focus areas reflect the EU's ambition to mainstream hydrogen across multiple sectors as a core component of a sustainable and resilient energy system.

10.3.2.3 Japan

Japan was the first nation worldwide to formally launch a comprehensive national strategy for hydrogen energy development, setting an early benchmark for global hydrogen policy. This pioneering approach is largely shaped by Japan's limited domestic energy resources and long-standing concerns over energy security. As a result, the country has pursued a clear, stable, and long-term policy orientation toward hydrogen, with its national strategy undergoing continuous refinement and upgrading as technologies and market conditions evolve. Central to Japan's approach is a strong emphasis on the advancement and commercialization of hydrogen-related technologies. This is reinforced through sustained government support mechanisms, including fiscal incentives, subsidies, and targeted tax measures aimed at accelerating innovation and market adoption. Policy assessments suggest that Japan deliberately positions hydrogen as a strategic energy carrier to enhance national energy security while simultaneously strengthening the competitiveness of its industrial sector. In line with this strategic vision, Japan's hydrogen deployment priorities focus on several key application areas. These include transportation particularly the development and large-scale adoption of hydrogen fuel cell vehicles alongside the expansion of hydrogen refueling infrastructure as well as hydrogen-based power generation and residential fuel cell combined heat and power (CHP) systems. Collectively, these initiatives reflect Japan's integrated and security-oriented approach to building a resilient hydrogen economy (Lebrouhi et al. 2022).

10.3.2.4 China

China has promulgated a series of national-level planning and policy frameworks aimed at systematically guiding and accelerating the development of its hydrogen energy industry. These policies are designed to foster coordinated and balanced growth across the entire hydrogen value chain, spanning hydrogen production, storage, transmission, and transportation infrastructure. Through this integrated policy approach, China seeks to create a mature and resilient hydrogen ecosystem capable of supporting large-scale deployment. Policy analyses indicate that China strategically positions hydrogen as a central pillar of its future national energy system and as a critical energy carrier for driving the green and low-carbon transformation of downstream sectors. At the same time, hydrogen is identified as a strategic emerging industry and a priority focus for long-term industrial upgrading and technological innovation. This multifaceted positioning reflects China's ambition to align hydrogen development with broader objectives related to energy security, carbon neutrality, and high-quality economic growth. Consistent with this strategic vision, China's hydrogen deployment roadmap prioritizes key application domains, including transportation, large-scale and distributed energy storage, hydrogen-based power generation, and industrial processes. Together, these focus areas underscore China's intent to integrate hydrogen across multiple sectors, supporting both decarbonization and industrial modernization (Chen et al. 2024).

10.4 Strategic Analysis of Hydrogen Energy Technology

10.4.1 Analysis of Global Hydrogen Energy Technology Layout

An analysis of more than 3000 hydrogen-related research projects conducted between 2017 and 2022 was carried out using data collected from government portals, leading research and development institutions, and widely recognized databases such as the CORDIS database, covering major countries worldwide. The global technological landscape of hydrogen research projects exhibits several defining characteristics. First, the scope of hydrogen-based power generation technologies incorporated into national hydrogen projects continues to expand across countries. Second, investment levels across diverse hydrogen technology pathways are steadily increasing. Third, hydrogen research and development efforts are predominantly concentrated in power generation applications, followed by electrolytic hydrogen production and hydrogen storage technologies. These trends highlight the growing diversification and intensification of global investment in hydrogen technologies (Sun et al. 2022).

10.4.1.1 United States

In the United States, both policy emphasis and financial commitments to hydrogen energy have expanded dramatically in recent years, reflecting the growing strategic importance of hydrogen within the national clean energy agenda. Public investment increased from approximately USD 13.06 million in 2018 to nearly USD 7.1 billion by 2022, signaling a decisive shift from exploratory support toward large-scale deployment and commercialization. This rapid escalation in funding underscores the federal government's intent to accelerate hydrogen innovation, infrastructure development, and market adoption. Between 2018 and 2021, federal research and development (R&D) funding for hydrogen was largely directed toward foundational technologies, including advanced hydrogen production methods, storage and transportation systems, fuel cell technologies, and hydrogen-capable gas turbines. By 2022, however, the focus of US hydrogen R&D had evolved in response to decarbonization priorities and industrial scale-up requirements. Greater emphasis was placed on clean hydrogen production pathways, the build-out of hydrogen transportation and storage infrastructure, and downstream applications such as combined heat and power (CHP) systems and hydrogen-based electricity generation. Recent trends in project deployment further illustrate this strategic realignment. Hydrogen projects in the United States have increasingly concentrated on electrolytic hydrogen production and hydrogen-fueled power generation, which together account for more than 70% of all US projects related to hydrogen for electricity generation (Zhang et al. 2024). This concentration reflects both technological maturity and policy incentives favoring low-carbon hydrogen and flexible power solutions. In terms of electrolysis technologies, US R&D efforts initially prioritized proton exchange membrane (PEM) electrolysis during the period from 2018 to 2020. From 2019 onward, attention progressively expanded to include solid oxide electrolysis cell (SOEC) technologies, driven by their potential for higher efficiency and integration with high-temperature industrial processes. Concurrently, frontier research initiatives launched in 2019 began targeting alkaline anion exchange membrane (AEM) electrolysis, with a particular focus on the development of low-cost, high-performance, and durable electrode materials. With respect to hydrogen-based power generation technologies, the United States pursued both proton exchange membrane fuel cell (PEMFC) and solid oxide fuel cell (SOFC) systems throughout the 2018–2022 period. More recently, R&D portfolios have broadened to include hydrogen-fired gas turbines and cogeneration technologies, with gas turbine research gaining momentum around 2021 and cogeneration applications emerging more prominently from 2022 onward. These efforts aim to enhance grid flexibility, support large-scale renewable energy integration, and decarbonize dispatchable power generation. Beyond production and utilization, 2022 marked a notable increase in attention to large-capacity and long-duration hydrogen storage technologies. Research and demonstration activities increasingly targeted solid-state hydrogen storage and geological storage options, recognizing their importance for seasonal energy storage, grid resilience, and large-scale hydrogen deployment. Collectively, these developments highlight the United States' strategic commitment to building an integrated, robust, and resilient hydrogen technology ecosystem encompassing hydrogen production, conversion, utilization, and storage across multiple sectors.

10.4.1.2 Germany

Within the European Union, hydrogen energy research and development activities in Germany are largely supported through EU research framework programs, notably Horizon 2020 and its successor, Horizon Europe. Funding allocated under these programs primarily targets hydrogen production pathways based on renewable energy sources, the development of large-capacity and long-duration hydrogen storage technologies, and the advancement of fuel cell technologies for transportation applications. In parallel, substantial policy and financial support is directed toward the integration of hydrogen across key end-use sectors, including industrial processes, transportation networks, and power generation systems. Recent trends in project deployment indicate that hydrogen-related projects in Germany with a focus on electricity applications are predominantly centered on hydrogen-based power generation technologies, which account for more than 50% of all projects in the hydrogen-for-electricity category. In the domain of electrolytic hydrogen production, Germany has demonstrated a clear and sustained preference for solid oxide electrolysis cell (SOEC) technology over the 4-year period from 2019 to 2022, reflecting its strategic interest in high-efficiency, high-temperature electrolysis pathways. Complementary research initiatives launched in 2020 also explored alkaline anion exchange membrane (AEM) electrolysis, with particular attention given to the development of cost-effective and durable electrode materials. In contrast, proton exchange membrane (PEM) electrolysis has not featured prominently within Germany's national hydrogen R&D portfolio. With regard to hydrogen storage technologies, German research efforts between 2018 and 2020 concentrated on cryogenic liquid hydrogen storage and metal-based solid-state storage materials. From 2021 onward, the focus expanded toward large-scale storage solutions, particularly underground salt cavern hydrogen storage and organic liquid hydrogen carrier (LOHC) technologies, which are well suited for seasonal and bulk hydrogen storage. In the area of hydrogen-based power generation, Germany consistently pursued both proton exchange membrane fuel cell (PEMFC) and solid oxide fuel cell (SOFC) technologies from 2018 to 2021. Beginning in 2021, R&D activities broadened to include hydrogen-based cogeneration systems, followed by an increasing emphasis on hydrogen gas turbine technologies during the 2021–2022 period. By 2022, Germany's hydrogen research and development landscape was primarily focused on solid oxide electrolysis for hydrogen production, proton exchange membrane fuel cell technologies, and hydrogen gas turbine applications. This evolution reflects a deliberate strategic emphasis on integrated hydrogen production and power generation systems capable of supporting large-scale decarbonization of the energy sector.

10.4.1.3 Japan

Japan's New Energy and Industrial Technology Development Organization (NEDO) has progressively and substantially expanded financial support for hydrogen energy research and development over recent years, with annual funding increasing from approximately ¥15.1 billion in 2018 to about ¥193.5 billion by 2022. This marked rise in investment reflects Japan's long-term strategic commitment to hydrogen as a core component of its future energy system. NEDO's funding has predominantly focused on strengthening the hydrogen supply chain, advancing fuel cell technologies, expanding hydrogen refueling infrastructure, and developing hydrogen-capable gas turbine systems. In 2022, NEDO further broadened its R&D portfolio to include hydrogen utilization in steelmaking processes, signaling an increasing emphasis on deploying hydrogen across a wide range of end-use sectors. These include transportation, industrial manufacturing, power generation, and the building sector, highlighting Japan's intention to embed hydrogen deeply within its national decarbonization strategy. Consistent with this approach, hydrogen-related projects in Japan that target electricity applications have, in recent years, been largely concentrated on hydrogen-based power generation technologies, which account for more than 60% of all projects in the hydrogen-for-electricity category. With respect to electrolytic hydrogen production, Japan initially explored both proton exchange membrane (PEM) electrolysis and solid oxide electrolysis cell (SOEC) technologies in 2018. Over time, however, research priorities shifted predominantly toward PEM electrolysis, reflecting a strategic preference for technologies with higher maturity levels and stronger prospects for near-term commercialization. In the field of hydrogen-based power generation, Japan has maintained sustained investment in proton exchange membrane fuel cells (PEMFCs), solid oxide fuel cells (SOFCs), and hydrogen gas turbine technologies throughout the 5-year period from 2018 to 2022. Notably, more than 70 individual projects have focused on PEMFC technology alone, underscoring the country's strong and consistent commitment to fuel cell innovation. Beyond hydrogen production and conversion technologies, Japan's hydrogen R&D landscape in 2022 also expanded to include organic liquid hydrogen carrier (LOHC) storage solutions and integrated renewable–hydrogen, or "power-to-hydrogen," systems. These additions further enhance Japan's comprehensive hydrogen technology ecosystem by addressing critical challenges related to hydrogen storage, system integration, and large-scale deployment, thereby reinforcing the country's leadership in hydrogen energy research and development.

10.4.1.4 China

China's national key hydrogen energy research programs have expanded steadily since 2018, reflecting a growing and sustained commitment to hydrogen as a strategic pillar of the country's energy transition. Over the past 5 years, research and development activities have been concentrated on several core technological domains, including proton exchange membrane (PEM) electrolysis for hydrogen production, cryogenic liquid hydrogen storage, PEM fuel cell technologies, hydrogen-based cogeneration systems, and a range of Power-to-X conversion pathways. In terms of application focus, hydrogen deployment has been directed primarily toward the power generation and transportation sectors, which are viewed as critical leverage points for large-scale decarbonization. A significant milestone in China's hydrogen R&D trajectory occurred during 2021 and 2022, when a more integrated and comprehensive research framework was established. This framework encompasses hydrogen production for power generation, advanced storage technologies, hydrogen utilization, and electric–hydrogen coupling, enabling closer coordination across different segments of the hydrogen value chain. In recent years, project development within the hydrogen-for-electricity domain has been dominated by hydrogen utilization technologies, which account for approximately 50% of all projects in this category, highlighting a strong emphasis on downstream applications. With respect to electrolytic hydrogen production, China initiated systematic research and development in PEM electrolysis technologies in 2021, followed by a substantial intensification of research efforts in 2022. In parallel, R&D activities in solid oxide electrolysis cell (SOEC) technologies were launched from 2022 onward, signaling an expansion toward high-efficiency, next-generation electrolysis pathways (Zhao et al. 2020; Zhang et al. 2022). In the area of hydrogen storage, China has maintained continuous research on low-temperature liquid hydrogen storage technologies since 2019, while research priorities broadened in 2022 to include metal-based solid-state hydrogen storage materials. Regarding hydrogen utilization technologies, China has consistently invested in both PEM and solid oxide fuel cell research since 2018, complemented by sustained R&D on hydrogen-based cogeneration technologies beginning in 2020. In the field of electric–hydrogen coupling, China initiated research on renewable-based hydrogen production—particularly wind- and solar-powered electrolysis—in 2018, with renewed emphasis in 2021. Power-to-X technologies have received focused and sustained attention since 2021, while hydrogen–grid interaction technologies were explored selectively in 2020. By 2022, China's hydrogen R&D portfolio had evolved into a comprehensive and increasingly coordinated national program, integrating electrolytic hydrogen production, storage, hydrogen-based power generation, and electric–hydrogen coupling to support large-scale and system-level hydrogen deployment (Yue et al. 2021).

10.5 Conclusions

In summary, while countries that have formulated hydrogen energy strategies are at different stages of development and progress at varying rates of implementation, their overarching strategic directions show a high degree of convergence. Most national strategies prioritize expanding the role of hydrogen across both energy production systems and end-use sectors, with the common aim of enabling a clean, low-carbon, and sustainable energy transition. This broad alignment underscores hydrogen's growing recognition as a key component of future energy systems worldwide. From the perspective of science and technology project deployment, however, national priorities exhibit notable differences. Compared with other countries, China and Japan place particular emphasis on hydrogen gas turbine technologies, reflecting their interest in hydrogen-based power generation and system flexibility. In addition, China assigns significant priority to hydrogen-based cogeneration systems, highlighting its focus on integrated energy utilization and efficiency improvements across industrial and power sectors. As renewable energy capacity continues to expand rapidly, hydrogen production based on renewable electricity is expected to emerge as a dominant pathway for future hydrogen development. This evolution will foster multi-level coupling interactions between hydrogen systems and power grids. Establishing a coordinated electric–hydrogen coupling mechanism characterized by clean electricity as the primary energy source and hydrogen as a complementary energy carrier and developing a new power system paradigm of "electricity-dominated systems with electric–hydrogen coupling" will enable the integrated deployment of hydrogen across energy sources, networks, loads, and storage within next-generation power systems. Such integrated electric–hydrogen systems can significantly enhance renewable energy utilization, provide large-capacity and long-duration energy storage, and promote coordinated optimization across electricity, heat, and fuel networks. Ultimately, this approach will strengthen the flexibility and regulation capability of power grids, while ensuring the safe, stable, and resilient operation of future low-carbon energy systems.

References

Al Humaidan FS, Absi Halabi M, Rana MS, Vinoba M (2023) Blue hydrogen: current status and future technologies. Energy Convers Manag 283:116840. https://doi.org/10.1016/j.enconman.2023.116840

Aldy JE, Gerarden TD, Sweeney RL (2023) Investment versus output subsidies: implications of alternative incentives for wind energy. J Assoc Environ Resour Econ 10:981–1018. https://doi.org/10.1086/723142

Al-tabatabaie KF, Hossain MB, Islam MK, Awual MR, TowfiqulIslam ARM, Hossain MA, Islam A (2022) Taking strides towards decarbonization: the viewpoint of Bangladesh. Energ Strat Rev 44:100948

Andeobu L, Wibowo S, Grandhi S (2024) Renewable hydrogen for the energy transition in Australia-current trends, challenges and future directions. Int J Hydrog Energy 87:1207–1223

Azadnia AH, McDaid C, Andwari AM, Hosseini SE (2023) Green hydrogen supply chain risk analysis: a European hard-to-abate sectors perspective. Renew Sust Energ Rev 182:113371. https://doi.org/10.1016/J.RSER.2023.113371

Bade SO, Tomomewo OS (2024) A review of governance strategies, policy measures, and regulatory framework for hydrogen energy in the United States. Int J Hydrog Energy 78:1363–1381

Bouacida I (2024) France's hydrogen strategy: focusing on domestic hydrogen production to decarbonise industry and mobility. In: The geopolitics of hydrogen: volume 1: European strategies in global perspective. Cham, Springer Nature Switzerland, pp 67–84

Brown A, Grünberg N (2022) China's nascent green hydrogen sector: how policy, research and business are forging. Mercato Inst China Stud 77:1–24

Brunner S, Flachsland C, Marschinski R (2012) Credible commitment in carbon policy. Clim Pol 12:255–271. https://doi.org/10.1080/14693062.2011.582327

Burgess J (2021) Hydrogen fever in EU puts 2024 target of 6-GW electrolyzer capacity in reach. S&P Global Platts. Available at: https://www.spglobal.com/platts/en/market-insights/latest-news/electric-power/070721-hydrogen-fever-in-eu-puts-2024-target-of-6-gw-electrolyzer-capacity-in-reach. Accessed Jan 2026

Carboni T. (2023) La strada per arrivare all'idrogeno verde non è affatto semplice. L'Espresso. Available at: https://lespresso.it/c/mondo/2023/8/3/la-strada-per-arrivare-allidrogeno-verde-non-e-affatto-semplice/2629

Chen P, Zhong S, Zheng S, Ullah S, Musa M (2024) Quantifying the influence of natural resources rent, financial development, and institutional quality on energy security risk. Energy Environ 0958305X241266526. https://doi.org/10.1177/0958305X241266526

Craen S (2023) Financing a world scale hydrogen export project. Oxford Institute for Energy Studies (OIES)

Edquist C, Hommen L (1999) Systems of innovation: theory and policy for the demand side. Technol Soc 21:63–79

European Commission (2020) A hydrogen strategy for a climate-neutral Europe, COM(2020) 301 final. European Commission, Brussels

European Commission (2022) REPowerEU Plan, COM (2022) 230 final. European Commission, Brussels

European Commission (2023) EU Hydrogen Bank: scaling up renewable hydrogen. Directorate-General for Energy, Brussels

Evro S, Oni BA, Tomomewo OS (2024) Carbon neutrality and hydrogen energy systems. Int J Hydrog Energy 78:1449–1467. https://doi.org/10.1016/j.ijhydene.2024.06

Federal Ministry for Economic Affairs and Climate Action (BMWK) (2020) The National Hydrogen Strategy. BMWK, Berlin

Federal Ministry for Economic Affairs and Climate Action (BMWK) (2023) Update of the National Hydrogen Strategy. BMWK, Berlin

García-Herrero A, Tagliapietra S, Vorsatz V (2021) Hydrogen development strategies: a global perspective. Bruegel

Hassan Q, Algburi S, Jaszczur M, Al-Jiboory AK, Al Musawi TJ, Ali BM et al (2024) Hydrogen role in energy transition: a comparative review. Process Saf Environ Prot 184:1069–1093. https://doi.org/10.1016/j.psep.2024.02.030

Hesel P, Braun S, Zimmermann F, Fichtner W (2022) Integrated modelling of European electricity and hydrogen markets. Appl Energy 328:120162

International Energy Agency (IEA) (2023) Global hydrogen review. IEA

Ken Research (2025) Netherlands green hydrogen production and export market 2024–2030. Ken Research. https://www.kenresearch.com/netherlands-green-hydrogen-production-and-export-market

Lebrouhi BE, Djoupo JJ, Lamrani B, Benabdelaziz K, Kousksou T (2022) Global hydrogen development-a technological and geopolitical overview. Int J Hydrog Energy 47(11):7016–7048

Lee J, Jun S-P, Lee C (2022) Does demand-side innovation policy drive lock-in? Global evidence from solar energy in 155 countries. Energy Res Soc Sci 89:102539. https://doi.org/10.1016/j.erss.2022.102539

Li J, Wei Y-M, Liu L, Li X, Yan R (2022a) The carbon footprint and cost of coal-based hydrogen production with and without carbon capture and storage technology in China. J Clean Prod 362:132514. https://doi.org/10.1016/j.jclepro.2022.132514

Li Y, Shi X, Phoumin H (2022b) A strategic roadmap for large-scale green hydrogen demonstration and commercialisation in China: a review and survey analysis. Int J Hydrog Energy 47(58):24592–24609

Liu W, Wan Y, Xiong Y, Gao P (2022) Green hydrogen standard in China: standard and evaluation of low-carbon hydrogen, clean hydrogen, and renewable hydrogen. Int J Hydrog Energy 47:24584–24591. https://doi.org/10.1016/j.ijhydene.2021

Lux B, Deac G, Kiefer CP, Kleinschmitt C, Bernath C, Franke K, Sensfuß F (2022) The role of hydrogen in a greenhouse gas-neutral energy supply system in Germany. Energy Convers Manag 270:11

Megía PJ, Vizcaíno AJ, Calles JA, Carrero A (2021) Hydrogen production technologies: from fossil fuels toward renewable sources. A mini review. Energy Fuels 35:16403–16415. https://doi.org/10.1021/acs.energyfuels.1c02501

Meng X, Chen M, Gu A, Wu X, Liu B, Zhou J et al (2022) China's hydrogen development strategy in the context of double carbon targets. Nat Gas Ind B 9:521–547. https://doi.org/10.1016/j.ngib.2022.11.00

Ministère de l'Économie. (2020). Présentation de la stratégie nationale pour le développement de l'hydrogène décarboné en France. https://www.economie.gouv.fr/presentation-strategie-nationale-developpement-hydrogene-decarbone-France

Nygrén NA, Kontio P, Lyytimäki J, Varho V, Tapio P (2015) Early adopters boosting the diffusion of sustainable small-scale energy solutions. Renew Sust Energ Rev 46:79–87. https://doi.org/10.1016/j.rser.2015.02.031

OECD (2023) World Energy Outlook 2023. OECD. https://doi.org/10.1787/827374a6-en

Postigo A (2022) Vaccine research and development (R&D) in the Asia-Pacific: the economics of vaccine R&D and policy recommendations to overcome market failures and promote R&D cooperation. ARTNeT, Bangkok

Qazi UY (2022) Future of hydrogen as an alternative fuel for next-generation industrial applications; challenges and expected opportunities. Energies 15:4741. https://doi.org/10.3390/en15134741

Quitzow R, Nunez A, Marian A (2024) Positioning Germany in an international hydrogen economy: a policy review. Energ Strat Rev 53:101361

Rehman MA, Hasan M, Pata UK, Chen XH, Kontoleon A (2024) Exploring the nexus of technology, environmental policy stringency, and political globalization: pathways to achieving sustainability. Energy Econ 107979. https://doi.org/10.1016/j.eneco.2024.107979

Shirizadeh B, Ailleret A, Guillon A, Bovari E, El Khatib N, Douguet S et al (2023) Towards a resilient and cost-competitive clean hydrogen economy: the future is green. Energy Environ Sci 16:6094–6109. https://doi.org/10.1039/d3ee0228

Słotwinski S (2022) The significance of the “power purchase agreement” for the development of local energy markets in the theoretical perspective of polish legal conditions. Energies 15:6691. https://doi.org/10.3390/en15186691

Somenzi E, Del Grosso F, Pontoni F, Vaglietti G (2021) A critical assessment of national hydrogen strategies. FEEM, Milan

Sun W, Harrison GP, Dodds PE (2022) A multi-model method to assess the value of power-to-gas using excess renewable. Int J Hydrog Energy 47(15):9103–9114. https://doi.org/10.1016/j.ijhydene.2021.12.248

Taylor M (2008) Beyond technology-push and demand-pull: lessons from California’s solar policy. Energy Econ 30:2829–2854. https://doi.org/10.1016/j.eneco.200

Ullah S, Sohail S, Majeed MT, Sohail MT (2025) Future of hydrogen energy: the role of environmental policy and political globalization. Int J Hydrog Energy 122:12–21

Yue M, Lambert H, Pahon E, Roche R, Jemei S, Hissel D (2021) Hydrogen energy systems: a critical review of technologies, applications, trends and challenges. Renew Sust Energ Rev 146:111180. https://doi.org/10.1016/j.rser.2021.111180

Zhang Q, Chen W, Ling W (2022) Policy optimization of hydrogen energy industry considering government policy preference in China. Sustainable Prod Consumption 33:890e902. https://doi.org/10.1016/j.spc.2022.08.017

Zhang L, Ji C, Bai F, Wang W, An S, Zhao K, Li Z, Li J, Sun H (2024) A comprehensive review of the promising clean energy carrier: hydrogen production, transportation, storage, and utilization (HPTSU) technologies. Fuel 355:129455. https://doi.org/10.1016/j.fuel.2023.129455

Zhao F, Mu Z, Hao H, Liu Z, He X, Victor Przesmitzki S, Ahmad Amer A (2020) Hydrogen fuel cell vehicle development in China: an industry chain perspective. Energ Technol 8(11):2000179

Zou C, Li J, Zhang X, Jin X, Xiong B, Yu H, Pan S (2022) Industrial status, technological progress, challenges, and prospects of hydrogen energy. Nat Gas Ind B 9(5):427–447

Zun MT, McLellan BC (2023) Cost projection of global green hydrogen production scenarios. Hydrogen 4:932–960. https://doi.org/10.3390/hydrogen4040055

Chapter 11
Evaluating Hydrogen Penetration Scenarios in Future Energy Systems: Insights from Case Studies

Abstract Hydrogen offers a promising pathway to decarbonize transport, industry, and space heating while providing energy storage for intermittent renewable sources. This chapter evaluates the future engineering potential of hydrogen and identifies research priorities for reducing economic barriers to its adoption. Using top-level system models based on end-use energy demands, four case studies are examined, covering global energy systems, climate–economy impacts, hydrogen-based transportation in the United States, and space heating in the United Kingdom. The results indicate substantial potential for hydrogen deployment, with possible contributions of about 3% to global energy demand by 2050, reduced rejected energy in US transport, and cost-effective decarbonization of heating in the UK. However, uncertainties related to nuclear power phase-out and the high costs associated with current hydrogen technologies remain significant challenges for a large-scale transition.

Keywords Hydrogen · Penetration · Energy model · Carbon emissions · Economics decarbonization

11.1 Introduction

To address climate change and align with globally defined decarbonization objectives, national energy systems must undergo profound transformations through pathways tailored to country-specific conditions. The Paris Agreement, adopted in December 2015, established a long-term objective of limiting the rise in global average temperature to well below 2 °C above pre-industrial levels, while pursuing efforts to cap warming at 1.5 °C (UNFCCC 2017). Achieving these targets requires each country to implement targeted measures aimed at substantially reducing greenhouse gas (GHG) emissions. A central challenge lies in identifying optimal mitigation strategies that simultaneously ensure economic growth, industrial competitiveness, and security of energy supply. At present, the energy sector accounts for approximately 32.2 Gt of global CO_2 emissions, with the power sector

D. M. Y. Mir et al., *Low-Carbon Hydrogen*, Clean Energy Production Technologies,
https://doi.org/10.1007/978-981-95-8142-9_11

alone contributing around 42% of this total. As a result, electricity plays a pivotal role in the decarbonization of the overall energy system. This transition is driven by two complementary trends: an increasing reliance on electrification across end-use sectors and the decarbonization of electricity generation itself. The latter has accelerated the deployment of renewable energy sources. However, the large-scale integration of variable renewables introduces new operational challenges, including increased risks of power system imbalances that can threaten grid stability. Consequently, system operators face growing complexity in maintaining real-time balance by procuring adequate reserves and managing fluctuations in supply and demand. Moreover, meeting the 2 °C—and especially the more ambitious 1.5 °C—climate targets necessitates a broader, system-wide perspective that extends beyond the power sector. Other major emitting sectors, particularly transport, which accounts for nearly 22.7% of global energy-related CO_2 emissions (UNFCCC 2017), must be integral to decarbonization strategies. The transport sector remains heavily reliant on fossil-fuel-based internal combustion engines, making its transition particularly challenging. In response, the European Union has implemented stringent emission standards for passenger vehicles, targeting a limit of 95 g CO_2/km by 2020. These targets are notably more ambitious than those projected for the United States, China, and Japan, with respective limits of 121, 117, and 105 g CO_2/km (Alanne and Cao 2017). Ensuring security of supply while achieving deep emissions reductions requires a fundamental rethinking of energy system management. Sector coupling—linking the power system with heating, transport, and gas networks through pathways such as power-to-heat and power-to-mobility, either directly via electricity or indirectly through synthetic fuels—has emerged as a promising approach. This integrated framework can enhance system flexibility, facilitate higher renewable energy penetration, and enable coordinated, multi-sector decarbonization (Moliner et al. 2016; Ishimoto et al. 2017). Within this context, hydrogen systems are increasingly recognized as key enablers of cross-sector integration. When produced using low-carbon technologies particularly water electrolysis powered by a decarbonized electricity mix, hydrogen offers a versatile energy carrier capable of providing both flexibility services and interconnection between energy sectors (Staffell et al. 2017). Hydrogen can be utilized for a wide range of applications, including industrial feedstocks, energy storage, and fuel for transport and power generation. In 2015, global hydrogen production reached approximately 61 Mt. per year, with around 96% derived from fossil-based processes such as natural gas and refinery gas reforming (48%), chemical processing (30%), and coal gasification (18%) (Midilli and Dincer 2008). Electrolysis accounted for only about 4% of total hydrogen production (IEA 2015). Currently, three principal electrolyser technologies are under development. Alkaline electrolysis is the most mature and widely commercialized technology. Proton exchange membrane (PEM) electrolysers are at an early commercial stage—particularly for large-scale applications—but are well suited for grid-related services due to their high operational flexibility and ability to handle variable loads. In contrast, solid oxide electrolysis cells (SOECs), which operate at high temperatures, remain largely in the research and development phase, offering promising efficiency gains but requiring further technological advancement before large-scale deployment (Michalski et al. 2018). This process presents significant potential due to its reversible operation and its ability to enable the co-electrolysis of water and carbon dioxide for the production of synthesis gas (syngas). In addition to hydrogen produced intentionally, substantial quantities of by-product hydrogen are generated across various industrial processes. One of the most significant sources of such by-product hydrogen is catalytic reforming in petroleum refineries, where hydrogen is typically recovered and reused internally to support other refinery operations (Moliner et al. 2016). At present, hydrogen is predominantly consumed as an industrial feedstock, with approximately 80% of global demand arising from refinery processes and ammonia production (Moliner et al. 2016). However, hydrogen demand in industrial applications is expected to expand considerably. Beyond its established role in fertilizer production and refining, hydrogen and hydrogen-based compounds can serve as key process agents in emerging low-carbon industrial pathways, such as hydrogen-assisted steelmaking (Michalski et al. 2018). Moreover, hydrogen has the potential to decarbonize industrial thermal energy requirements and to contribute to emissions reductions in other end-use sectors, including buildings and transport. Consequently, the role of hydrogen as a versatile energy carrier is anticipated to grow significantly in the coming decades (Ahmed et al. 2016). Despite this potential, hydrogen is currently utilized only to a limited extent in energy applications. Existing uses include blending hydrogen into natural gas networks and deployment in transport, heating, and power generation. Through these applications, hydrogen could contribute to the decarbonization of a broad range of sectors, including hard-to-abate domains such as transport. However, its large-scale market penetration remains uncertain. The pace and extent of hydrogen adoption will strongly depend on regional conditions, particularly the policy and regulatory frameworks that can either enable or constrain deployment. In this context, the present study conducts a multi-regional assessment of the feasibility of hydrogen market penetration in light of recently announced policies and targets. The analysis focuses on four regions the United States, Europe, China, and Japan each characterized by distinct energy systems and policy environments, thereby allowing hydrogen deployment to be evaluated under diverse conditions. Two end-use markets are examined: the mobility sector, through fuel cell electric vehicles (FCEVs) for light-duty passenger transport, and the direct injection of hydrogen into existing natural gas networks. Studies adopting a multi-regional perspective on future hydrogen deployment while simultaneously considering multiple applications remain relatively limited in the literature. Existing contributions (Colbertaldo et al. 2019) typically focus either on normative scenarios characterized by stringent CO_2 constraints and strong policy or industrial support, or on prospective hydrogen cost assessments without explicitly benchmarking these costs against the economic thresholds required for market entry. The objective of this paper is to address this gap by proposing an alternative approach to evaluate hydrogen market penetration feasibility through a combined assessment of hydrogen production and supply costs and the corresponding market entry costs. The economic analysis is conducted within the framework of current and announced governmental policies to determine whether existing policy measures are sufficient to stimulate hydrogen

Table 11.1 Hydrogen penetration scenarios in future energy systems

Scenario type	Hydrogen share in final energy mix	Key characteristics	Primary application sectors	References
Low penetration (incremental transition)	<5–10% by 2050	Limited adoption due to high costs and infrastructure constraints; mainly fossil-based hydrogen with CCS	Oil refining, ammonia production, niche industrial heat	IEA (2019)
Moderate penetration (policy-driven expansion)	10–20% by 2050	Strong climate policies enable blue and green hydrogen scale-up; regional hydrogen hubs	Steelmaking (DRI), heavy-duty transport, power system flexibility	IEA (2023)
High penetration (net-zero alignment)	20–30% by 2050	Rapid cost reductions; large-scale green hydrogen deployment; expanded infrastructure	Industry, long-haul transport, seasonal energy storage	IRENA (2021)
Transformational (hydrogen economy)	>30% by 2050	Hydrogen as a core energy carrier; integrated hydrogen-electricity systems	Industry, transport, buildings, synthetic fuels	Staffell et al. (2019a, b)

deployment. The first part of the study presents a prospective analysis aimed at identifying future market entry costs for the selected applications. These costs represent competitiveness benchmarks, defined relative to incumbent technologies and alternative energy carriers. The second part evaluates current and projected hydrogen costs along different production and supply pathways, extending from production to end use. By comparing these costs with the identified target costs, the study assesses market penetration feasibility based on the magnitude of the cost gap. A larger gap indicates greater barriers to market entry, whereas a smaller gap suggests more favourable conditions for hydrogen adoption. As shown in Table 11.1, higher hydrogen penetration levels are associated with stronger climate policy frameworks, accelerated deployment of renewable-based hydrogen, and expanded infrastructure development.

11.2 Case Studies

11.2.1 Case Study 1: Global Energy Model

To assess the potential penetration of hydrogen within future global energy systems, this study employs an integrated global energy system model originally developed by Komiyama and Fujii and further extended by Hosoya and Fujii (Hosoya and Fujii 2011). The model provides a detailed representation of the global energy landscape by disaggregating the world into 82 distinct regions and embedding a comprehensive set of techno-economic, environmental, and policy-relevant parameters. These include constraints on carbon dioxide emissions, the availability and cost of primary energy resources, infrastructure and capital investment requirements, energy conversion and storage technologies, sectoral end-use energy demands, capacity utilization factors, and the variability and intermittency associated with renewable energy generation. A key strength of the modelling framework lies in its explicit treatment of interregional energy trade. Imports and exports of energy carriers are modelled endogenously, enabling a realistic depiction of cross-border energy flows and global market interactions. This feature allows the model to capture regional disparities in resource endowments and infrastructure development, as well as the implications of international energy cooperation and trade on system-wide outcomes. Within this integrated modelling environment, the global energy system model is applied to estimate the extent to which hydrogen could be adopted across different sectors of the economy, including industry, transport, power generation, and energy storage. To facilitate a robust evaluation of hydrogen's role, the model incorporates a series of targeted assumptions, constraints, and structural modifications, which are detailed in the following section. These adjustments ensure that hydrogen production pathways, distribution infrastructures, and end-use technologies are consistently represented within the broader energy system context. The application of hydrogen is constrained to a limited set of end uses, namely, fuel cell vehicles (FCEVs) for both passenger and freight transport, substitution for city gas in urban distribution networks, co-firing in natural gas–fired power plants, and use as a feedstock for methanol synthesis.

1. Greenhouse gas (GHG) emission trajectories are aligned with Representative Concentration Pathway 2.6 (RCP 2.6) (IPCC 2014), corresponding to the objective of limiting global warming to 2 °C or below, as stipulated in the 2015 Paris Agreement (UNFCCC 2017). Under this scenario, Organization for Economic Co-operation and Development (OECD) countries are required to achieve an 80% reduction in CO_2 emissions, while a global emission reduction target of 65% is imposed by 2050.
2. Constraints are placed on the expansion of nuclear power generation capacity, with no new nuclear deployment permitted in Japan from 2010 onward and in other regions from 2030 onward, in accordance with projections from the World Nuclear Association (2017). It is important to note that the long-term phase-out of nuclear energy not only necessitates the accelerated deployment of alternative zero-carbon technologies but also removes one of the established carbon-free options for large-scale hydrogen production.
3. Carbon dioxide captured through mitigation technologies is assumed to be stored in depleted natural gas reservoirs and deep saline aquifers or utilized for enhanced oil recovery. Transport of captured CO_2 is permitted between geographically connected regions via land-based infrastructure or pipelines, while maritime transport is allowed only to a limited extent.

The model optimizes the global energy system by minimizing total system costs, accounting for both fixed and variable components, subject to the constraints described above. The analysis spans the period from 2000 to 2050, with results generated at ten-year intervals. The principal outputs relevant to this study include changes in total energy system costs, patterns of energy supply and demand, carbon capture and storage (CCS) requirements, and the evolving structure of the energy supply system. With specific regard to hydrogen, the model quantifies the volume of hydrogen introduced into the energy system and identifies the associated production pathways, primary energy feedstocks, geographic origins, and end-use applications. The cost-optimal outcomes can be examined at multiple spatial scales, including global, regional (e.g., OECD, Asia, Europe), and national levels. Under the assumed constraints, both total energy demand and electricity consumption are projected to increase steadily through 2050. This growth is met primarily by a substantial expansion of renewable energy generation, while the contributions of natural gas and oil remain relatively stable over time. In contrast, electricity generation from nuclear power declines continuously toward 2050, consistent with the imposed deployment restrictions. As a result, significant CCS capacity is required to achieve compliance with the global two-degree CO_2 emissions target. By 2050, hydrogen is projected to account for approximately 450 million tonnes of oil equivalent (Mtoe), corresponding to about 3% of total final energy consumption. The majority of this hydrogen is used as a gaseous fuel, largely substituting for city gas, while only a small share is allocated to passenger transport applications by mid-century. This outcome reflects the assumption that hydrogen is most economically viable as a replacement for existing city gas infrastructure, although such a transition would require substantial capital investment, which is not explicitly quantified in this study. Advances in hydrogen-based transport technologies could significantly alter this distribution, suggesting that hydrogen demand in mobility applications may be underestimated under the current assumptions. From a supply perspective, the cost-minimization framework and present-day cost assumptions indicate that hydrogen production in 2050 is dominated by fossil-based and biomass pathways primarily natural gas, coal, oil, and biomass with CCS applied where necessary. Only a marginal fraction of hydrogen is produced via electrolysis by 2050. Stringent CO_2 emission constraints necessitate a rapid scale-up of renewable energy deployment alongside extensive CCS infrastructure development. These combined requirements result in a pronounced and sustained increase in total global energy system costs. Under the cost-minimization scenario, total system expenditures rise by a factor of approximately 2.4 between 2020 and 2050. Specifically, global energy system costs increase from about USD 5.5 trillion in 2020 to roughly USD 13.2 trillion by 2050. Over this period, the share of variable costs relative to fixed costs increases substantially, from around 13% to 36%, driven largely by the growing contribution of CCS-related operating costs. Figure 11.1 highlights the decoupling of economic growth from energy consumption at the global level between 1990 and 2021. Figure 11.2 illustrates the evolution of global demand for pure hydrogen from 1975 to 2018. The figure highlights a steady and sustained increase in hydrogen consumption over the period, driven primarily by growing demand from industrial sectors such as oil refining, ammonia production, and chemical manufacturing. The long-term trend

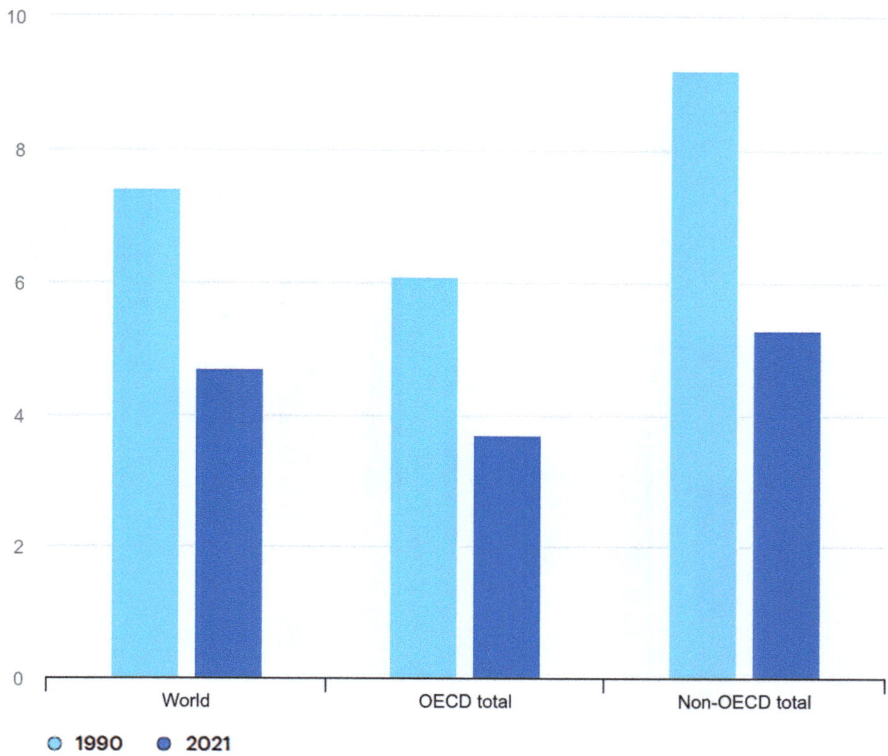

Fig. 11.1 Comparison of global energy intensity levels between 1990 and 2021 (IEA 2023)

reflects the expanding role of hydrogen as a critical industrial feedstock, while also indicating its increasing relevance within broader energy system transitions.

11.2.2 Case Study 2: Social Welfare Economics Evaluation

While Case Study 1 examined the financial implications of global carbon mitigation, the present case study shifts the focus toward the social cost of carbon (SCC) associated with atmospheric emissions. The SCC is commonly estimated using a range of methodological approaches; in this analysis, it is derived using integrated assessment models (IAMs) (Stanton et al. 2009). Specifically, the study evaluates climate change impacts through a welfare optimization-based IAM, which offers both computational simplicity and strong suitability for conceptual and policy-oriented analysis. The assessment employs the dynamic integrated climate–ecosystem–economics model (CEEM) developed by Tamaki et al. (2017) to estimate the impacts arising from the diffusion of hydrogen energy. The CEEM is an extension of the well-established Dynamic Integrated model of Climate and the Economy

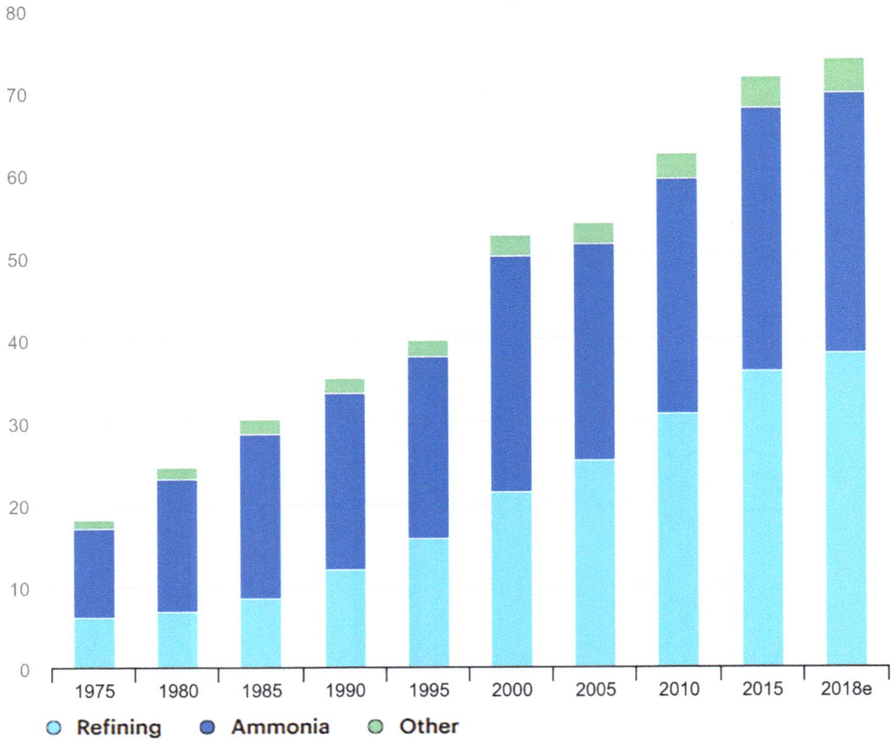

Fig. 11.2 Global Pure Hydrogen Demand Over the Period 1975–2018 (IEA 2023)

(DICE) and is designed to incorporate multiple energy technologies, enabling a more comprehensive evaluation of technological transitions (Alanne and Cao 2017). Within this framework, the backstop technology cost (expressed in US\$/tCO_2) is treated as an exogenous variable. Backstop technologies represent alternative solutions capable of alleviating resource scarcity and limiting emissions growth. Although hydrogen is not a primary energy source, its use as a fuel does not result in direct CO_2 emissions. Looking ahead, if hydrogen can be produced efficiently using renewable energy, nuclear power, or fossil fuels combined with carbon capture and storage (CCS), it could enable the provision of low-carbon and economically competitive energy. Such developments would effectively reduce the cost of backstop technologies. Consequently, changes in the diffusion of hydrogen energy are reasonably modelled as changes in the backstop cost parameter (Moliner et al. 2016). In the CEEM framework, the backstop cost in period t, denoted as $p(t)$, is specified as a declining function over time, reflecting technological learning and innovation. Enhanced research and development efforts are expected to accelerate this decline. For example, Kalinci et al. (2015) demonstrate that hybrid PV–wind–hydrogen systems using hydrogen for energy storage can substantially reduce net present costs. Similarly, Barreto et al. (2003) argue that hydrogen technologies are likely to experience significant, yet realistic, improvements in performance and cost, enabling widespread diffusion. Based on these insights, the model assumes an increasing rate of cost reduction over time and defines three future scenarios accordingly. If large-scale, efficient hydrogen production from renewable sources is realized, backstop costs could decline more rapidly than anticipated. In such a case, the present analysis may underestimate hydrogen adoption, and hydrogen-based systems would gain a competitive advantage over greenhouse-gas-emitting technologies. The GOR is defined as the ratio of gross output under each scenario relative to the baseline gross output, which corresponds to the optimal solution of the DICE model. Higher GOR values indicate improved economic productivity relative to the baseline. The vertical axis illustrates the GOR for each scenario. Although the GOR for Scenario 3 exhibits a slight decline between 2055 and 2060, it remains above unity throughout the entire period, indicating sustained productivity gains compared to the baseline case. This period corresponds to the transition point at which alternative technologies begin to replace the backstop technology, resulting in net-zero CO_2 emissions. In contrast, under the baseline scenario, CO_2 emissions do not decline to zero even by 2070. Although gross output increases from 2015 onward in both the baseline and scenario cases, the evolution of the gross output ratio (GOR) is governed by the relative growth rates of gross output in each case. The observed temporary decline in the GOR can be attributed to the transition point in Scenario 3, where backstop technologies are superseded by other low-carbon technologies and emissions fall to zero. Despite this short-term reduction, the magnitude of the GOR decline is modest, and the ratio remains consistently above unity throughout the study period. This outcome indicates that the diffusion of hydrogen energy can support higher economic output when climate change impacts are taken into account. It is important to note that the DICE model, which forms the foundation of the present analysis, adopts a relatively optimistic view of climate change impacts and may therefore underestimate the full benefits of mitigation measures. As a result, the findings presented here can be interpreted as lower-bound estimates, suggesting that the true positive effects of hydrogen energy diffusion on climate outcomes may be even greater. Moreover, this study focuses on impacts that can be explicitly quantified within the modelling framework. Declining backstop technology costs enable emissions reductions to be achieved at lower abatement costs. At the same time, by limiting temperature increases, climate-related damage costs are also reduced (Staffell et al. 2019a, b). Consequently, both abatement expenditures and damage costs decline as a result of reduced CO_2 emissions and suppressed temperature rise. Under Scenarios 1, 2, and 3, gross output in 2070 increases by approximately 0.09%, 0.25%, and 0.30%, respectively, relative to the baseline. When expressed in monetary terms, these gains correspond to increases of roughly US$0.4 trillion, US$1.1 trillion, and US$1.5 trillion, respectively, above the baseline gross output of US$461 trillion in 2070.

11.2.3 Case Study 3: The United States: Hydrogen as an Energy Carrier for Road Vehicles

The United States is selected as a representative case study to examine the feasibility of hydrogen as an energy carrier capable of supplying the energy requirements of the on-road transportation sector. The analysis estimates the quantity of hydrogen necessary to replace the energy currently consumed by road vehicles and evaluates the corresponding electricity demand assuming hydrogen is produced via water electrolysis powered by renewable energy sources. Throughout the assessment, efficiency losses along the entire energy supply chain from electricity generation to hydrogen production, compression, and end use are explicitly incorporated. Given that water is a critical input for electrolysis-based hydrogen production, the associated water requirements are also quantified. However, the analysis assumes that water availability does not pose a limiting constraint on hydrogen deployment. In addition, an order-of-magnitude estimate of hydrogen production costs is developed based on projected hydrogen demand in the transportation sector. The assessment builds upon the national energy flow analysis conducted by Lawrence Livermore National Laboratory, which provides a comprehensive depiction of current energy sources and end-use consumption patterns in the United States. Under this framework, hydrogen produced at scale could replace approximately 22 quads of petroleum-derived fuels that met the energy needs of on-road vehicles primarily passenger cars and trucks—in the United States in 2016. Energy consumption associated with other transport modes, including maritime transport, aviation, rail, off-road equipment (such as agricultural, construction, and mining machinery), and miscellaneous vehicle categories, is excluded from the scope of this analysis. Because hydrogen fuel cell vehicles operate at substantially higher efficiencies than conventional internal combustion engine vehicles, the energy content of hydrogen required to displace petroleum fuels is not assumed to be equivalent on a one-to-one basis. To account for this efficiency advantage, average fuel economies for gasoline-powered vehicle classes including light-duty vehicles, buses, and commercial trucks are scaled by a factor of 2.5. For diesel-powered vehicles, such as heavy freight trucks, a lower scaling factor of 1.2 is applied to reflect more modest efficiency gains. These adjustments yield a notional estimate of vehicle energy demand under hydrogen fuel cell operation. Hydrogen production via water electrolysis is assumed to require 49 kWh of electricity per kilogram of hydrogen, with compression to 700 bar for onboard storage in vehicles. Assuming a capacity factor of 25%, characteristic of utility-scale solar installations in the US desert southwest, generating the additional electricity required approximately 16 quads would necessitate roughly 2.1 TW of new solar generation capacity. For comparison, total installed electricity generation capacity in the United States in 2016 was approximately 1.2 TW, including about 108 GW of non-hydropower renewable capacity. The total generation capacity required is highly sensitive to the composition of the renewable energy mix. While solar power typically exhibits lower capacity factors, modern wind installations routinely achieve capacity factors between 35% and 50%, and geothermal power plants can exceed 80%. Under a diversified generation portfolio comprising 25% solar, 50% wind, and 25% geothermal power, the total dedicated capacity required for hydrogen production would decline to approximately 1.4 TW. By comparison, the US petroleum industry currently delivers the equivalent of about 0.7 TW of liquid fuel capacity to satisfy road transportation energy demand. Consequently, a transition to hydrogen-based road transport would require replacing 0.7 TW of liquid fuel supply with approximately 1.4–2.1 TW of renewable electricity capacity, depending on the renewable technology mix. Although this pathway entails substantial investment in new electricity generation infrastructure, it delivers significant system-level efficiency gains. The deployment of renewable hydrogen for road transportation could reduce total transportation-sector energy demand by roughly 40%, decrease rejected energy across the US economy by nearly 10%, and lower carbon dioxide emissions by more than 1500 million metric tons equivalent to an approximately 30% reduction in total US CO_2 emissions. Beyond energy and emissions considerations, the analysis also evaluates water requirements associated with large-scale hydrogen production. Electrolysis-based hydrogen production requires approximately 15.5 litres of water per kilogram of hydrogen. At the scale necessary to produce nearly 11 quads of hydrogen annually, total water demand would reach about 1.5 trillion litres per year, corresponding to roughly 0.4% of total freshwater withdrawals in the United States. It is important to note that overall water consumption and withdrawal levels depend on the electricity generation technologies used to supply electrolysis, as different power sources exhibit distinct water-use profiles. The levelized cost of hydrogen produced via electrolysis is estimated at approximately US$5.60/kg for distributed systems and US$3.20/kg for centralized systems. At an annual production level of roughly 95 billion kilograms of hydrogen, total production costs would amount to approximately US$530 billion for distributed electrolysis and US$300 billion for centralized electrolysis, excluding the capital costs of renewable electricity generation infrastructure. Looking ahead, the US Department of Energy has established cost targets of US$4/kg for distributed electrolysis and US$2/kg for centralized electrolysis, with continued technological progress, manufacturing scale-up, and market penetration expected to further reduce hydrogen costs over time. While this analysis focuses exclusively on electrolysis-based hydrogen production for on-road transportation, hydrogen can be produced through multiple pathways and deployed across a wide range of sectors within the US energy system. Emerging system-level studies indicate that under high-renewable scenarios, electrolysis becomes the dominant hydrogen supply pathway, revealing strong cross-sectoral synergies and reinforcing hydrogen's potential to deliver deep decarbonization across the broader energy economy.

11.2.4 Case Study 4: Decarbonizing the Gas Grid Using Hydrogen

In the United Kingdom, heat demand within buildings constitutes a dominant share of national energy use, accounting for approximately 40% of total final energy consumption and nearly one-fifth of overall greenhouse gas (GHG) emissions. When transport-related energy use is excluded, heating requirements represent almost 80% of domestic energy consumption. At present, natural gas remains the backbone of residential heating, supplying space and water heating needs in roughly 85% of UK households. Despite its central role in the national energy system and its substantial contribution to emissions, the decarbonization of heat has historically received considerably less policy and research attention than the decarbonization of electricity generation. Most system-level energy modelling studies in the UK have therefore focused primarily on electrification-based pathways for low-carbon heating. These include the deployment of air-source and ground-source heat pumps, district heating networks supplied by low-carbon energy sources, solar thermal systems, and biomass-based heating technologies. While these options offer meaningful emissions reduction potential, their large-scale deployment faces particular challenges in high-latitude countries such as the UK, where heat demand is both high and strongly seasonal. As a result, many electrification-based solutions require substantial capital investment while experiencing relatively low utilization during warmer periods, which can undermine their overall economic efficiency. Furthermore, the widespread adoption of electrified heating technologies would necessitate not only significant upgrades to electricity infrastructure but also substantial behavioural and lifestyle changes by end users. These requirements may render electrification pathways more disruptive to consumers than alternatives that leverage the existing natural gas infrastructure. In this context, adapting the current gas grid to distribute hydrogen has emerged as a potentially less intrusive and more socially acceptable decarbonization strategy. Consequently, the role of hydrogen and fuel cell technologies in low-carbon heating has attracted increasing attention in recent years. Two major UK research institutions the Hydrogen and Fuel Cell (H2FC) SUPERGEN Research Hub and the Sustainable Gas Institute have produced influential White Papers that systematically assess hydrogen's potential contribution to heat decarbonization. The first focuses on the broader role of hydrogen and fuel cells within future UK energy systems, while the second evaluates pathways toward a lower-carbon gas network. Both studies synthesize extensive academic, industrial, and policy literature, with a strong emphasis on UK-specific technical, economic, and institutional conditions. The H2FC SUPERGEN White Paper employs the UK TIMES Model (UKTM), the energy system model used by the UK Government, to explore six distinct scenarios for hydrogen and fuel cell deployment under alternative future energy pathways. All scenarios are consistent with the UK's statutory commitment to achieving an 80% reduction in GHG emissions, but they differ in terms of technology choices and system configurations. One particularly illustrative pathway—the Gas Conversion scenario—examines the large-scale conversion of existing gas networks to hydrogen, with household uptake of low-carbon heating and transport technologies driven primarily by cost optimization. The modelling results indicate that hydrogen-based heating could play a substantial role in achieving cost-optimal decarbonization of the UK energy system. Findings from both White Papers suggest that repurposing the existing gas grid to deliver hydrogen for space and water heating may be more cost-effective than many currently available low-carbon heat alternatives. These conclusions are reinforced by evidence from other industrial and demonstration studies. The Sustainable Gas Institute analysis provides detailed estimates of network conversion costs, consumer prices, and household-level expenditures associated with different decarbonization pathways. The additional cost of converting a household to a hydrogen-based gas network is estimated at approximately US$2200 per dwelling. By comparison, the installation costs for air-source heat pumps range from roughly US$3000 to US$8200, while ground-source heat pumps require substantially higher upfront investments of approximately US$9600 to US$14,800. Although ground-source heat pumps can achieve high operational efficiencies up to 250% and thereby reduce ongoing energy costs, their high capital costs remain a significant barrier to widespread adoption without strong policy support or financial incentives. In contrast, the cost of converting a home to hydrogen-based heating is broadly comparable to that of replacing a conventional gas boiler. This cost similarity, combined with minimal required changes to household heating practices, suggests that hydrogen-based heating may represent a more economically attractive and less disruptive option for consumers. Overall, these findings underscore hydrogen's potential to play a critical role in the decarbonization of heat in the UK, particularly in scenarios that prioritize cost-effectiveness, infrastructure reuse, and consumer acceptability.

11.3 Potential Areas for Research and Development

The growing interest in hydrogen as a solution to global energy challenges is accompanied by a range of complex and interrelated obstacles. Economic feasibility remains one of the most significant barriers, as hydrogen production, storage, and distribution are currently more costly than conventional fossil fuel based systems. This persistent cost differential highlights a critical research need for the development of low-cost catalysts and high-efficiency electrolyzers capable of narrowing the economic gap and improving overall system competitiveness (Dewangan et al. 2022). In parallel, the limited maturity of hydrogen infrastructure including refuelling stations, transport networks, and large-scale storage facilities poses a major challenge to large-scale deployment. Addressing these infrastructure gaps is essential for enabling a fully integrated and functional hydrogen economy (Olabi et al. 2021). Safety considerations further complicate hydrogen adoption, as the fuel's high flammability and the risks associated with high-pressure storage and transport necessitate stringent safety standards, advanced containment systems, and robust regulatory frameworks (Agaton et al. 2022; Raad et al. 2022). Moreover, despite hydrogen's high gravimetric energy density, its low volumetric density imposes substantial spatial requirements, underscoring the need for innovative and compact storage solutions. The environmental implications of hydrogen production also warrant careful consideration. Hydrogen derived from fossil fuels can carry a substantial carbon footprint, reinforcing the importance of prioritizing renewable-powered electrolysis and biogas-based production pathways to ensure genuine emissions reductions (Hassan et al. 2023). Looking ahead, the integration of renewable energy sources into hydrogen production represents a promising avenue for establishing a sustainable and flexible energy system. By leveraging the intermittency and cyclical nature of solar and wind power, hydrogen can function as both an energy carrier and a long-term storage medium, thereby supporting grid stability while minimizing greenhouse gas emissions. Advancements in fuel cell technologies, particularly within the transportation sector, further enhance hydrogen's potential. The transition toward hydrogen fuel cell vehicles offers a low-emission alternative to conventional gasoline-powered transport and could play a transformative role in decarbonizing mobility systems (Baeyens et al. 2020; Hassan et al. 2023). This comparative perspective not only clarifies the current technological landscape but also sheds light on anticipated pathways of innovation and market adoption, which are critical for strategic planning in sustainable energy transitions. Collectively, these data emphasize the importance of accelerating the development of advanced, low-emission hydrogen production technologies. Although renewable-powered water electrolysis currently entails higher costs, it remains the most environmentally sustainable option and represents a rapidly expanding sector with the potential to significantly reshape the global energy market. In summary, hydrogen energy stands at a pivotal juncture: while its challenges are substantial, the opportunities it offers are equally compelling. As global energy systems increasingly prioritize sustainability, hydrogen emerges as a promising cornerstone of a low-carbon future. Successfully transitioning hydrogen into a central energy carrier will require overcoming economic, technological, and infrastructural barriers, but the long-term rewards include a cleaner, more resilient, and diversified energy system.

11.4 Directions for Future Research

Future research on hydrogen energy systems is critical for improving their efficiency, sustainability, and economic competitiveness, as well as for overcoming the technological and infrastructural barriers that currently limit large-scale deployment. A key research priority lies in the development of more efficient and cost-effective electrolysis technologies. Existing electrolysis methods typically achieve energy efficiencies in the range of 60–80%; however, advances in catalyst design particularly those enabling hydrogen production from alternative feedstocks such as biomass offer significant potential to reduce production costs while simultaneously lowering environmental impacts. At present, the cost of producing green hydrogen is approximately US$5 per kilogram, underscoring the urgent need for technological breakthroughs that can drive substantial cost reductions. Equally important are innovations in hydrogen storage and infrastructure. The development of lightweight, durable, and affordable storage materials would help address the limitations of current high-pressure and cryogenic storage systems, which remain costly and energy intensive. In parallel, significant expansion of hydrogen distribution infrastructure is required, including pipelines and refuelling stations. To meet the anticipated demand, the global number of hydrogen refuelling stations will need to increase by roughly an order of magnitude by 2030, highlighting the scale of infrastructure investment required. Addressing safety and regulatory challenges is also essential for the successful integration of hydrogen into existing energy systems. This includes the establishment of comprehensive safety standards, advanced monitoring systems, and robust containment strategies to ensure system reliability and build public confidence. Future research should further examine the design of effective regulatory frameworks capable of governing hydrogen production, storage, and transportation while balancing safety, efficiency, and innovation. In addition, the integration of hydrogen with other energy systems, such as electricity and natural gas networks, represents a promising pathway to enhance energy system flexibility and resilience. Research efforts should focus on the development of hybrid energy systems that combine hydrogen with renewable electricity and gas infrastructure, enabling improved load balancing, long-term energy storage, and sector coupling across power, heat, and transport. Finally, improving the economic viability of hydrogen energy systems extends beyond technical cost reductions. It also requires the exploration of innovative business models, market mechanisms, and financing strategies that can accelerate adoption and support early-stage infrastructure deployment. Targeted subsidies, policy incentives, and novel investment frameworks may play a crucial role in mitigating high upfront costs and fostering market growth. Overall, realizing hydrogen's full potential as a clean and sustainable energy carrier will depend on addressing these challenges through coordinated research and development efforts involving policymakers, industry stakeholders, and the academic community. Such collaboration is essential to ensure that hydrogen energy systems evolve into a cornerstone of a resilient, low-carbon global energy landscape.

11.5 Conclusion

This research demonstrates that expanding the role of hydrogen within the energy system can deliver substantial energetic efficiency gains while supporting deep decarbonization at both national and global scales. In the United States, the use of hydrogen for on-road transportation is shown to reduce rejected energy by nearly 10% and lower national CO_2 emissions by approximately 30%. In the United Kingdom, the analysis indicates that hydrogen may represent a least-cost pathway for decarbonizing space heating, particularly when compared with more capital-intensive electrification-based alternatives. A key cross-cutting finding is that, despite its technical promise, cost remains the primary barrier to the widespread adoption of hydrogen energy under current levels of technological maturity. These costs arise predominantly from two sources. First, the production, storage, and transmission of hydrogen remain expensive. Although these aspects were not explicitly optimized in the present analysis, there is strong evidence that substantial cost reductions are achievable through continued, targeted research and development. If such cost declines are realized, hydrogen's potential to emerge as a least-cost decarbonization option could exceed the estimates presented in this study. Second, the transition to a hydrogen-based energy system will require significant investment in dedicated infrastructure, including production facilities, pipelines, storage systems, and refuelling networks. While ongoing research aims to reduce infrastructure costs, managing and containing capital expenditures remains a persistent challenge across large-scale infrastructure projects worldwide. A further important insight from this work concerns the implications of a widespread phase-out of nuclear power. Eliminating a major source of low-carbon baseload electricity would necessitate substantially higher investment in alternative energy sources, with uncertain consequences for overall energy system costs and the ability of nations to limit global temperature rise. Nuclear energy has also been proposed as a viable option for large-scale hydrogen production, given its high and stable capacity factors. In Case Study 3, hydrogen production was assumed to rely on renewable energy sources with capacity factors ranging from 25% to 80%. By comparison, nuclear power plants can operate at capacity factors exceeding 90%, which could significantly reduce the amount of installed capacity required for hydrogen generation. Finally, results from the global energy system model suggest that, under the assumed constraints, hydrogen could account for approximately 3% of total final energy consumption by 2050, primarily through the deployment of fuel cell vehicles and the replacement of city gas with hydrogen, particularly in OECD countries. While this market share is non-trivial, it reflects conservative assumptions and multiple constraints that may be relaxed over time. Sustained research and development, alongside reductions in the costs of hydrogen production, storage, distribution, and end-use technologies, could enable hydrogen to capture a substantially larger share of the future energy mix.

References

Agaton CB, Batac KIT, Reyes EM Jr (2022) Prospects and challenges for green hydrogen production and utilization in The Philippines. Int J Hydrog Energy 47:17859e70. https://doi.org/10.1016/j.ijhydene.2022.04.101

Ahmed A, Al-Amin A, Ambrose A, Saidur R (2016) Hydrogen fuel and transport system: a sustainable and environmental future. Int J Hydrog Energy 41(3):1369e80

Alanne K, Cao S (2017) Zero-energy hydrogen economy (ZEH2E) for buildings and communities including personal mobility. Renew Sust Energ Rev 71:697e711

Baeyens J, Zhang H, Nie J, Appels L, Dewil R, Ansart R, Deng Y (2020) Reviewing the potential of bio-hydrogen production by fermentation. Renew Sustain Energy Rev 131:110023

Barreto L, Makihira A, Riahi K (2003) The hydrogen economy in the 21st century: a sustainable development scenario. Int J Hydrog Energy 28(3):267e84

Colbertaldo P, Agustin S, Campanari S, Brouwer J (2019) Impact of hydrogen energy storage on California electric power system: towards 100% renewable electricity. Int J Hydrogen Energy. https://doi.org/10.1016/j.ijhydene.2018.11.062

Dewangan SK, Mohan M, Kumar V, Sharma A, Ahn B (2022) A comprehensive review of the prospects for future hydrogen storage in materials-application and outstanding issues. Int J Energy Res 46(12):16150–16177

Hassan Q, Sameen AZ, Salman HM (2023) Hydrogen energy horizon: balancing opportunities and challenges. Energy Harvesting and Systems 11(1):Art, 20220132. https://doi.org/10.1515/ehs-2022-0132

Hosoya Y, Fujii Y (2011) Analysis of energy strategies to halve CO2 emissions by the year 2050 with a regionally disaggregated world energy model. Energy Procedia 43:5853e60

IEA (2015) Technology roadmap: hydrogen and fuel cells. IEA Publications, Paris

IEA (2019) Global demand for pure hydrogen, 1975–2018. IEA, Paris. https://www.iea.org/data-and-statistics/charts/global-demand-for-pure-hydrogen-1975-2018, Licence: CC BY 4.0

IEA (2023) Global energy intensity, 1990 versus 2021, IEA, Paris. https://www.iea.org/data-and-statistics/charts/global-energy-intensity-1990-versus-2021, Licence: CC BY 4.0

International Renewable Energy Agency (IRENA) (2021) World Energy Transitions Outlook: 1.5°C Pathway. IRENA, Abu Dhabi

IPCC (2014) Climate change 2014: synthesis report. Contribution of working groups I, II and III to the fifth assessment report of the intergovernmental panel on climate change. IPCC, Geneva, Switzerland

Ishimoto Y, Kurosawa A, Sasakura M, Sakata K (2017) Significance of CO2-free hydrogen globally and for Japan using a longterm global energy system analysis. Int J Hydrog Energy 42(19):13357e67

Kalinci Y, Hepbasli A, Dincer I (2015) Techno-economic analysis of a stand-alone hybrid renewable energy system with hydrogen production and storage options. Int J Hydrog Energy 40(24):7652e64

Michalski J, Poltrum M, Bunger U (2018) The role of renewable fuel supply in the transport sector in a future decarbonized energy system. Int J Hydrog Energy. https://doi.org/10.1016/j.ijhydene.2018.10.110

Midilli A, Dincer I (2008) Hydrogen as a renewable and sustainable solution in reducing global fossil fuel consumption. Int J Hydrog Energy 33(16):4209e22

Moliner R, Lazaro MJ, Suelves I (2016) Analysis of the strategies for bridging the gap towards the hydrogen economy. Int J Hydrog Energy 41(43):19500e8

Olabi AG, Abdelghafar AA, Baroutaji A, Sayed ET, Alami AH, Rezk H, Abdelkareem MA (2021) Large-vscale hydrogen production and storage technologies: current status and future directions. Int J Hydrog Energy 46(45):23498–23528

Raad SMJ, Leonenko Y, Hassanzadeh H (2022) Hydrogen storage in saline aquifers: opportunities and challenges. Renew Sustain Energy Rev 168:112846

Staffell I, Dodds PE, Scamman D, Abad AV, Mac Dowell N, Ward KR, ... & Ekins P (2017) H2FC supergen-the role of hydrogen and fuel cells in future energy systems

Staffell I, Scamman D, Abad AV, Balcombe P, Dodds PE, Ekins P et al (2019a) The role of hydrogen and fuel cells in the global energy system. Energy Environ Sci 12(2):463–491

Staffell I, Scamman D, Abad AV, Balcombe P, Dodds PE, Ekins P, Shah N, Ward KR (2019b) The role of hydrogen and fuel cells in the global energy system. Energy Environ Sci 12(2):463–491

Stanton EA, Ackerman F, Kartha S (2009) Inside the integrated assessment models: four issues in climate economics. Clim Dev 1(2):166e84

Tamaki T, Nozawa W, Managi S (2017) Evaluation of the ocean ecosystem: climate change modelling with backstop technologies. Appl Energy 205:428e39

United Nations Framework Convention on Climate Change (UNFCCC) (2017) The Paris agreement. http://unfccc.int/paris_agreement/items/9485.php

World Nuclear Association (2017) Nuclear power in the European Union. http://www.world-nuclear.org/informationlibrary/country-profiles/others/european-union.aspx

Chapter 12
Techno-Economic Analysis of Hydrogen Production Pathways: Status, Challenges, and Future Outlook

Abstract This chapter examines the role of hydrogen in enabling a low-carbon energy transition through a techno-economic analysis of hydrogen production pathways and the broader supply chain, including storage, transport, and end-use distribution. It evaluates key economic, technological, and policy challenges, with emphasis on cost structures, infrastructure requirements, and environmental impacts. The analysis incorporates global and regional policy frameworks and financial instruments such as contracts for difference, power purchase agreements, and regulated asset base models to assess their effectiveness in reducing investment risk and accelerating market adoption. Cost projections indicate a significant decline in green hydrogen prices, while storage options reveal trade-offs between economic feasibility and geographical constraints. Overall, this chapter highlights the need for coordinated policy interventions, innovative financing mechanisms, and large-scale infrastructure investments to support a sustainable and economically viable hydrogen economy.

Keywords Hydrogen economics · Risk assessment · Financial risks · Mitigation strategies · Refueling networks

12.1 Introduction

The concept of a hydrogen economy envisions a future energy system in which hydrogen emerges as a central alternative to conventional fossil fuels, acting as a clean, sustainable, and highly versatile energy carrier across multiple sectors. Owing to its flexibility and low-carbon potential, hydrogen is widely regarded as a critical enabler of deep decarbonization, particularly in energy-intensive and hard-to-abate sectors, and is closely aligned with broader global sustainable development objectives. Early scenario-based analyses by Barreto et al. (2003) outline a long-term vision in which hydrogen becomes a dominant energy carrier over the course

D. M. Y. Mir et al., *Low-Carbon Hydrogen*, Clean Energy Production Technologies, https://doi.org/10.1007/978-981-95-8142-9_12

of the twenty-first century, driven by advances in renewable energy technologies and strengthened international commitments to climate change mitigation. Their work highlights hydrogen's capacity to substantially reduce carbon emissions across key domains such as transportation, heavy industry, and power generation, while emphasizing that the realization of such a system would require significant upfront investments in infrastructure, continuous technological innovation—especially in renewable hydrogen production—and the implementation of coherent and stable policy frameworks. Building on this foundation, Crabtree et al. (2004) explore both the opportunities and challenges associated with transitioning toward a hydrogen-based energy system. They underscore hydrogen's role as a flexible and clean energy vector that can be produced via multiple pathways, including renewable electricity–driven water electrolysis and fossil-fuel-based processes integrated with carbon capture and storage (CCS). However, their analysis also draws attention to persistent technical, economic, and systemic barriers that must be overcome, such as the development of efficient hydrogen storage technologies, safe and cost-effective transport networks, and durable, high-performance fuel cells. These challenges, they argue, represent critical bottlenecks that will shape the pace and scale of hydrogen deployment. Further contributions by Muradov and Veziroğlu (2005) examine the broader transition from a hydrocarbon-dominated energy system toward one centered on hydrogen. They argue that sustainably produced hydrogen has the potential to substantially reduce fossil fuel dependence and lower greenhouse gas (GHG) emissions, provided that carbon-neutral production pathways are successfully developed. Without such pathways, the environmental advantages of hydrogen would remain limited. The authors position hydrogen as a transitional energy carrier capable of bridging existing fossil-based systems and a future low-carbon energy paradigm, while emphasizing the necessity of sustained research and development to improve production efficiency, reduce costs, and enhance system integration. Similarly, Marbán and Valdés-Solís (2007) assess the practical feasibility of a hydrogen economy through a detailed examination of technological and economic constraints. By reviewing a range of hydrogen production routes, including electrolysis, thermochemical cycles, and fossil-derived processes with CCS, they conclude that while hydrogen holds considerable promise for emissions reduction, its widespread adoption is likely to be gradual rather than immediate. Achieving large-scale deployment will require coordinated efforts among governments, industry stakeholders, and research institutions to address issues related to cost competitiveness, scalability, and infrastructure readiness. Expanding on these perspectives, Muradov and Veziroğlu (2008) present a comprehensive overview of carbon-neutral technologies capable of enabling a hydrogen-based energy system. Their analysis emphasizes the pivotal role of green hydrogen production, particularly renewable-energy-powered electrolysis, in minimizing environmental impacts. They also identify complementary low-carbon pathways, such as biomass gasification and nuclear-assisted hydrogen production, as potentially important contributors. Collectively, their work proposes a strategic and phased roadmap for transitioning toward a hydrogen economy, in which reliance on fossil fuels is progressively reduced and hydrogen becomes a core pillar of a sustainable global energy system.

Taken together, these studies highlight both the transformative potential of hydrogen in addressing climate and energy challenges and the substantial technical, economic, and policy-related barriers that must be resolved for this vision to materialize. This evolving discourse unfolds against the backdrop of a broader global energy transition, as countries intensify efforts to reduce fossil fuel dependence and curb greenhouse gas emissions. These efforts are driven largely by ambitious climate targets, most notably the objective of achieving net-zero emissions by mid-century, as articulated in international frameworks such as the Paris Agreement and the United Nations Sustainable Development Goals (UNFCCC 2015). The Paris Agreement, in particular, commits signatory nations to limiting global temperature rise to well below 2 °C above pre-industrial levels, thereby reinforcing the urgency of deploying low-carbon energy technologies. Within this global context, hydrogen has gained increasing strategic importance due to its ability to decarbonize multiple sectors while also providing large-scale and long-duration energy storage. As a result, the hydrogen economy is now widely viewed as an essential component of international climate mitigation strategies. At the regional level, the European Union has placed hydrogen at the core of its climate and energy agenda. The EU hydrogen strategy, as highlighted in European Parliament briefings and European Commission communications, positions hydrogen as a key pillar of the European Green Deal, outlining a comprehensive roadmap for scaling up production, expanding infrastructure, and fostering market development in pursuit of a climate-neutral energy system. Hydrogen is particularly valued for its role in decarbonizing "hard-to-electrify" or no-regret sectors, including energy-intensive industries such as steel and cement production, as well as long-haul transportation modes such as aviation, shipping, and freight. Often described as the "fuel of the future," hydrogen can function as both a clean fuel and a flexible energy carrier across a wide range of applications. In the power sector, it can be utilized in fuel cells, co-fired with natural gas, or combusted in turbines and internal combustion engines to generate electricity with minimal or zero carbon emissions (Abdalla et al. 2018). In transportation, hydrogen-powered fuel cell vehicles offer advantages such as long driving ranges and rapid refueling, making them particularly suitable for heavy-duty applications. Beyond industry and transport, hydrogen also holds potential in residential and commercial heating, where it could partially or fully replace natural gas, thereby further reducing emissions. Moreover, assessments such as those by the Royal Society highlight hydrogen's critical role in large-scale electricity storage, particularly in energy systems with high shares of variable renewable energy. By enabling the storage of surplus renewable electricity over long durations, hydrogen enhances grid flexibility, reliability, and resilience. At the global level, the International Energy Agency's net-zero roadmap underscores hydrogen's central role in achieving climate neutrality by 2050, particularly in sectors that are difficult to decarbonize through direct electrification (IEA 2021). Industry-led initiatives, such as the Hydrogen Council's Path to Hydrogen Competitiveness, further demonstrate how hydrogen could become cost-competitive by around 2030 through economies of scale, technological learning, and coordinated investment. Similarly, IRENA emphasizes renewable hydrogen as a cornerstone of the global energy transition, especially for decarbonizing high-temperature industrial processes and chemical production (IRENA 2019). Despite its significant promise, large-scale hydrogen deployment continues to face challenges related to high costs, limited infrastructure, and the scalability of green hydrogen production. Nevertheless, rapid technological progress, supportive policy frameworks, and increasing international collaboration are accelerating hydrogen development worldwide. As investment in hydrogen production and infrastructure expands, hydrogen is increasingly positioned to play a decisive role in shaping a cleaner, more resilient, and sustainable global energy system.

12.2 Hydrogen Economics and Risk Assessment

This section provides an integrated evaluation of the economic factors and financial risks across the hydrogen value chain. In Sect. 12.2.1, supply chain costs, including production, storage, and transport are discussed, followed by a discussion of hydrogen's broader economic impacts and key financing instruments. This chapter continues in Sects. 12.2.2 and 12.2.3 with an assessment of investment risks and mitigation strategies, offering insights into how targeted policies and financial tools can enhance the economic viability of hydrogen deployment.

12.2.1 Hydrogen Production Cost

Primary fuels refer to naturally occurring energy resources that can be directly utilized or converted into other usable forms of energy. Common examples include coal, crude oil, natural gas, wind energy, and solar radiation. In contrast, secondary fuels are energy carriers generated through the transformation of primary energy sources. Hydrogen, electricity, and refined petroleum products such as gasoline and diesel belong to this category, as their production inherently depends on energy inputs derived from primary fuels. Because hydrogen is a secondary energy carrier, its production cost is strongly influenced by both the selected production pathway and the characteristics and price of the primary energy source employed. As a result, hydrogen production costs reflect not only the cost of supplying primary energy but also the expenses associated with converting that energy into hydrogen. These costs are typically classified into two main components: capital costs and operating costs. This section first outlines the key elements of these cost categories and subsequently compares the cost characteristics of major hydrogen production technologies. Capital costs include all expenditures associated with the initial development and commissioning of a hydrogen production facility. Major contributors generally comprise electrical and mechanical equipment, core process technologies, civil construction, site preparation, engineering design, and installation activities (Yukesh et al. 2021). Given the long operational lifetimes of hydrogen production systems, capital investments are commonly annualized over the expected plant lifetime using capital recovery factors that depend on the discount rate and project duration. Consequently, while high capital costs may appear less prominent in the short term, their financial implications become increasingly significant over medium- and long-term time horizons. Operating costs encompass all recurring expenditures required to ensure continuous plant operation and hydrogen output. Key determinants include feedstock or raw material costs, water consumption, electricity use, labor, and maintenance requirements (Yukesh et al. 2021). Operating costs are typically divided into fixed and variable components. Fixed operating costs are incurred annually per unit of installed capacity and are largely independent of utilization rates, whereas variable operating costs scale with operational intensity and are commonly expressed per unit of energy produced, energy consumed, or distance traveled in end-use applications. Among commercially mature hydrogen production pathways, steam methane reforming (SMR) is associated with the lowest capital cost, reflecting its technological maturity and widespread deployment. Moradpoor et al. (2023) report that capital investments for coal gasification plants are marginally lower than those required for alkaline and proton-exchange membrane (PEM) electrolysis systems, while being significantly lower than the capital costs of biomass gasification facilities. In electrolysis-based hydrogen production, electrolyzer efficiency and, by extension, material selection and manufacturing processes play a decisive role in determining overall system costs. This explains why solid oxide electrolyzers (SOECs), which operate at high temperatures and require advanced materials, currently represent the most capital-intensive electrolysis technology. The operating cost of hydrogen is largely governed by the price of the primary energy input. For example, natural gas prices dominate operating costs in SMR, coal prices are critical for coal gasification, and electricity prices are the primary cost driver for electrolysis-based pathways. While the absolute price levels of coal, natural gas, and renewable electricity vary considerably, the sensitivity of hydrogen production costs to changes in primary energy prices differs across technologies. Comparative analyses indicate that coal gasification exhibits the highest sensitivity to fluctuations in primary fuel prices, whereas solid oxide electrolysis shows the lowest sensitivity. This implies that relatively small changes in coal or electricity prices can result in disproportionately larger increases in hydrogen costs for coal gasification compared with SOEC-based hydrogen production. These differences are primarily attributable to variations in conversion efficiencies across technologies. Economies of scale constitute another critical factor influencing hydrogen production costs. As plant capacity increases, unit production costs—and consequently the delivered hydrogen price—tend to decline. Evidence suggests that economies of scale exert a stronger influence on SMR facilities than on water electrolysis systems. In electrolysis-based production, scale effects predominantly affect balance-of-plant (BOP) components, such as power electronics, water treatment systems, and cooling infrastructure, rather than the electrolyzer stack itself (Zhou et al. 2022). Cost breakdown studies indicate that, on average, the BOP accounts for approximately 55% of total electrolyzer system costs, while the electrolyzer stack contributes around 45% (IRENA

2019). From a strategic standpoint, economies of scale play a central role in decisions regarding centralized versus decentralized hydrogen production. Large, centralized facilities can benefit substantially from scale-related cost reductions but often incur higher costs associated with hydrogen storage and transportation. In contrast, decentralized or onsite production systems reduce distribution requirements but may forgo some scale-related cost advantages. Recent studies suggest, however, that both centralized and onsite hydrogen supply configurations can deliver hydrogen to end users at comparable overall costs when evaluated across the full value chain (Zhou et al. 2022). Accordingly, the subsequent section focuses on hydrogen storage and distribution costs, which represent a substantial share of total hydrogen supply chain expenditures. Figure 12.1 illustrates the projected levelized cost of hydrogen production across selected technologies in Northwest Europe under the Announced Pledges Scenario for the period 2023–2030. Costs are expressed in USD kg^{-1} H_2, enabling a direct comparison of the economic performance of different production pathways and highlighting expected cost trajectories as technologies mature and scale up.

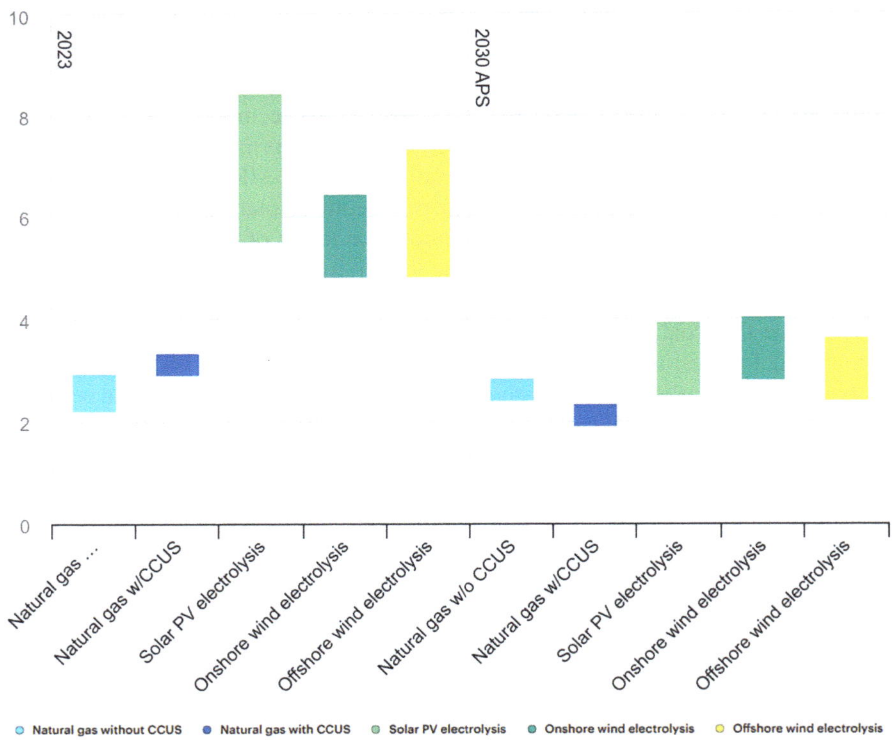

Fig. 12.1 Comparative levelized costs of hydrogen production technologies in northwest Europe within the announced pledges scenario, 2023–2030 (USD kg^{-1} H_2) (IEA 2024)

12.2.2 Hydrogen Storage Cost

Hydrogen storage constitutes a critical element in the design and optimization of hydrogen supply chain architectures, as it is shaped by a range of technical, economic, and system-level considerations. These include the nature of the primary energy source, particularly whether hydrogen production is continuous or intermittent; the demand characteristics of end users; requirements for system flexibility, resilience, and emergency preparedness; the need to balance temporal mismatches between supply and demand; and the geographical distance between hydrogen production sites and points of consumption. Consequently, effective hydrogen storage plays a central role in enhancing the stability, operational flexibility, and overall reliability of hydrogen supply systems. Despite its strategic importance, direct hydrogen storage presents significant technical challenges stemming from hydrogen's intrinsic physical properties, including its low volumetric energy density, high diffusivity, and high flammability (Prakash and Rohit 2023). Direct hydrogen storage technologies are generally classified into three main categories: compressed hydrogen, liquefied hydrogen, and cryo-compressed hydrogen (Ghorbani et al. 2023). Among these options, compressed hydrogen storage is the most technologically mature and widely deployed. It encompasses both pressurized surface storage systems such as tanks and cylinders—and underground storage solutions, including salt caverns and aquifers (Patonia and Poudineh 2023). Pressurized hydrogen tanks and cylinders are commonly categorized into four types based on their structural configuration and material composition, which directly affect their cost, weight, and performance. Type I tanks are fully metallic, typically fabricated from steel or aluminum, making them the heaviest but also the most cost-effective option; they are predominantly used in stationary and industrial applications. Type II tanks consist of a metallic liner reinforced with composite materials, offering moderate weight reduction and suitability for select transport applications. Type III tanks feature a metallic liner fully wrapped in composite materials, resulting in substantial weight savings and widespread adoption in hydrogen vehicles and aerospace applications. Type IV tanks represent the most advanced and lightweight design, comprising a fully composite structure with a polymer liner, and are widely used in modern fuel cell vehicles (FCVs) (Su et al. 2021). Underground hydrogen storage offers the potential for comparatively low capital expenditure relative to other storage options; however, its deployment is severely constrained by geological availability, which limits its widespread applicability (Ghorbani et al. 2023). While underground storage generally exhibits the lowest capital costs among storage technologies, it is often associated with significantly higher operating costs than pressurized tank or liquefied hydrogen storage systems. Furthermore, literature indicates that the operating costs of liquid hydrogen storage—largely driven by the high energy demand of the liquefaction process—are substantially higher than those associated with compressed hydrogen storage. In addition to capital investment requirements, direct hydrogen storage is characterized by considerable energy penalties. These arise primarily from hydrogen's inherently low volumetric energy density and its extremely low boiling point at atmospheric pressure (−252.8 °C) (Ghorbani et al. 2023). In compressed hydrogen systems, hydrogen is typically stored at pressures ranging from 350 to 700 bar, requiring an energy input of approximately 5.9–6.8 kWh kg^{-1} H_2 for compression. Despite this energy expenditure, the volumetric energy density achieved remains relatively low, in the range of 1.3–3.4 MJ L^{-1}, compared to approximately 32.4 MJ L^{-1} for conventional gasoline storage (Graf 2021). For liquid hydrogen storage, the liquefaction stage constitutes the most energy-intensive step, with energy requirements of roughly 10–13 kWh kg^{-1} H_2, equivalent to approximately 30%–40% of hydrogen's lower heating value. As a result, the liquefaction process alone can account for 30%–80% of total storage costs in liquid hydrogen systems (Cipriani et al. 2014). Additional challenges arise from boil-off losses, typically on the order of 0.2%–0.3% per day, which are particularly problematic for long-term storage applications (Collis and Schomacker 2022). The high capital costs and substantial energy penalties associated with direct hydrogen storage have stimulated extensive research into alternative indirect storage technologies. Unlike direct storage, which relies on hydrogen in gaseous or liquid form under high-pressure or cryogenic conditions, indirect storage approaches involve chemically binding hydrogen within stable compounds. These methods can be broadly categorized into five groups: synthetic hydrocarbons, chemical hydrides, metal hydrides, liquid organic hydrogen carriers (LOHCs), and porous materials (Patonia and Poudineh 2023). A key advantage of indirect storage is that hydrogen bound in chemically stable carriers is generally easier and potentially less costly to store, transport, and handle than pure hydrogen. Moreover, several indirect storage pathways—particularly synthetic hydrocarbons such as synthetic natural gas (SNG), as well as hydrogen derivatives including methanol (CH_3OH) and ammonia (NH_3)—can leverage existing infrastructure developed for conventional fuels such as natural gas, gasoline, and diesel. This infrastructure compatibility can significantly reduce the capital investment required for hydrogen supply chain deployment compared with systems relying exclusively on direct hydrogen storage. Nevertheless, indirect hydrogen storage technologies also face substantial limitations. Many remain at relatively low technology readiness levels, with hydrogenation and dehydrogenation processes that are energy intensive and economically challenging (Patonia and Poudineh 2023). Additional method-specific issues—such as toxicity and corrosiveness in ammonia and methanol systems, low gravimetric hydrogen density in LOHCs, and material availability constraints—can further increase costs and complicate deployment. Auxiliary processes, including carbon management for synthetic hydrocarbons and LOHCs, ammonia cracking, and downstream hydrogen purification following dehydrogenation, introduce further cost and efficiency penalties (Patonia and Poudineh 2023). In summary, indirect hydrogen storage technologies offer considerable potential to reduce the overall cost and complexity of hydrogen supply chains, particularly by enabling infrastructure integration and improved handling characteristics. However, their large-scale deployment remains constrained by low technological maturity, limited commercial experience, and significant cost uncertainties. Among indirect storage options, metal hydrides exhibit both the highest costs and the greatest uncertainty, primarily due to their substantial

Table 12.1 Global average annual investment in hydrogen infrastructure, net zero scenario (2022–2050)

S. no	Hydrogen infrastructure	2022–25	2026–30	2031–40	2041–50
1.	H2 refueling stations	4 USD bn	4 USD bn	6 USD bn	6 USD bn
2.	H2 pipelines	1 USD bn	8 USD bn	15 USD bn	34 USD bn
3.	H2 underground storage	1 USD bn	11 USD bn	19 USD bn	28 USD bn

capital investment requirements. Finally, Table 12.1 presents the average annual global investment in hydrogen infrastructure under the Net Zero Scenario for the period 2022–2050, illustrating the long-term capital requirements associated with hydrogen production, transport, storage, and end-use infrastructure necessary to support a global transition toward a hydrogen-based energy system (IEA 2023a, b).

12.2.3 Hydrogen Transportation Cost

Although hydrogen transportation costs depend on multiple factors such as demand volume, transport distance, and geographical conditions (Yang et al. 2023), the method used to preserve hydrogen during transport is a dominant determinant of overall transportation cost. Similar to storage technologies, hydrogen transportation pathways can be broadly categorized into direct and indirect methods. In direct transport methods, hydrogen is delivered in its elemental form without being converted into another compound, whereas indirect methods involve chemically converting hydrogen into a more stable carrier and reconverting it back to hydrogen at the destination. Compressed gaseous hydrogen and liquid hydrogen are the two most common direct transport options. Due to its low volumetric energy density, transporting compressed hydrogen over long distances, particularly overseas, is economically impractical. As a result, pipelines and road transport via trucks are generally considered viable options for this mode of transport (Yang et al. 2023). Liquid hydrogen, on the other hand, is more suitable for maritime shipping; however, hydrogen losses due to boil-off increase with transport duration and distance, which significantly raises costs over long routes (Ghorbani et al. 2023). Liquid hydrogen transport via pipelines is not technically feasible, and tanker-based transport is typically recommended only for medium-distance applications, generally exceeding 500 km (Graf 2021). Indirect transportation methods involve converting hydrogen into a chemically stable carrier that is easier to handle during transport, followed by reconversion to hydrogen at the point of use. Among these options, ammonia and liquid organic hydrogen carriers (LOHCs) are the most extensively studied. Ammonia offers flexibility in transportation, as it can be transported via pipelines, trucks, and ships, while LOHCs are typically transported by truck or ship only (Nunez-Jimenez and De Blasio 2022). For short-distance transport, pipelines generally represent the most economical option; however, the high capital expenditure required to develop hydrogen pipeline networks makes this option attractive primarily in scenarios with sufficiently high and stable hydrogen demand (Collis and Schomacker 2022). Transporting hydrogen as a compressed gas via trucks or pipelines tends to be more expensive than other options. Nevertheless, when hydrogenation and dehydrogenation costs are included, compressed gas hydrogen emerges as the most cost-effective option for distances below approximately 100 km by truck and 1000 km by pipeline. To reduce capital costs associated with new hydrogen pipeline infrastructure, several studies have examined the feasibility of repurposing existing natural gas pipelines for hydrogen transport. While direct reuse without significant modification is generally discouraged due to safety and material compatibility issues (Abdalla et al. 2023), upgrading existing pipelines has been estimated to cost around 20%–40% of the expense required to build new dedicated hydrogen pipelines. When hydrogenation and dehydrogenation costs are excluded, strong competition exists among liquid hydrogen, ammonia, and LOHC transport for distances below 500 km. Beyond this range, liquid hydrogen transport costs increase sharply relative to ammonia and LOHC, primarily due to higher boil-off losses during long-distance transport. However, when conversion and reconversion costs are considered LOHC-based transport becomes significantly more economical than both liquid hydrogen and ammonia. This cost advantage arises because ammonia cracking requires high temperatures typically in the range of 600–900 °C and specialized catalysts, making the process energy intensive and costly. In contrast, LOHC dehydrogenation occurs at lower temperatures with comparatively higher efficiency, resulting in lower overall conversion costs. Beyond cost considerations, ammonia and LOHC offer additional advantages related to chemical stability and ease of handling during transport. Preservation conditions for these carriers are generally less demanding than those required for compressed or liquid hydrogen, improving safety and logistical convenience (Toyota 2023). Nevertheless, ammonia's toxicity introduces safety and regulatory challenges that must be carefully managed. LOHCs, while safer to handle, suffer from higher energy losses typically in the range of 26%–43% compared to ammonia (11%–19%), compressed hydrogen (5%–15%), and liquid hydrogen (9%–22%) (Collis and Schomacker 2022). Another drawback of indirect transport methods is the requirement to return the depleted carrier material to the point of origin for rehydrogenation, which can increase logistical complexity and transportation costs (Graf 2021). While this is often cited as a disadvantage of indirect methods, it should be noted that direct transport methods also face inherent inefficiencies, such as the need to maintain minimum pressure levels in compressed hydrogen systems or cryogenic temperatures in liquid hydrogen transport, which can similarly limit deliverable hydrogen quantities and increase operational costs (Yang et al. 2023).

12.3 Financial Risks and Mitigation Strategies

This section examines the key financial risks associated with the implementation of hydrogen policies and discusses strategies that can be used to mitigate or hedge against these risks. In the context of hydrogen policy, financial risks refer to events or actions that may arise across the hydrogen value chain and potentially hinder successful policy execution. Such risks differ in both their likelihood of occurrence and the magnitude of their potential impacts, making a systematic and comprehensive risk assessment essential. Although hydrogen has gained prominence as a critical decarbonization tool in sectors such as industry and transport, it is not without inherent risks. Many of these risks are linked to hydrogen's fundamental physical properties, particularly its high flammability, which can pose safety hazards (Azadnia et al. 2023). As a result, hydrogen policy implementation carries the risk of accidents, including fires and explosions (Groth and Hecht 2017). Nevertheless, hydrogen is often considered safer than conventional fuels such as gasoline, owing to its low density, high diffusion rate, and rapid dispersion in open environments, which significantly reduces the likelihood of explosive concentrations forming (Sharma and Ghoshal 2015). To minimize accident risks, it is strongly recommended that hydrogen production and handling facilities be located away from densely populated or vulnerable areas (Scipioni et al. 2023). In addition, the adoption of stringent safety standards, regulations, and operational protocols can further reduce the probability and severity of accidents (Scipioni et al. 2023). Hydrogen production and utilization also rely heavily on advanced technologies and involve a complex, international value chain with multiple stakeholders operating across different regions (Ball and Weeda 2015). This complexity introduces additional risks related to technological performance and market coordination, which may disrupt the effective implementation of hydrogen policies (OIES 2023). Consequently, a robust risk assessment framework must account for both technological and market dynamics along the hydrogen value chain. In this context, the following discussion focuses on the main risks arising from these two dimensions. From a technological perspective, limited technological readiness represents a major barrier to successful hydrogen policy deployment (IRENA 2022). This risk emerges when key technologies are either unavailable at commercial scale or remain immature and insufficiently tested (Madadi et al. 2022). Mitigating such risks requires the development and enforcement of comprehensive safety standards, adequate workforce training, and rigorous testing and certification of hydrogen technologies before large-scale deployment (Madadi et al. 2022). Continued investment in research and development, together with enhanced collaboration among industry, research institutions, and policymakers, is also essential to accelerate technological maturation and reduce associated risks (Scipioni et al. 2023). Moreover, heavy reliance on imported hydrogen technologies exposes countries to supply risks arising from geopolitical tensions or changes in international trade relations (Scipioni et al. 2023). In this regard, fostering domestic hydrogen technology development and production capacity is a crucial strategy for enhancing resilience. In parallel, market-related risks within the hydrogen value chain can significantly affect policy effectiveness. In particular, uncertainty and volatility in hydrogen demand may disrupt the smooth functioning of the entire value chain (Almaraz et al. 2022). Such risks arise when demand growth fails to keep pace with supply expansion or when inadequate infrastructure limits the ability to deliver hydrogen from production sites to end-use sectors (Erdoğan and Güler 2023). To address these challenges, improving the cost competitiveness of hydrogen is essential for stimulating sustained market demand (Dincer and Acar 2017). This can be achieved through scaling up production to realize economies of scale and through continuous technological innovation aimed at cost reduction (Lebrouhi et al. 2022). However, as these structural changes require time, short-term government intervention remains necessary. Targeted policy measures such as subsidies, tax incentives, and financial support for hydrogen-powered technologies (e.g., fuel cell vehicles) and hydrogen production systems can play a critical role in bridging the gap between current market conditions and long-term economic viability. Moreover, the use of long-term contractual arrangements with fixed hydrogen prices can effectively reduce exposure to price volatility, thereby enhancing demand stability and investment confidence in hydrogen markets (Lebrouhi et al. 2022). In parallel, substantial upfront investments in hydrogen production, storage, and distribution infrastructure can help address risks associated with insufficient or underdeveloped facilities (Sharma and Ghoshal 2015). Strategic planning and close cooperation between public authorities and private stakeholders are therefore essential for establishing a robust and reliable hydrogen infrastructure network. The geographical configuration of the hydrogen supply chain also represents a significant source of risk. When hydrogen production or critical supply chain segments are concentrated in a limited number of countries, or when hydrogen is heavily reliant on imports, supply disruptions may occur, undermining the overall security and resilience of the hydrogen system (Nunez-Jimenez and De Blasio 2022). These risks can be mitigated by fostering a diversified portfolio of hydrogen suppliers, supported by long-term contractual agreements that enhance supply security. At the same time, market competition and geopolitical tensions present additional challenges to the successful integration of hydrogen into the energy mix (IRENA 2022). The emerging hydrogen market is characterized by intense competition, as incumbent firms and new entrants seek to secure strategic positions and capture future value. Established players, particularly from the oil and gas sector, are increasingly diversifying their portfolios to include hydrogen technologies, while start-ups are driving innovation to gain market share. In this highly competitive environment, promoting collaboration and partnerships among stakeholders is essential to unlock synergies and support the sustained growth of hydrogen markets (Azadnia et al. 2023). Beyond technological and market-related risks, hydrogen's status as a secondary energy carrier introduces important environmental considerations. The emissions footprint of hydrogen production is determined entirely by the primary energy source used (Feitosa and Costa 2021). Consequently, blue and gray hydrogen pathways pose environmental risks due to their associated carbon emissions. In situations where renewable hydrogen production cannot be scaled up because of limited renewable electricity capacity, there is also a risk of long-term lock-in to fossil fuel–based hydrogen systems (Kemfert et al. 2022). To address these challenges, policy instruments such as feed-in tariffs (FITs) and contracts for difference (CfDs) can be deployed to incentivize renewable-based hydrogen production, alongside investments in cross-border electricity interconnections to compensate for domestic supply shortfalls. Even green hydrogen is not entirely risk-free from an environmental perspective. The manufacture of electrolysis systems and related infrastructure requires significant quantities of critical raw materials, which may have adverse implications for human health and biodiversity (Arvesen et al. 2018). In addition, green hydrogen production through electrolysis entails considerable water consumption, raising concerns about water scarcity, particularly in regions already facing severe water stress, such as parts of Africa (Azadnia et al. 2023). Advancing electrolysis technologies to improve water efficiency and reduce overall resource intensity represents an effective strategy for mitigating these risks (Scipioni et al. 2023). Finally, hydrogen projects are also exposed to broader macroeconomic and political risks, including interest rate fluctuations, high inflation, and political instability, similar to other large-scale energy investments (OIES 2023). A detailed examination of these risks and their mitigation strategies lie beyond the scope of this study, and interested readers are referred to the relevant literature for further discussion. Increasing public awareness of hydrogen's potential benefits and actively engaging citizens as participants and co-creators within the hydrogen value chain can further support social acceptance and facilitate the successful and widespread deployment of hydrogen technologies (Madadi et al. 2022).

12.4 Technological Challenges of Hydrogen Use

12.4.1 Infrastructure Challenges

Although hydrogen is widely recognized as a cornerstone of the global decarbonization transition, major infrastructure-related challenges persist, particularly with respect to storage systems, distribution networks, and refueling infrastructure (Ball and Wietschel 2009). Addressing these bottlenecks is essential to enable the large-scale deployment of green hydrogen across key sectors such as transportation, industry, and power generation. Without coordinated investment in these enabling infrastructures, the practical integration of hydrogen into existing energy systems will remain limited. A major barrier to scaling up sustainable hydrogen production is the continued predominance of gray hydrogen, which is derived from natural gas through carbon-intensive processes such as steam methane reforming (SMR). The International Energy Agency reports that approximately 95% of global hydrogen production currently falls into this category, resulting in substantial carbon dioxide emissions (IEA 2023a, b). Replacing gray hydrogen with green hydrogen produced via water electrolysis using renewable electricity is therefore critical to meeting global climate and decarbonization targets. However, the transition to green hydrogen is constrained by several economic and technological challenges, most notably its high levelized cost of hydrogen (LCOH). Achieving cost competitiveness with gray hydrogen will require large-scale investments in renewable power generation as well as significant expansion of electrolyzer manufacturing and deployment. In this context, continued technological progress in electrolyzer efficiency and durability, sustained reductions in renewable electricity costs, and the realization of economies of scale through mass production and deployment are all essential. Together, these developments are necessary to drive down costs and enable green hydrogen to become a commercially viable and widely adopted energy carrier in a low-carbon future.

12.4.2 Electrolysis Technologies

Electrolysis represents the core technological pathway for green hydrogen production, in which water molecules are split into hydrogen and oxygen using electricity generated from renewable energy sources (Kumar and Lim 2022). Several electrolysis technologies are currently available, each characterized by specific advantages and limitations that influence their suitability for large-scale green hydrogen deployment. Alkaline water electrolysis (AWE) is the most mature and commercially established technology (Zeng and Zhang 2010). It is valued for its long operational lifetime, robustness, and relatively low capital cost. Nevertheless, AWE generally operates at lower current densities and exhibits reduced efficiency compared to more advanced electrolyzer technologies, which can limit its attractiveness in applications where high efficiency and dynamic operation are required. Proton exchange membrane water electrolysis (PEMWE) offers higher efficiency, compact system design, and superior operational flexibility, particularly under variable power inputs from intermittent renewable sources such as wind and solar (Ayers 2019). Despite these advantages, PEM electrolyzers remain costly, primarily due to their reliance on scarce and expensive noble metal catalysts, including iridium and platinum, which constrains large-scale deployment. Solid oxide electrolyzer cells (SOECs) operate at elevated temperatures, typically in the range of 700–1000 °C, enabling higher electrical efficiency through the utilization of thermal energy (Ni et al. 2008). However, the demanding operating conditions, material degradation, and system complexity significantly increase capital costs and pose challenges for long-term durability and large-scale commercialization. Overall, advancing electrolysis technologies by improving efficiency, reducing reliance on critical materials, enhancing durability, and achieving cost reductions through scale-up is crucial for accelerating green hydrogen production and enabling it to compete economically with conventional fossil fuel–based hydrogen pathways.

12.4.3 Challenges in Hydrogen Distribution

Hydrogen transportation is particularly challenging due to its low volumetric energy density and the technical complexity associated with maintaining it in either gaseous or liquid form during transit. Several transportation pathways have been proposed, each presenting distinct technical and economic constraints (Ishaq et al. 2022). One common approach involves compressing hydrogen to high pressures, typically up to 700 bar, enabling transport in gaseous form. However, compression is highly energy-intensive and necessitates specialized high-pressure storage vessels and transport tanks, which substantially increase capital and operational costs. Alternatively, hydrogen can be transported in liquid form by cooling it to cryogenic temperatures of approximately −253 °C (20 K). Liquid hydrogen offers a higher volumetric energy density than compressed gas, but the liquefaction process is costly and energy-intensive, consuming roughly 30% of the hydrogen's intrinsic energy content. In addition, boil-off losses during transportation and storage further reduce overall efficiency. An indirect transportation route involves converting hydrogen into ammonia (NH_3), which is easier to liquefy, store, and transport using existing infrastructure (Klerke et al. 2008). At the point of use, ammonia can be cracked to release hydrogen; however, this reconversion step introduces additional system complexity and energy penalties, thereby lowering the overall efficiency of the supply chain. Material-based hydrogen storage options have also been explored in research and development efforts (Ren et al. 2017), but these technologies remain at low technology readiness levels and are unlikely to contribute significantly to large-scale decarbonization targets by 2050. Infrastructure limitations further complicate hydrogen transport. Existing natural gas pipelines cannot be directly repurposed for pure hydrogen without substantial modification, as hydrogen's small molecular size can cause metal embrittlement and leakage, raising safety and durability concerns. While blending hydrogen into natural gas pipelines may serve as a transitional solution, the long-term establishment of a hydrogen-based energy system will ultimately require the development of dedicated hydrogen pipeline infrastructure.

12.4.4 Storage and Utilization

Hydrogen storage poses significant technical and economic challenges, particularly when considered at large scale (Usman 2022). As discussed earlier, the main hydrogen storage options include compressed gas, cryogenic liquid, and hybrid cold/pressurized gas storage, each with inherent limitations. Compressed gas storage (CGS) involves storing hydrogen at very high pressures in steel or composite tanks. Although this method is relatively mature and well understood, it requires expensive, high-strength materials to safely withstand the extreme pressures involved, which restricts its economic feasibility for large-scale or long-duration storage applications. Cryogenic liquid storage requires hydrogen to be cooled and maintained at extremely low temperatures (approximately −253 °C). While liquefaction significantly increases volumetric energy density, it necessitates advanced insulation systems and continuous energy input to limit boil-off losses. These requirements add considerable cost and operational complexity, particularly for long-term storage. Cold or pressurized gas storage represents a hybrid approach that combines elevated pressures with reduced temperatures to achieve higher energy densities than conventional compressed gas storage. Despite its technical advantages, this method relies on specialized storage vessels and sophisticated control systems, leading to high capital costs and stringent safety requirements. Beyond aboveground storage tanks, underground salt caverns have been widely explored as an option for long-term and large-scale hydrogen storage. These caverns can accommodate very large quantities of hydrogen at relatively low cost per unit of storage. However, their deployment is constrained by geological availability, as suitable salt formations are limited to specific regions, which restricts their universal applicability.

12.4.5 Hydrogen Refueling Networks

The establishment of a robust hydrogen refueling network is essential for facilitating the large-scale deployment of hydrogen fuel cell vehicles (FCVs) and hydrogen-powered transportation systems (Rose and Neumann 2020). At present, hydrogen refueling infrastructure remains limited, with only a few hundred stations worldwide, predominantly concentrated in countries such as Japan, Germany, and the United States. Expanding this network requires substantial capital investment, coordinated planning, and strong governmental support to overcome early-stage market barriers. Furthermore, hydrogen refueling is inherently more complex than conventional fueling processes, as it relies on high-pressure storage, specialized compressors, and advanced safety systems. These technical requirements significantly increase both capital and operational costs, posing an additional challenge to the rapid scaling of hydrogen refueling infrastructure.

12.4.6 Hydrogen for Transport

In the transportation sector, hydrogen is utilized primarily through fuel cells and modified combustion technologies. Hydrogen fuel cells convert chemical energy directly into electricity with high efficiency and near-zero tailpipe emissions, making them particularly attractive for hydrogen fuel cell vehicles (FCVs) and buses. In addition, fuel cell technology is increasingly viewed as a promising solution for decarbonizing heavy-duty transport modes, including trucks, maritime vessels, and rail systems (Singla et al. 2021). Alternatively, hydrogen can be combusted in adapted internal combustion engines (ICEs). Although hydrogen combustion is generally less efficient than fuel cell systems and may still generate pollutants such as nitrogen oxides (NO_x), it offers benefits such as a high power-to-weight ratio and compatibility with existing engine architectures. As a result, hydrogen combustion engines may serve as a transitional option in applications where fuel cell deployment remains technically or economically constrained. Furthermore, hydrogen is being explored as a fuel for modified gas turbine engines in aviation, presenting a potential pathway toward reducing emissions in the aerospace sector (Boretti 2020).

12.4.7 Hydrogen for Power Generation

Hydrogen offers considerable promise for power generation, particularly as a means of improving the flexibility, resilience, and sustainability of modern energy systems. It can be deployed through both fuel cell technologies and combustion-based applications. Hydrogen fuel cells generate electricity by directly converting chemical energy into electrical energy, achieving high efficiencies and minimal emissions. While fuel cells are already in use in hydrogen fuel cell vehicles (FCVs) and buses, they are also attracting growing interest for their potential to decarbonize heavy-duty transport sectors, including trucks, shipping, and rail systems. In addition to fuel cells, hydrogen can be utilized in modified internal combustion engines (ICEs). Although hydrogen combustion is generally less efficient than fuel cell operation and may lead to the formation of nitrogen oxides (NO_x), it provides advantages such as a high power-to-weight ratio and compatibility with existing engine technologies. Consequently, hydrogen-fueled ICEs may act as a transitional solution in applications where fuel cell technologies are not yet technically mature or economically viable (Cummins 2023). Hydrogen also holds significant potential in gas turbine (GT) and combined cycle gas turbine (CCGT) power plants. Existing gas turbines can operate on hydrogen–natural gas blends, enabling a gradual and pragmatic transition toward hydrogen-based power generation. Several industrial players are actively developing hydrogen-capable turbines to reduce the carbon intensity of electricity production. In the longer term, CCGT plants operating on high hydrogen shares—or even pure hydrogen—could substantially decarbonize large-scale power generation while maintaining system reliability and dispatchability. To fully unlock hydrogen's role in a clean energy future, however, substantial investments are required to address persistent infrastructure challenges. Advancements in hydrogen production capacity, cost-effective and safe storage technologies, efficient distribution networks, and widespread refueling and utilization infrastructure are essential. Overcoming these barriers is critical for enabling hydrogen to contribute meaningfully to the decarbonization of power generation, transportation, and industrial sectors (Toyota 2023).

12.5 Conclusion

The hydrogen economy holds the potential to fundamentally transform global energy systems and to play a pivotal role in decarbonizing hard-to-electrify sectors, including heavy industry, aviation, maritime transport, and long-distance freight. However, translating this potential into reality requires overcoming substantial technical, economic, and institutional challenges across the entire hydrogen value chain, encompassing production, storage, transportation, and regulatory governance. Among the available pathways, green hydrogen produced using renewable electricity offers a particularly compelling advantage, as it enables large-scale, long-duration energy storage and mitigates the inherent intermittency of wind and solar generation. In this regard, green hydrogen is increasingly recognized as a critical enabler of a resilient and low-carbon energy system. Despite these advantages, the high cost of green hydrogen production remains a major impediment to large-scale deployment. Electrolysis-based hydrogen currently remains significantly more expensive than hydrogen produced from fossil fuels, even when carbon capture technologies are applied. Achieving cost competitiveness will depend on multiple interrelated factors, including rapid scale-up of production capacity, technological innovation in electrolyzer design and manufacturing, and continued declines in the cost of renewable electricity. Without sustained cost reductions, green hydrogen is unlikely to achieve widespread adoption across price-sensitive industrial and transport applications. In parallel, hydrogen storage and transportation represent critical technical and economic bottlenecks. Hydrogen's low volumetric energy density and high flammability make it more challenging and costly to handle than conventional fuels, necessitating substantial investments in dedicated pipelines, storage facilities, refueling infrastructure, and advanced safety systems. These infrastructure requirements underscore the need for coordinated planning and long-term investment strategies to avoid fragmented or inefficient hydrogen networks. Given these challenges, strong and coherent policy support at both national and international levels is indispensable. In the long term, predictable policy frameworks are essential to de-risk private investment, accelerate research and development, and provide market certainty. Policy instruments such as carbon pricing, production subsidies, contracts for difference, and tax incentives—already implemented in leading economies such as China—must be expanded, refined, and harmonized across regions to support global hydrogen market development. Equally important is international cooperation to establish common standards for hydrogen production methods, safety protocols, and cross-border trade, thereby enabling the formation of efficient and transparent global value chains. The strategic value of hydrogen is further enhanced through its integration with renewable energy systems. By absorbing surplus renewable electricity during periods of excess generation and supplying energy during times of scarcity, hydrogen can play a key role in balancing power systems and maintaining grid stability in countries pursuing rapid renewable energy expansion. This system-level flexibility positions hydrogen not merely as a fuel, but as an essential component of future energy system architecture. The successful development of a hydrogen economy will ultimately depend on close collaboration among governments, industry stakeholders, research institutions, and international organizations. Public–private partnerships will be critical for mobilizing the capital required for large-scale infrastructure deployment, while continuous innovation across the hydrogen value chain must be supported by clear, robust, and adaptive regulatory frameworks. International organizations such as the International Energy Agency (IEA) and the Hydrogen Council have consistently emphasized the importance of coordinated global action to ensure that hydrogen markets develop in a cost-effective and climate-aligned manner. Cross-border cooperation will be particularly important for scaling hydrogen production and establishing global supply chains, especially for regions endowed with abundant renewable resources but limited domestic hydrogen infrastructure. In this context, China's state-led and infrastructure-focused approach has positioned it at the forefront of hydrogen development, while many Western economies risk falling behind due to slower policy implementation and fragmented investment strategies. To remain competitive, Western countries must move beyond ambition and signaling toward decisive policy execution, large-scale infrastructure investment, and accelerated market deployment. In conclusion, hydrogen has the potential to emerge as a cornerstone of the global energy transition, offering a versatile and low-carbon solution for sectors where direct electrification is impractical. Unlocking this potential will require decisive and sustained policy action, substantial infrastructure investment, and deep international cooperation. With the appropriate strategies in place, hydrogen can serve as a critical bridge between today's fossil-fuel-dependent energy systems and a sustainable, low-carbon future, contributing meaningfully to global climate objectives, including those enshrined in the Paris Agreement.

References

Abdalla AM, Hossain S, Nisfindy OB, Azad AT, Dawood M, Azad AK (2018) Hydrogen production, storage, transportation and key challenges with applications: a review. Energy Convers Manag 165:602–627

Abdalla AM, Osama A, Bo W, Xinzhi W, Azad Abul K, Dawood MK (2023) Hydrogen production technologies: conventional processes. In: Hydrogen economy. Elsevier. https://doi.org/10.1016/B978-0-323-99514-6.000

Almaraz SDL, Rácz V, Azzaro-Pantel C, Szántó ZO (2022) Multiobjective and social cost-benefit optimisation for a sustainable hydrogen supply chain: application to Hungary. Appl Energy 325:119882. https://doi.org/10.1016/J.APENERGY.2022

Arvesen A, Luderer G, Pehl M, Bodirsky BL, Hertwich EG (2018) Deriving life cycle assessment coefficients for application in integrated assessment modelling. Environ Model Softw 99:111–125. https://doi.org/10.1016/J.ENVSOFT.2017.09.01

Ayers K (2019) The potential of proton exchange membrane–based electrolysis technology. Curr Opin Electrochem 18:9–15

Azadnia AH, McDaid C, Andwari AM, Hosseini SE (2023) Green hydrogen supply chain risk analysis: a European hard-to-abate sectors perspective. Renew Sust Energ Rev 182:113371. https://doi.org/10.1016/J.RSER.2023.113371

Ball M, Weeda M (2015) The hydrogen economy—vision or reality? Int J Hydrog Energy 40:7903–7919. https://doi.org/10.1016/J.IJHYDENE.2015.04.032

Ball M, Wietschel M (2009) The future of hydrogen–opportunities and challenges. Int J Hydrog Energy 34(2):615–627
Barreto L, Makihira A, Riahi K (2003) The hydrogen economy in the 21st century: a sustainable development scenario. Int J Hydrog Energy 28(3):267–284
Boretti A (2020) Hydrogen internal combustion engines to 2030. Int J Hydrog Energy 45(43):23692–23703
Cipriani G, Di Dio V, Genduso F, La Cascia D, Liga R, Miceli R et al (2014) Perspective on hydrogen energy carrier and its automotive applications. Int J Hydrog Energy 39:8482–8494. https://doi.org/10.1016/J.IJHYDENE.2014.03.174
Collis J, Schomacker R (2022) Determining the production and transport cost for H2 on a global scale. Front Energy Res 10:909298. https://doi.org/10.3389/fenrg.2022.90
Crabtree GW, Dresselhaus MS, Buchanan MV (2004) The hydrogen economy. Phys Today 57(12):39–44
Cummins (2023) Hydrogen engines for heavy-duty applications. Available at: www.cummins.com/engines/hydrogen
Dincer I, Acar C (2017) Innovation in hydrogen production. Int J Hydrog Energy 42:14843–14864. https://doi.org/10.1016/j.ijhydene.2017.04.107
Erdoğan A, Güler MG (2023) Optimization and analysis of a hydrogen supply chain in terms of cost, CO2 emissions, and risk: the case of Turkey. Int J Hydrog Energy 48:22752–22765. https://doi.org/10.1016/j.ijhydene.2023.04.300
Feitosa FEB, Costa AL (2021) Application of a multicriteria methodology for evaluation of energy alternatives for hydrogen production for the automotive sector—case study. Int J Hydrog Energy 46:20799–20814. https://doi.org/10.1016/J.IJHYDENE.2021.03.209
Ghorbani B, Zendehboudi S, Saady NMC, Dusseault MB (2023) Hydrogen storage in North America: status, prospects, and challenges. J Environ Chem Eng 11:109957. https://doi.org/10.1016/J.JECE.2023.109957
Graf M (2021) Costs and energy efficiency of long-distance hydrogen transport options. University of Natural Resources and Life Sciences, Vienna
Groth KM, Hecht ES (2017) HyRAM: a methodology and toolkit for quantitative risk assessment of hydrogen systems. Int J Hydrog Energy 42:7485–7493. https://doi.org/10.1016/J.IJHYDENE.2016.07.002
IEA (2021) Net zero by 2050: a roadmap for the global energy sector. OECD Publishing. Available at: www.iea.org/reports/net-zero-by-2050
IEA (2023a) Average annual global investment in hydrogen and natural gas infrastructure in the net zero scenario, 2016–2050. IEA, Paris. https://www.iea.org/data-and-statistics/charts/average-annual-global-investment-in-hydrogen-and-natural-gas-infrastructure-in-the-net-zero-scenario-2016-2050
IEA (2023b) Hydrogen production and infrastructure projects database. IEA. Available at: www.iea.org/data-and-statistics/data-product/hydrogen-production-and-infrastructure-projects-database
IEA (2024) Levelised cost of hydrogen via selected technologies in Northwest Europe in the announced pledges scenario, 2023–2030. IEA, Paris. https://www.iea.org/data-and-statistics/charts/levelised-cost-of-hydrogen-via-selected-technologies-in-northwest-europe-in-the-announced-pledges-scenario-2023-2030
IRENA (2019) Hydrogen: a renewable energy perspective. www.irena.org/-/media/Files/IRENA/Agency/Publication/2019/Sep/IRENA_Hydrogen_2019.pdf
IRENA (2022) Geopolitics of the energy transformation: the hydrogen factor. IRENA
Ishaq H, Dincer I, Crawford C (2022) A review on hydrogen production and utilization: challenges and opportunities. Int J Hydrog Energy 47(62):26238–26264
Kemfert C, Präger F, Braunger I, Hoffart FM, Brauers H (2022) The expansion of natural gas infrastructure puts energy transitions at risk. Nat Energy 7:582–587. https://doi.org/10.1038/s41560-022-01060-3
Klerke A, Christensen CH, Nørskov JK, Vegge T (2008) Ammonia for hydrogen storage: challenges and opportunities. J Mater Chem 18(20):2304–2310
Kumar SS, Lim H (2022) An overview of water electrolysis technologies for green hydrogen production. Energy Rep 8:1

Lebrouhi BE, Djoupo JJ, Lamrani B, Benabdelaziz K, Kousksou T (2022) Global hydrogen development—a technological and geopolitical overview. Int J Hydrog Energy 47:7016–7048. https://doi.org/10.1016/J.IJHYDENE.2021.12.076

Madadi AV, Zendehboudi S, Cata Saady NM, Dusseault MB (2022) A comprehensive review on hydrogen production and utilization in North America: prospects and challenges. Energy Convers Manag 269:115927. https://doi.org/10.1016/J.ENCONMAN.2022.115927

Marbán G, Valdés-Solís T (2007) Towards the hydrogen economy? Int J Hydrog Energy 32(12):1625–1637

Moradpoor I, Syri S, Santasalo-Aarnio A (2023) Green hydrogen production for oil refining—Finnish case. Renew Sust Energ Rev 175:113159. https://doi.org/10.1016/J.RSER.2023.113159

Muradov NZ, Veziroğlu TN (2005) From hydrocarbon to hydrogen–carbon to hydrogen economy. Int J Hydrog Energy 30(3):225–237

Muradov NZ, Veziroğlu TN (2008) "Green" path from fossil-based to hydrogen economy: an overview of carbon-neutral technologies. Int J Hydrog Energy 33(23):6804–6839

Ni M, Leung MK, Leung DY (2008) Technological development of hydrogen production by solid oxide electrolyzer cell (SOEC). Int J Hydrog Energy 33(9):2337–2354

Nunez-Jimenez A, De Blasio N (2022) Competitive and secure renewable hydrogen markets: three strategic scenarios for the European Union. Int J Hydrog Energy 47:35553–35570. https://doi.org/10.1016/J.IJHYDENE.2022.08.170

Patonia A, Poudineh R (2023) Hydrogen storage for a net-zero carbon future (No. 23). OIES Paper: ET

Prakash C, Rohit S (2023) Chapter 21: Approach toward economical hydrogen storage. In: Solar-driven green hydrogen generation and storage. Elsevier. https://doi.org/10.1016/B978-0-323-99580-1.00003-0

Ren J, Musyoka NM, Langmi HW, Mathe M, Liao S (2017) Current research trends and perspectives on materials-based hydrogen storage solutions: a critical review. Int J Hydrog Energy 42(1):289–311

Rose PK, Neumann F (2020) Hydrogen refueling station networks for heavy-duty vehicles in future power systems. Transp Res Part D Transp Environ 83:102358

Scipioni A, Manzardo A, Ren J (2023) Hydrogen economy: processes, supply chain, life cycle analysis and energy transition for sustainability. Elsevier

Sharma S, Ghoshal SK (2015) Hydrogen the future transportation fuel: from production to applications. Renew Sust Energ Rev 43:1151–1158. https://doi.org/10.1016/J.RSER.2014.1

Singla MK, Nijhawan P, Oberoi AS (2021) Hydrogen fuel and fuel cell technology for cleaner future: a review. Environ Sci Pollut Res Int 28(13):15607–15626

Su Y, Lv H, Zhou W, Zhang C (2021) Review of the hydrogen permeability of the liner material of type IV on-board hydrogen storage tank. World Electr Veh J 12:130. https://doi.org/10.3390/WEVJ12030130

Toyota (2023) Evolution of hydrogen technology and taking on new challenges in Australia. Available at: global.toyota/en/newsroom/corporate/40081554

UNFCCC (2015) The Paris Agreement. Available at: unfccc.int/process-and-meetings/the-paris-agreement

Usman MR (2022) Hydrogen storage methods: review and current status. Renew Sust Energ Rev 167:112743

Yang M, Hunger R, Berrettoni S, Sprecher B, Wang B (2023) A review of hydrogen storage and transport technologies. Clean Energy 7(1):190–216. https://doi.org/10.1093/ce/zkad021

Yukesh KR, Kavitha S, Preethi Parthiba Karthikeyan O, Kumar G, DaiViet NV et al (2021) Techno-economic assessment of various hydrogen production methods—a review. Bioresour Technol 319:124175. https://doi.org/10.1016/J.BIORTECH.2020.124

Zeng K, Zhang D (2010) Recent progress in alkaline water electrolysis for hydrogen production and applications. Prog Energy Combust Sci 36(3):307–326

Zhou Y, Searle S, Baldino C (2022) Cost of renewable hydrogen produced onsite at hydrogen refueling stations in Europe. International Council on Clean Transportation, Washington, DC, p 17

MIX
Papier aus verantwortungsvollen Quellen
Paper from responsible sources
FSC® C105338

If you have any concerns about our products,
you can contact us on
ProductSafety@springernature.com

In case Publisher is established outside the EU,
the EU authorized representative is:
Springer Nature Customer Service Center GmbH
Europaplatz 3, 69115 Heidelberg, Germany

Printed by Libri Plureos GmbH
in Hamburg, Germany